COAL COMBUSTION AND GASIFICATION

The Plenum Chemical Engineering Series

Series Editor
Dan Luss, *University of Houston, Houston, Texas*

ENGINEERING FLOW AND HEAT EXCHANGE
Octave Levenspiel

COAL COMBUSTION AND GASIFICATION
L. Douglas Smoot and Philip J. Smith

A Continuation Order Plan is available for this series. A continuation order will bring delivery of each new volume immediately upon publication. Volumes are billed only upon actual shipment. For further information please contact the publisher.

COAL COMBUSTION AND GASIFICATION

L. Douglas Smoot
and
Philip J. Smith

Brigham Young University
Provo, Utah

PLENUM PRESS • NEW YORK AND LONDON

Library of Congress Cataloging in Publication Data

Smoot, L. Douglas (Leon Douglas)
  Coal combustion and gasification.

  (The Plenum chemical engineering series)
  Bibliography: p.
  Includes index.
  1. Coal—Combustion. 2. Coal gasification. I. Smith, Philip J. II. Title. III. Series.
TP325.S57   1985   662.6′2   84-17937
ISBN 0-306-41750-2

This limited facsimile edition has been issued
for the purpose of keeping this title available
to the scientific community.

© 1985 Plenum Press, New York
A Division of Plenum Publishing Corporation
233 Spring Street, New York, N.Y. 10013

Printed in the United States of America

DEDICATION

Recognizing its several limitations, we unpretentiously dedicate this book to our Creator, Who, through holy writ, has inspired us to seek light and understanding where darkness exists, and Who, through these same records, has taught us to place our professional pursuits in perspective with greater obligations of service to family and to society.

PREFACE

The use of coal is required to help satisfy the world's energy needs. Yet coal is a difficult fossil fuel to consume efficiently and cleanly. We believe that its clean and efficient use can be increased through improved technology based on a thorough understanding of fundamental physical and chemical processes that occur during consumption. The principal objective of this book is to provide a current summary of this technology.

The past technology for describing and analyzing coal furnaces and combustors has relied largely on empirical inputs for the complex flow and chemical reactions that occur while more formally treating the heat-transfer effects. Growing concern over control of combustion-generated air pollutants revealed a lack of understanding of the relevant fundamental physical and chemical mechanisms. Recent technical advances in computer speed and storage capacity, and in numerical prediction of recirculating turbulent flows, two-phase flows, and flows with chemical reaction have opened new opportunities for describing and modeling such complex combustion systems in greater detail. We believe that most of the requisite component models to permit a more fundamental description of coal combustion processes are available. At the same time there is worldwide interest in the use of coal, and progress in modeling of coal reaction processes has been steady.

We have been working during the past ten years on coal combustion and gasification processes. This work has included both basic measurements and development of analytical models. In our modeling work, we have attempted to develop fundamental models based upon the general equations of conservation and to use the most appropriate literature information available as sources for model components and parameters. One of the major purposes of this new book is to document our general modeling approach for pulverized-coal systems. This model emphasizes processes using finely pulverized coal which is entrained in a gaseous phase. Problems of particular interest include combustion in pulverized-coal furnaces and power generators, entrained coal gasifiers, coal-fired MHD power generators, and flame propagation in laminar and turbulent coal dust/gas mixtures. In addition, we have reviewed work by others in modeling of combustion or gasification in entrained, fluidized, and fixed beds. We have also briefly considered other combustion problems associated with oil shale, tar sands, and other fossil fuels.

This book is the second published by the authors (with others in the first book, Plenum Publishing Co., New York, 1979) which relates to coal combustion and gasification. For the case of pulverized-coal models, the authors (with others) provided detailed fundamentals on which modeling is based in the earlier book (Smoot and Pratt, 1979). This earlier book emphasized general principles for reacting, turbulent or laminar, multiphase systems. General conservation equations were developed and summarized. The basis for computing thermochemical equilibrium in complex, heterogeneous mixtures was presented, together with techniques for rapid computation and reference to required input data. Rate processes were discussed, including pertinent aspects of turbulence, chemical kinetics, radiative heat transfer, and gas–particle convective–diffusive interactions.

This book differs from the first in several significant ways. Most of the material in this book is different from that in the first book. The scope of the new book has been expanded to include material on fixed- and fluidized-bed process characteristics. The details of interactions between turbulence and chemistry for coal-laden systems is a major new emphasis in this book, which has been prepared for use both as an advanced textbook and as a reference and source book for those who work in fossil-fuel combustion and gasification. Technical problems for solution have been included at the end of each chapter. A single nomenclature and list of references is used for the entire book.

The book is divided into five major topic areas:

1. Chapters 1 and 2 consider general characteristics of coal processes and properties.
2. Chapters 3–5 treat basic reaction processes of coal particles, including coal devolatilization, char oxidation, and volatiles combustion. Some of these topics were treated in the first book but have been updated or expanded significantly. New sections on heatup and ignition have been added.
3. Chapters 6–8 deal with practical fossil combustion flames. Processes are classified by flame type and then each is considered in some detail, with particular emphasis on pulverized-coal flames. Recent laboratory data for combustion and gasification are used to illustrate flame characteristics. A review of methods for modeling fixed, fluidized, and entrained beds is included together with a summary of data for evaluation of predictive methods.

4. Chapters 9, 10, and 14 provide the fundamental equations and back-ground for turbulent combustion systems.
5. Chapters 11–13 and 15 document in some detail the approach and theory for the interactions between chemistry and turbulence in reacting systems encompassing gaseous flames, particle-laden systems, and pollutant formation in these systems.

This work has formed the basis for model development at our Combustion Laboratory. These fundamentals were treated generally in the first book. The treatment in this new book for governing equations and for radiation is not greatly changed. However, significant new material is added for chemically reacting, turbulent, heterogeneous systems and for formation of nitrogen oxide pollutants.

Included in several of these chapters are comparisons of measurements with predictions from a comprehensive model developed at the Combustion Laboratory. These comparisons demonstrate the state of development of the comprehensive code which, in the view of the authors, is suitable for some practical application to pulverized-coal systems. While this method has been developed principally for pulverized-coal combustion and gasification, this and similar methods have far more general applicability. The code is directly applicable to nonreactive gaseous and particle-laden flows and to reactive gaseous systems. Progress is being made for application to coal slurries and only minor work would be required for use with liquid fuel flames.

The authors recognize the several uncertainties in constructing and applying models of this nature. Coal is a very complex heterogeneous substance whose structure and behavior are highly variable and are not well known. Pyrolysis and oxidation of coal are dependent upon coal type, size, size distribution, temperature history, etc., thus making generalization difficult. In addition, the complexities of turbulent recirculating flows, turbulent reacting flows, and turbulent two-phase flows are not fully resolved. Only recently has formal computation of recirculating flows been possible. Suitable generalized parameters for the turbulence models are not well-established. The very important interactions of turbulence with chemical reaction are even less well-developed. Significant uncertainties also pertain to the effects of gas turbulence on the motion of the particles. It is known that the random gas fluctuations are a major force in dispersing particles, but treatments of this effect are in the formative stages.

Finally, there has been little previous attempt to develop models which consider several of these aspects jointly. For this reason, models of pulverized-coal systems must be validated by comparison with experimental measurements. Measurements of outlet gas or char composition or temperature alone are not suitable for this model validation. Profiles of at least time-mean, local properties from within the reactor provide a more acceptable basis for the model evaluation.

As a part of the research program being conducted by the authors and colleagues such measurements are being obtained, some of which are presented and compared with model predictions in this book. These measurements include data for laminar, premixed coal dust/air flames and methane–air flames, as well as spatially resolved measurements of gas and char composition from inside entrained gasifiers and combustors. Further, profile measurements have been and are being made for particle-laden, recirculating jet flows without chemical reaction. Comparison of these measurements and those of several other investigators with model predictions will permit evaluation of particle and gas-flow effects in the absence of chemical reaction complications.

Much of the work upon which this book is based has been supported at our laboratory by federal and industrial organizations. We especially express our appreciation to The United States Department of Energy (Morgantown Energy Technology Center and Pittsburgh Energy Technology Center), the Electric Power Research Institute (Fossil Fuels Division), the National Science Foundation, the Tennessee Valley Authority, and the Utah Power and Light Co., for contract or grant support related to measurements and model development for pulverized-coal systems. We further acknowledge the College of Engineering and Technology and Research Division of Brigham Young University for financial research support and for assistance in preparation of the manuscript. We are also grateful to Elaine Alger for typing and editorial work on the manuscript and to Mr. Daniel Gleason for preparation of the illustrations. Appreciation is also due Exxon Corporation, Los Alamos National Laboratory, and Morgantown Energy Technology Center for financial support for preparation of parts of these materials which were used in technical lectures given by the author(s) at each of these institutions. Much of the materials of Chapters 1–5 were initially prepared at the request of Exxon Corporation, who kindly gave permission for inclusion of these materials in this book. We gratefully acknowledge our faculty colleagues and graduate students at the Combustion Laboratory for significant research contributions related to much of the material in this book.

Contents

Behold, all ye that kindle a fire, that compass
yourselves about with sparks: walk in the light of
your fire, and in the sparks that ye have kindled.
                                    Isaiah 50:11

INTRODUCTION

1.1. OBJECTIVES

This book deals with reaction processes involving coal, char, coal–water mixtures, and other solid fossil fuels. Properties and uses of these solid fossil fuels are treated; physical and reaction processes of coal particles are also considered and modeled. Then, these results are applied to the description of coal processes.

Key objectives for preparation of this book have been to:

1. Provide a review of the existing and potential uses of coal and the processes most commonly applied.
2. Identify the general chemical and physical properties of solid fossil fuels, emphasizing the complexity and variability of these natural materials.
3. Summarize major issues being addressed in the increasing uses of coal and other solid fossil fuels.
4. Characterize effects of key variables such as particle size, heating rate, temperature, pressure, and oxidizer type on coal particle reaction rates.
5. Outline useful existing methods for modeling of coal particle reactions in a turbulent environment including ignition, devolatilization, gas-phase reaction of volatiles, and heterogeneous oxidation processes.
6. Identify the nature and controlling processes of coal dust flames in various coal processes, including direct combustion and coal gasification.
7. Provide a fundamental foundation for a description of complex, reacting particle-laden systems through treatment of turbulence and its interaction with chemical reactions, radiation and related basic topics.

1

8. Outline general methods for modeling of turbulent coal reaction processes, and illustrate by application of one-dimensional and two-dimensional models.
9. Provide example problems that illustrate application of the computational methods.
10. Apply results to various coal processes.

1.2. SCOPE

The entire area of coal reaction processes is very extensive. This field of study includes or interacts with such topics as:

1. The origin and geologic nature of coal.
2. The chemical and physical properties and classification of coal.
3. The relationship of coal structure and composition to coal reaction processes.
4. The relationship of coal to other solid and solid-derived fossil fuels, such as oil shale or solvent-refined coal.
5. Thermal devolatilization of coal and its dependence on coal type, particle size, heating rate, temperature, etc.
6. The nature and chemical composition of coal volatiles and their dependence on coal type, heating rate, temperature, etc.
7. The chemical reaction of coal volatiles in the turbulent gas phase, including formation of soot, cracking of hydrocarbons, etc.
8. The formation of char during devolatilization, including swelling, softening, cracking, formation of internal surfaces, etc.
9. The reaction of char particles, including oxidizer diffusional processes internal and external to the particle, effects of volatiles on transpiration, surface reaction, and product diffusion.
10. The turbulent flow, dispersion, vaporization, and reaction of coal slurries.
11. Formation and control in a turbulent environment of a variety of pollutant species, including oxides of nitrogen and their precursors, oxides of sulfur, oxides of carbon, potentially carcinogenic hydrocarbons, carbon dioxide, volatile trace metals, small particulates, etc.
12. Radiative processes of coal and its solid products (i.e., soot, ash, slag, and char) and gaseous products (e.g., CO_2 and H_2O).
13. Formation of ash and slag particles, their change in particle sizes, and their control and removal.
14. Interaction with walls and surfaces, including formation of ash or slag layers.

15. Particle–gas interactions including convective and radiative heat transfer, reactant and product diffusion, and particle motion in the turbulent gas media.
16. Design and optimization of coal reaction processes.
17. Others.

In this book, solid fossil fuels and coal–water mixtures (CWM) are emphasized, and emphasis is further placed on finely pulverized coal reaction processes. This form of coal is dominant in existing coal processes. Treatment is also given to processing of larger coal particles. Reactions of coal are considered in some detail. The application of these reaction sequences to modeling of coal reaction processes is treated and illustrated with one- and two-dimensional models.

1.3. APPROACH

An effort is made to apply the foundations of combustion to the treatment of coal reaction processes. An attempt is also made to identify the relationships among coal properties, basic coal particle reactions, and the behavior of coal flames in turbulent flows common to practical processes.

Selected exeperimental data are presented to illustrate typical coal behavior and reaction rates. Then, existing methods for correlating these data are developed and generalized where possible. These correlative and predictive methods are integrated into methods for describing coal reaction processes. Sample problems are included to illustrate the application of coal modeling methods.

1.4. GENERAL REFERENCES

The multivolume series *Chemistry of Coal Utilization*, edited by Lowry, published about two decades ago, still provides a good description of coal origins, properties and reactions, and coal processes, including storage, handling, processing, and products. More recently, Elliott (1981) has edited the second supplementary volume of *Chemistry of Coal Utilization*. The book, *Pulverized Coal Combustion*, by Field *et al.*, published in England in 1967, provides a good general description of the modeling of coal reaction. The book *Pulverized Coal Combustion and Gasification*, edited by Smoot and Pratt and published in 1979, provides an advanced summary of the foundations and methods for describing and modeling pulverized-coal flames and reaction processes. In addition to the above books, several review articles have been prepared on various aspects of coal combustion. Table 1.1 summarizes several of these recent reviews, which by themselves

TABLE 1.1. Summary of Selected Surveys in Coal Combustion in the Past Decade

Author(s)	Date	Topic	Subject area(s)
Grumer	1974	Explosions	Review of practices for extinguishment of coal dust explosions, including rock dust, water, and chemicals; characteristics of explosions (63 refs.).
Anson	1976	Fluidized beds	Descriptive review of the state of development of fluidized beds for generating power from coal, including aspects of combustion, emissions, and cycles (50 refs.).
Anthony and Howard	1976	Coal devolatilization	Basic review of thermal decomposition of coal in inert gases and in hydrogen, including product yield, and pyrolysis models (180 refs.).
Beér	1976	Fluidized beds	Basic and applied aspects of fluid-bed coal combustion, including theory, related kinetic data, sulfur and nitrogen emissions, and heat transfer (69 refs.).
Breen	1977	Boiler	Impact of furnace/boiler modifications (e.g., reduced excess air) on efficiency and pollutant level (15 refs.).
Essenhigh	1977	Particle, dust reactions	Basic discussion of coal particle reactions: coal dust flame propagation (160 refs.).
Godridge and Read	1976	Large furnace heat transfer	Combustion and heat transfer in large coal and oil boilers, including boiler characteristics, burner designs, heat-transfer models, and furnace measurements (58 refs.).
Sarofim and Flagen	1976	NO_x pollution control	Summary of NO_x pollutant formation processes and control methods for sources (95 refs.).
Littlewood	1977	Gasification	Theory and applications for gasification of coal and other fossil fuels, including a description of several commercial and development systems (119 refs.).
Smoot and Horton	1977	Dust flames	Review of measurements and theory for propagation of premixed, laminar coal dust flames (86 refs.).
Belt and Bissett	1978	Flash and hydropyrolysis	Review of past 15-years work in pyrolysis; eight heating methods reviewed; key variables identified; need for additional data recognized (35 refs.).

Gibson	1978	Coal properties	The constitution of coal and its relationship to pyrolysis, liquefaction, gasification, and hydrogenation processes (46 refs.).
Laurendeau	1978	Heterogeneous kinetics	Detailed review of char combustion and gasification, including coal characteristics, surface phenomena, and kinetics (214 refs.).
Macek	1978	Pollution from combustor types	Stoker, cyclone, pulverized combustion comparisons for NO_x, fly-ash, SO_x, and coal/oil dispersions (34 refs.).
Sarofim and Hottel	1978	Radiation	Summary of optical properties of coal and its products: ash, char, and soot: application to solid-fossil fuels; state of prediction of radiation in coal flames (77 refs.).
Krazinski et al.	1979	Coal dust flames	Review of measurements and models in dust flames; development of a model for dust flame prediction (92 refs.).
Wall et al.	1979	Coal mineral matter	Mineral matter in coal and its effect on large-boiler performance: characteristics of coal ash release during grinding, mining, heating, and combustion; wall deposits and radiative properties (85 refs.).
Wendt	1980	NO_x pollution	A review of coal combustion mechanisms and pollutant formation in furnaces. Furnace-flame types are illustrated. Effects of staging and other variables on NO_x levels; basic coal particle reaction processes (61 refs.).
Smoot	1980	Theory, modeling	Review of foundations and submodels for pulverized-coal turbulent flames: summary and comparison of pulverized-coal models developed in past decade; identification of major model needs (82 refs.).
Howard et al.	1981	Coal devolatilization	Review of experimental methods, data, and modeling of coal devolatilization, applied particularly to gasification (312 refs.).
Smoot and Hill	1983	Critical needs in combustion	An extensive review of combustion research work over the past five years, including coal combustion, with recommendations for further work (258 refs.).
Smith et al.	1982	Char oxidation	Review of data correlation methods, and experimental data (non-U.S. coals) for oxygen reaction heterogeneously with chars and cokes (55 refs.).

contain nearly 2000 references to recent studies of coal combustion and related topics.

A review of recent CWM work can be found in the *Fourth through Sixth International Symposia on Coal Slurry Combustion*, which were held in 1982, 1983, and 1984.

Where no wood is, there the fire goeth out:
Proverbs 26 : 20

SOLID FOSSIL FUELS (MOSTLY COAL): PROCESSES AND PROPERTIES

2.1. COAL AVAILABILITY AND USES

Coal is the most abundant fossil fuel known to exist in the United States. Present recoverable reserves are estimated to be nearly 300 billion tons (DTA, 1979), with potential total reserves far in excess of this amount. Deposits of coal are located in most regions in the United States as illustrated in Figure 2.1. Estimated production of coal by the year 2000 will be nearly two billion tons per year with the bulk being consumed through combustion processes as illustrated in Figure 2.2. Thus, present recoverable reserves are adequate to meet the nation's coal needs for many decades and potentially much longer.

Most of the consumption of coal by the year 2000 will be for electric power generation as illustrated in Figure 2.2, with industrial consumption of coal for steam and heat and for metallurgical processes being other major uses. Areas for potential expanding uses of coal, while hard to project in magnitude at the present time, include the following:

1. Fluidized-bed combustion
2. Coal exports
3. Combined cycle power generation
4. Direct gasification for medium or high BTU gas
5. Liquefaction
6. Chemical feedstocks
7. Improved solid fuels (e.g., solvent-refined coal)
8. MHD (magnetohydrodyamics) power generation
9. Fuel cells

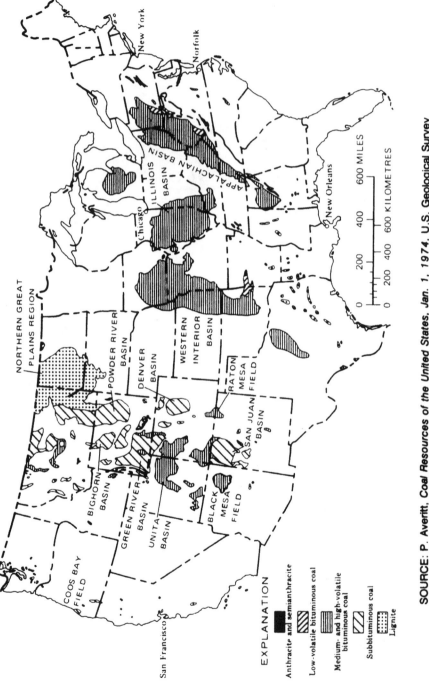

EXPLANATION

Anthracite and semianthracite

Low-volatile bituminous coal

Medium- and high-volatile bituminous coal

Subbituminous coal

Lignite

SOURCE: P. Averitt, *Coal Resources of the United States, Jan. 1, 1974*, U.S. Geological Survey Bull, 1412, at 5 (1975).

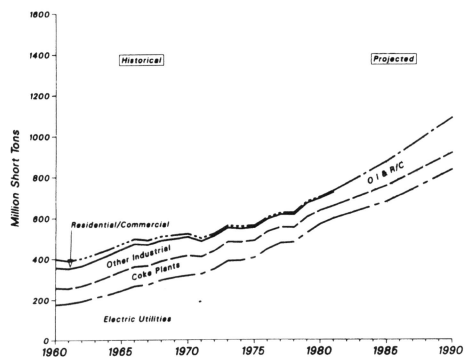

Figure 2.2 Combustion of coal and lignite by end-use sector [from Department of Energy Report DOE/EIA-0121 (82/4Q), 1982].

2.2. COAL PROCESSES

Most of the coal presently being consumed is by direct combustion of finely pulverized coal in large-scale utility furnaces for generation of electric power, and is likely to remain this way through this century (see Figure 2.2). However, there are a large number of other processes for the conversion of coal into other products or for the direct combustion of coal (Lowry, 1963; IGT, 1979; Pitt and Millward, 1979; Elliott, 1981). These processes can be classified by process type (or end product) and by coal particle size or temperature, etc. For example:

Direct Combustion

Pulverized coal combustors (smallest particles)
Fluidized bed (medium-sized particles)
Fixed bed (e.g., stokers) (larger particles)

<u>Gasification</u>

Entrained bed (smallest particles)
Fluidized bed (medium-sized particles)
Fixed or moving bed (larger particles)
Others (molten bath)

<u>Carbonization and Coking</u>

Low temperature (750–975 K)
Medium temperature (1025–1175 K)
High temperature (1175–1325 K)

<u>Liquefaction</u>

Pyrolysis
Extraction

<u>Other</u>

MHD
Fuel cell

This is not intended to be a complete review of coal processes, but to illustrate their wide variability. Table 2.1 summarizes some of the characteristics of these coal processes, including extent of use, scale size, and the coal particle sizes employed.

2.3. COAL SLURRIES

Interest in the use of coal slurries, and particularly coal–water mixtures (CWM), centers largely around use of coal slurries as a replacement fuel for existing oil-fired boilers. In addition, CWM may also be useful in ignition of pulverized coal in existing furnace systems, and as a substitute fuel in gas turbines. Research results for coal–oil mixtures (COM) have shown little economic incentive for COM as an oil replacement fuel. Further, full-scale utility boiler test results have uncovered unexpected problems in handling of ash fouling with the use of coal–oil slurries. Efforts to develop coal–water mixtures are much more recent and a suitable data base for this fuel does not yet exist. However, early test results have been encouraging. deLesdernier *et al.* (1982) show for a specific 200-MW$_e$ oil-fired boiler that CWM may have some economic incentive. They note that the critical technical issues include development and integration of advanced coal cleaning methods for up-front reduction of sulfur

TABLE 2.1. Summary of Selected Coal Processes

Process type	Description	Extent of coal use in U.S. (% of total used)[a]	Commercial use	Scale size (TPD)[d]	Coal types	Coal size
Direct combustion	Burning coal to produce electricity, heat, and steam.	73[3]–80[4]%				
Power station	Commercial electrical production.					
Pulverized	Rapid burning of finely grained coal.		Common	1000–10,000[5]	All	0.01–0.025 mm[5]
Fluidized bed.	Well-stirred combustion.		Pilot plant	2000–8000[5]	All	0.15–0.6 cm[5]
Stoker	Mechanically fed, fixed-bed.		Small	100[5]	Noncaking	1–5 cm[5]
MHD	Combustion energy capture by magnetic fields.		Laboratory	800–4000[6]		
Coal/Oil mixture (COM)	Burning coal/oil mixtures in oil furnaces.		Demonstration	TPC ce[*]		
Industrial heat/steam	Industrial-plant power providers.	8–11%				
Pulverized			Small	1–100		Same as above
Fluidized bed			Pilot	1–100		
Stoker			Common	1–100		
COM			Demonstration	1–100		
Domestic/commercial	Hand-stoked space heating.	1%		0.005–0.05	Noncaking	3–10 cm
Transportation	Fuel for railroads.	0.01–0.02%		0.1–1	Noncaking	

TABLE 2.1. (*Cont.*)

Process type	Description	Extent of coal use in U.S. (% of total used)*	Commercial use	Scale size (TPD)^d	Coal types	Coal size
Gasification[1]	Converting coal to low, medium, or high BTU gas for use as fuel or feedstock.	0				
Fixed bed[f]	Reacting coal in a fixed bed.					
Single stage						
Dry ash			*Common	60–800	Noncaking	0.3–0.5 cm[7]
Two stage						
Dry ash			Common	100	Noncaking	0.3–0.5 cm[7]
Slagging			Test units	20	Noncaking	0.3–0.5 cm[7]
Fluidized bed	Reacting coal in a well-mixed reactor.		Common		All	0.8 cm[7]
Entrained flow	Gasifying coal rapidly.					
Single stage			Common	600	All	0.01–0.02 mm
Two stage			Pilot plant	5–60	All	0.01–0.02 mm
Tumbling bed	Mechanically mixing the bed.		Pilot plant	44	All	0.3–0.2 cm
Molten bath	Gasifying coal in a hot liquid.					
In situ	Reacting coal in place, underground.		Pilot tests			
Liquefaction[2]	Converting coal to liquid fuel.	0				
Pyrolysis	Removal of volatile compounds by heating.					
Gentle			Pilot plant[a]	25	1–2 cm	
Flash			Commercial[b]			0.01–0.025 mm

Process	Description	Status		Size
Direct liquefaction	Hydrogenating, dissolving, and heating the coal to derive a liquid.			
Solvent extraction		Pilot plant	50–250	0.01–0.025 mm
Catalytic		Pilot plant	10–600	0.01–0.026 mm
Indirect liquefaction	The gasification of coal followed by catalytic recombination.	Commercial[c]	(see gasification)	
Coking	Carbonizing coal for use in metallurgical processes.	10^4–15%[b]	Common	All	0.15–0.6 cm[8]

*Coal equivalent.
[a]Was once widespread as a feedstock for chemicals.
[b]Commercial application is only found in Europe.
[c]Commercial application is only found in South Africa.
[d]TPD = tons/day.
[e]Very adaptable, exhibits high heat transfer in bed as well as low-level pollutant products.
[f]Used commonly in other parts of the world for the production of BTU gas.

1. Data from Bodle and Schora (1979).
2. Data from Nowacki (1979).
3. Data from Wilson (1980).
4. Data from McNair (1980).
5. Data from Essenhigh et al. (1979).
6. Data from Schweiger (1979).
7. Data from OTA (1979).
8. Data from Lowry (1945).

TABLE 2.2.   Technical Issues Relating to CWM
Development

Coal preparation
  Coal cleaning (S, ash)
  Particle size distribution
  Ultrafine grinding
CWM Characterization
  Coal loading
  Slurry stability
  Slurry rheology
Burner design
  Slurry dispersion
  Droplet size distibution
  Flame stability
  NO_x formation
  Nozzle erosion
Full-scale utility demonstration
  Slurry supply/stability
  Burner reliability
  Boiler derating
  Nozzle erosion
  Pollutant emissions
  Combustion efficiency
  Fouling, ash removal
  Boiler availability

---

and ash levels and demonstration of high combustion efficiency, reliability, and performance in a utility boiler.

The total maximum market for retrofitting of oil-fired utility boilers is about 50,000 MW_e in the United States, excluding world markets. Major market areas are the northeast and southeast parts of the country. Some of these oil units are in states (e.g., California) where coal cannot be burned and others are too old to be considered for retrofit. Several companies are engaged in development of CWM slurries, with pilot plants operating or nearing operation with production capacities of 25–120 tons/day. Some of the issues related to development of CWM are summarized in Table 2.2.

2.4.   ISSUES IN INCREASING USE OF COAL

Increasing use of coal in the United States presents many technical problems. Table 2.3 summarizes, without regard to any priority, related technical problem areas that are of concern. In addition to these technical problems, a host of other problems have a direct bearing on the consumption of coal, including

TABLE 2.3. Some Technical Issues and Problems Arising from the Expanding Use of Coal

	Problem area	General description	Related references
1.	Environmental concerns	Need for control of emissions from coal processes, including oxides of nitrogen, oxides of sulfur, carbon dioxide, fine particulates soot, heavy hydrocarbons, trace elements, etc.	OTA (1979); Hansen et al. (1978); Wendt (1980); Sarofim and Flagen (1976); Macek (1978)
2.	Safety and explosions	Need for methods to control explosions and fires in mines, utility and industrial boilers, and in coal storage facilities.	Grumer (1975); Burgess et al. (1979)
3.	Combustion/conversion efficiency	Need to maintain high combustion efficiency as process variables change for pollutant control, coal variability, etc.	Johnson and Sommer (1980)
4.	Ash/slag	Need to control ash and slag deposits and removal, and associated maintenance of surfaces, corrosion, etc., particularly for the variety of coals being considered, some with very high ash levels.	Wall et al. (1979)
5.	Process design/optimization	Need for development and verification of more efficient methods for designing, scaling, and optimizing coal conversion processes.	Smoot (1980)
6.	Basic process data	Need for relating coal characteristics (composition, structure, etc.) to conversion/combustion characteristics, including radiative properties, pyrolysis rates, char oxidation rates.	Gibson (1978); Spackman (1980); IGT (1979)
7.	Use of alternate solid fossil fuels	Basic information on conversion and combustion an alternate solid fuels and fuel products such as oil shale, char, solvent-refined coal, peat, tar sands, etc.	McCann et al. (1977); McRanie (1979); Granoff and Nuttal (1977)
8.	Coal/water mixtures	For retrofitting existing oil-fired furnaces and for direct coal ignition systems. Needs for coal preparation, CWM slurry characteristics, and full-scale boiler tests.	Fourth and fifth International Symposia on Coal Slurry Combustion (1982, 1983).

mine safety, labor availability and relations, water availability, feedstock trans-
portation, mine equipment manufacture, capital availability, and control of solid
and liquid wastes (OTA, 1979). While these issues are well beyond the scope
of this book, they will continue to have a major impact on the consumption of
coal.

2.5. COAL CHARACTERISTICS

2.5a. Formation and Variation

Coal is an inhomogeneous organic fuel, formed largely from partially
decomposed and metamorphosed plant materials. Formation has occurred over
long time periods, often under high pressures of overburden and at elevated
temperatures. Differences in plant materials and in their extent of decay influences the components present in coals, such as vitrain (from "vitro" meaning
glass), clarain (from "clare" meaning clear or bright), durain (from "dur" meaning
hard or tough), and fusain (meaning charcoal). Descriptions of such coal
components are part of the science of petrography (Hendrickson, 1975).

Coals vary greatly in their composition. Of 1200 coals categorized by the
Bituminous Coal Research Institute, no two had exactly the same composition
(Hendrickson, 1975). Typical compositions (mass percentages) of coal include
65–95% carbon, 2–7% hydrogen, up to 25% oxygen and 10% sulfur, and 1–2%
nitrogen (Essenhigh, 1977). Inorganic mineral matter (ash) as high as 50% has
been observed, but 5–15% is more typical. Moisture levels commonly vary from
2 to 20%, but values as high as 70% have been observed.

The process of conversion of plant materials, such as peat, to coal is called
"coalification" and takes place in stages producing a variety of coal products.
Hendrickson (1975) provides the following description of some of these coal
types:

> 'Lignite, the lowest rank of coal, was formed from peat which was compacted
> and altered. Its color has become brown to black and it is composed of recognizable
> woody materials imbedded in pulverized (macerated) and partially decomposed
> vegetable matter. Lignite displays jointing, banding, a high moisture content, and a
> low heating value when compared with the higher coals.
>
> Subbituminous coal is difficult to distinguish from bituminous and is dull, black
> colored, shows little woody material, is banded, and has developed bedding planes.
> The coal usually splits parallel to the bedding. It has lost some moisture content,
> but is still of relatively low heating value.
>
> Bituminous coal is dense, compacted, banded, brittle, and displays columnar
> cleavage and a dark black color. It is more resistant to disintegration in air than are
> subbituminous and lignite coals. Its moisture content is low, volatile matter content
> is variable from high to medium, and its heating value is high. Several varieties of
> bituminous coal are recognizable.

Anthracite is the highly metamorphosed coal, is jet black in color, is hard and brittle, breaks with a conchoidal fracture, and displays a high luster. Its moisture content is low and its carbon content is high.

Neither peat nor graphite are coal, but they are the initial and end products of the progressive coalification process.'

Lowry (1963), Hendrickson (1975), Given (1964), Neavil (1979), Spackman (1980), Hamblen *et al.* (1980), IGT (1979), and Elliott (1981) discuss further the origin and/or characteristics of coal.

2.5b. Coal Classification

Efforts have been made to classify the almost limitless number of coals into broad classifications, and to relate similarities among coals to their potential behavior in coal conversion processes. Possibly, the most common of these is the ASTM (American Society of Testing Materials) Classification, which is based upon fixed carbon level and heating value. Figure 2.3 illustrates the general characteristics of 12 such coal groups, ranging from soft lignite to very hard meta-anthracite.

Classifications have also been based on petrographic parameters. Lowry (1963) and Elliott (1981) summarize several other systems for general classification of coal including:

1. English National Coal Board System, based upon percent of proximate volatiles and Gray–King coking properties.
2. International System for hard coals and brown coals.
3. Mott System, based upon volatile matter, heating values, and ultimate (O–H) analysis.

None of these approximate systems is able to deal with the complex structural and compositional differences in coals. Studies by Spackman (1980) and Hamblen *et al.* (1980) are attempting to provide methods to predict the behavior of coals during conversion processes, from a knowledge of coal composition.

2.5c. Physical and Chemical Properties of Coal

Modeling of coal conversion processes requires data for physical properties of coal, such as thermal conductivity, specific heat, density, etc. Table 2.4 summarizes selected values of common physical properties of coal, including specific heat, specific gravity, thermal conductivity, and swelling index. Heating value data are shown in Figure 2.3. These properties will vary among coals, even of common rank, and will often be related to temperature and moisture content. Lowry (1963), Hendrickson (1975), IGT (1979), Spackman (1980), Hamblen (1980), and Elliott (1981) discuss several other physical, mechanical,

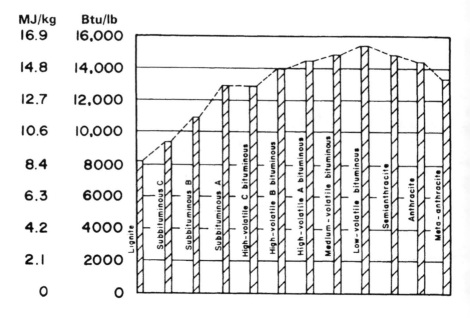

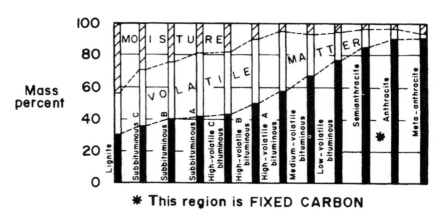

*** This region is FIXED CARBON**

Figure 2.3. Energy content and composition of coals according to ASTM coal rank: (a) heating value; (b) proximate analysis. (Figure used with permission from Hendrickson, 1975.)

thermal, and chemical properties of coal, including fate of trace elements, grindability, friability, compressive strength, dustiness, electrical resistivity, plasticity, optical density, indices of refraction, reflection, and absorption, magnetic susceptibility, electrical conductivity, dielectric constants, forms of sulfur, etc.

Properties of char are even more variable, since such properties are a function of the nature of the conversion process from which they were produced, in addition to being related to the coal from which they were derived. Chars are generally richer in carbon and leaner in hydrogen than the parent coal and are often porous and more regular in shape, having softened during the conversion process.

The composition of coal is traditionally characterized by ASTM proximate analysis or ASTM ultimate analysis. The former determines only the moisture content (by drying), percent volatiles (from inert devolatilization at about 1200 K), ash (residual after complete combustion in air), and fixed carbon (by difference). Coal rank vs. proximate analysis was shown in Figure 2.3. Proximate analysis for a selected variety of coals is shown in Table 2.5. In these coals alone, the percent of volatiles varies from 8.8 to 45.5% by weight. The char shown has only 2.4% proximate volatiles.

TABLE 2.4.   Typical Values for Selected Physical Properties of Coal[a]

A. Specific Heats of Air-Dried Coals
1. Proximate analysis

Coal sample source	Moisture (%)	Volatile matter (%)	Carbon (%)	Ash (%)
West Virginia	1.8	20.4	72.4	5.4
Pennsylvania (bituminous)	1.2	34.5	58.4	5.9
Illinois	8.4	35.0	48.2	8.4
Wyoming	11.0	38.6	40.2	10.2
Pennsylvania (anthracite)	0.0	16.0	79.3	4.7

2. Mean specific heat for °C temperature ranges
Temperature range (°C)

Coal sample size	28–65	25–130	25-177	25–227
West Virginia	0.261	0.288	0.301	0.314
Pennsylvania (bituminous)	0.286	0.308	0.320	0.323
Illinois	0.334			
Wyoming	0.350			
Pennsylvania (anthracite)	0.269			

B. Specific Gravity[b] of Coal

Probable rank	Specific gravity
Anthracite	1.7
Semianthracite	1.6
Bituminous	1.4
Subbituminous	1.3
Lignite	1.2

TABLE 2.4. *(Cont.)*

Coal type	C. Thermal Conductivity[c] Temperature (K)	$k\,J\,s^{-1}\,m^{-1}\,K^{-1}$
Monolithic anthracite	303	0.2–0.4
Monolithic bituminous	303	0.17–0.3
Pulverized bituminous	Ambient	0.10–0.15

D. Average Free-Swelling Index Values for Illinois and Eastern Bituminous Coals Rank	Coals	ASTM free-swelling index
High-volatile C	Illinois No. 6	3.5
High-volatile B	Illinois No. 6	4.5
High-volatile B	Illinois No. 5	3.0
High-volatile A	Illinois No. 5	5.5
High-volatile A	Eastern	6.0–7.5
Medium volatile	Eastern	8.5
Low volatile	Eastern	8.5–9.0

[a]Data from Hendrickson (1975). (Table from Smoot and Pratt, 1979.)
[b]Specific gravity has been shown by Lowry (1963) to be a function of hydrogen content.
[c]Thermal conductivity usually increases with increasing apparent density, volatile matter content, ash content, temperature, and probably with moisture content. Coal is thermally anisotropic with k greater perpendicular to bed.

ASTM ultimate analysis gives elemental analyses for carbon, hydrogen, nitrogen, sulfur, and oxygen, the latter often determined by difference. The residual mineral matter is shown as ash. Ultimate analyses for selected coals and a char are also shown in Table 2.5. Ash in coals has been shown to contain significant amounts of some elements, together with trace amounts of several elements, as shown in Table 2.6 for selected coals. Sarofim et al. (1977) discuss particle sizes of ash from pulverized coals and note that some ash components are volatile at higher temperatures.

Particle sizes of coal dust vary greatly, depending upon grinding technique and desired application. Typical size distributions for a fluid-bed gasifier application and a utility boiler application are shown in Table 2.7. Size distribution for one specific char is also shown in Table 2.7.

2.5d. Structural Characteristics of Coal

It is thought that coal structure is highly planar and layered with a pore volume of 8–20%. Hendrickson (1975), Anthony and Howard (1976), Lowry (1963), Given (1960), Solomon (1980), Hamblen et al. (1980), Spackman (1980), IGT (1979), and Elliott (1981) discuss chemical properties of coal, including details of proximate and ultimate analyses, plastic properties of coal, coal hydrogenation and halogenation, solvent extraction of coal components, and properties of minerals and coal chemical structure.

A recent idealized "model" for the chemical organic structure of coal is illustrated in Figure 2.4. The model shown from Solomon (1980) is similar to that prepared by other recent investigators (Wiser, 1975; Heredy and Wender, 1980) and is based on information from infrared measurements, nuclear magnetic resonance, ultimate and proximate analyses, and pyrolysis data. Mineral matter can exist as occlusions within the base organic-framework structure. This is not the specific structure of any particular coal, but only a model for interpreting and correlating coal conversion data. Coal structure is undoubtedly more varied and complex than even this structure suggests.

Infrared (Fourier transform) measurements provide quantitative concentrations for the following constituents in the raw coal or its components: hydroxyl, aliphatic or hydroaromatic hydrogen, aromatic hydrogen, aliphatic carbon, and aromatic carbon. From this and other information, the infrared spectrum (IR) of the parent coal or its constituents can be synthesized from the basic organic constituents, as illustrated in Figure 2.5. Figure 2.5 compares the measured spectra for a high-volatile bituminous (Pittsburgh) coal with that synthesized from the concentrations of the chemical constituents. The method for synthesizing spectra is discussed by Solomon (1980).

Presently, methods for design and analysis of existing coal conversion processes do not make significant use of this kind of information on the structure of coals. However, with increased interest in the use of coal in the United States, more attention is being given on the relationship of coal structure to coal reaction and conversion processes. Gibson (1978) recently discussed the relevance of coal constitution to coal conversion processes, including carbonization, liquefaction, gasification, pyrolysis, extraction, and hydrogenation. Given and Biswas (1979) have conducted an investigation of the relation of coal characteristics to liquefaction of coal, while Spackman (1982) has considered this issue with respect to production of clean energy fuels. Solomon (1980) has developed a method for predicting the rate of thermal decomposition of coal and the distribution of products based on measured IR coal structure. This method will be discussed in greater detail in the next subsection in order to illustrate the potential results that might be realized from relating the chemical structure of coal to its conversion.

2.5e. Mineral Matter Removal

Environmental constraints together with development of coal-based fuels for oil substitution have increased interest in cleaning of coal. Singh *et al.* (1982) and Liu (1982) have provided recent reviews of physical and chemical methods for cleaning of coal. Reduction in mineral and sulfur content of coals is of particular concern.

TABLE 2.5. Typical Proximate and Ultimate Analyses of Coals and Char[a] (Principally from Smoot and Pratt, 1979)[b]

Coal I.D. rank	Utah Church Mine bituminous	Pittsburgh[b] bituminous (high volatile)	Pittsburgh[b] bituminous	Sewell bituminous (medium volatile)	Pocahantas[b] bituminous (low volatile)
Moisture (%)	2.5–2.7	2.0	1.0	—	1.9
Proximate (%)					
Volatiles	44.1–45.5	36.6	28.9	29.2	16.3
Fixed carbon	42.6–44.2	55.4	63.2	63.9	75.6
Ash	9.2–9.5	6.0	6.9	6.9	6.2
Ultimate (%)					
Carbon	69.8–71.5	77.5	80.6	81.4	84.2
Hydrogen	5.5–5.6	5.3	4.9	4.8	4.3
Nitrogen	1.4–1.5	1.5	1.5	1.6	1.2
Sulfur	0.4–0.7	1.2	0.7	0.7	0.7
Oxygen	11.2–13.2	8.5	5.4	4.6	3.4
Ash	9.2–9.5	6.0	6.9	6.9	6.2

Coal I.D. rank	Anthracite[b] (low volatile)	Illinois coal bituminous	Illinois coal char	North Dakota lignite	Wyoming subbituminous	Kentucky bituminous
Moisture (%)	1.3	10.1	0.9	29.9	27.8	8.6
Proximate (%)						
Volatiles	8.8	35.9	2.4	29.5	32.9	35.2
Fixed carbon	71.8	46.7	76.1	33.4	34.3	41.5
Ash	18.1	7.3	20.6	7.2	5.0	23.3
Ultimate (%)						
Carbon	73.2	68.3	74.0	69.7	76.3	61.0
Hydrogen	3.1	5.0	0.7	3.8	4.4	4.4
Nitrogen	0.9	1.3	1.0	1.9	1.1	1.4
Sulfur	0.9	3.5	3.3	1.1	0.5	4.3
Oxygen	3.8	13.8	0.2	13.2	10.8	5.6
Ash	18.1	8.1	20.8	10.3	6.9	23.3

[a] Data courtesy of Bureau of Mines, Brigham Young University, and ERDA.
[b] Data principally from Smoot and Pratt (1979).
[c] From U.S. Bureau of Mines, Pittsburgh, Pa.
[d] From COED Gasification Process, supplied by ERDA, Pittsburgh, Pa. (4th stage product).

TABLE 2.6. Compositions of Typical Ashes[a]

A. Variations in Coal Ash Compositions with Rank

Rank	SiO$_2$ (%)	Al$_2$O$_3$ (%)	Fe$_2$O$_3$ (%)	TiO$_2$ (%)	CaO (%)	MgO (%)	Na$_2$O (%)	K$_2$O (%)	SO$_3$ (%)	P$_2$O$_5$ (%)
Anthracite	48–68	25–44	2–10	1.0–2	0.2–4	0.2–1	—	—	0.1–1	—
Bituminous	7–68	4–39	2–44	0.5–4	0.7–36	0.1–4	0.2–3	0.2–4	0.1–32	—
Subbituminous	17–58	4–35	3–19	0.6–2	2.2–52	0.5–8	—	—	3.0–16	—
Lignite	6–40	4–26	1–34	0.0–0.8	12.4–52	2.8–14	0.2–28	0.1–1.3	8.3–32	—
Utah bituminous[b]	43–48	16–19	3.8–4.2	0.0–1.0	6.5–8.1	0.9–1.1	4.3–4.9	0.4–4.7	3.5–4.1	1.0

B. Range of Amount of Trace Elements Present in Coal Ashes (ppm on Ash Basis)

Element	Anthracites	High-volatiles bituminous	Low-volatiles bituminous	Medium-volatiles bituminous	Lignites and subbituminous	Utah[b] bituminous
Ag	1	1–3	1–1.4	1	1–50	—
B	63–130	90–2800	76–180	74–780	320–1900	700–1500
Ba	540–1340	210–4660	96–2700	230–1800	550–13,900	700–1500
Be	6–11	4–60	6–40	4–31	1–28	5–7
Co	10–165	12–305	26–440	10–290	11–310	7–15
Cr	210–395	74–315	120–490	36–230	11–140	70–100

Cu	96–540	30–770	76–850	130–560	58–3020	62–68
Ga	30–71	17–98	10–135	10–52	10–30	30–70
Ge	20	20–285	20	20	20–100	—
La	115–220	29–270	56–180	19–140	34–90	70–100
Mn	58–220	31–700	40–780	125–4400	310–1030	400
Ni	125–320	45–610	61–350	20–440	20–420	15–30
Pb	41–120	32–1500	23–170	52–210	20–165	35–45
Sc	50–82	7–78	15–155	7–110	2–58	15
Sn	19–4250	10–825	10–230	29–160	10–660	—
Sr	80–340	170–9600	66–2500	40–1600	230–8000	1000–1500
V	210–310	60–840	115–480	170–860	20–250	150
Y	70–120	29–285	37–460	37–340	21–120	50–70
Yb	5–12	3–15	4–23	4–13	2–10	7
Zn	155–350	50–1200	62–550	50–460	50–320	58–64
Zr	370–1200	115–1450	220–620	180–540	100–490	200–300
Cd	—	—	—	—	—	1
Li	—	—	—	—	—	133–155
Nb	—	—	—	—	—	20–30

a Table used with permission from Hendrickson (1975)
b *Utah Power and Light Coal, Church Mine*. Analyses by U.S. Geological Survey, Denver, Co.

TABLE 2.7. Typical Size Distributions for Pulverized Coal[a]

| | A. Fluid-Bed Gasifier[b] | |
Tyler screen	Illinois bituminous	Illinois char
14	15.2	3.3
28	47.0	28.7
48	69.0	54.0
100	81.9	72.1
200	80.2	84.2
325	94.2	91.2
Pan	99.7	99.8

| B. Utility Boiler[c] (Utah Bituminous[d]) | |
Increment size (μm)	Percentage in increment[e]
2.85	0.3–0.5
3.59	0.4–0.6
4.52	0.5–0.7
5.70	0.7–1.0
7.18	1.2–1.6
9.04	2.0–2.5
11.39	3.0–3.7
14.35	4.1–5.1
18.10	5.5–6.9
22.80	7.1–9.0
28.70	9.0–11.1
36.15	10.9–13.3
45.55	10.7–12.4
57.40	11.5–12.1
72.30	9.2–12.4
85.30	5.3–8.3
90+	7.4–12.9

[a]Data from Smoot and Pratt (1979).
[b]Courtesy of U.S. Department of Energy.
[c]Termed 70% through 200 mesh.
[d]Church Mine, BYU, (Brigham Young University) data.
[e]Mass-mean diameter is about 50 μm (Coulter counter measurement)

Ultrafine grinding (micronization) of coal to an average particle diameter of about 10 μm separates much of the mineral matter from the organic coal. Subsequent separation of the coal and mineral matter by differences in specific gravity has led to coals with mineral matter levels of less than 1% and with sulfur levels about half the original value. Reduction in sulfur level by physical methods depends strongly upon the relative proportions of sulfur in the mineral matter (pyritic, sulfates) and that bound in the organic structure. Work is

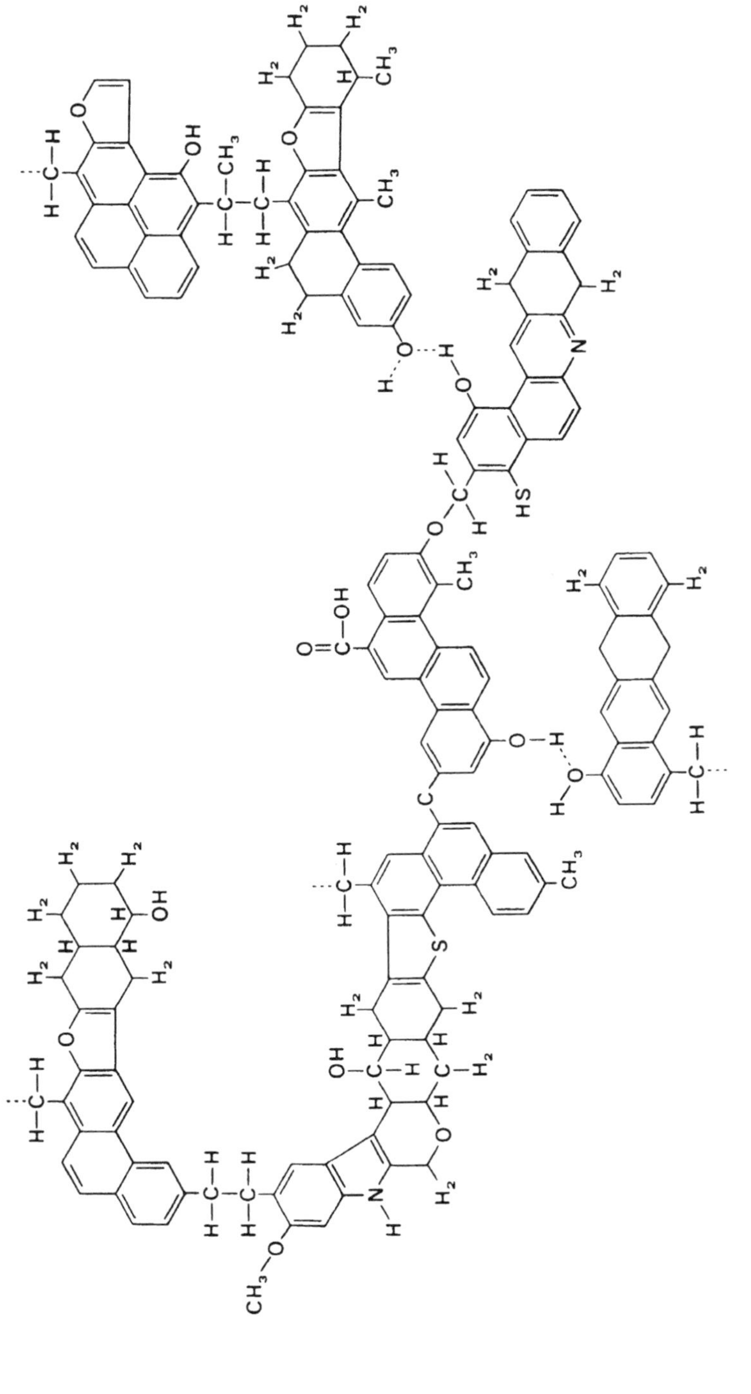

Figure 2.4. Summary of coal structure information in a hypothetical coal molecule. (Figure used with permission from Solomon, 1980.)

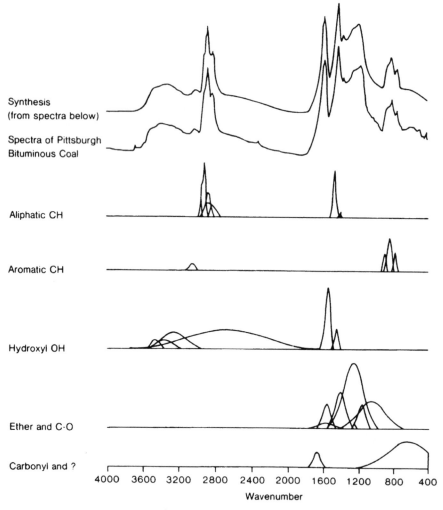

Figure 2.5. Synthesis of infrared spectrum. (Figure used with permission from Solomon, 1980.)

continuing on several physical and chemical coal cleaning processes (Singh *et al.*, 1982; Liu, 1982) in an effort to demonstrate economically viable processes.

2.6. PROPERTIES OF COAL SLURRIES

Coal slurries are more complex than coal in many respects. In addition to the variable physical and chemical characteristics of the coal in the mixture, the properties of the mixture are important. COM and CWM are non-Newtonian fluids whose rheological properties must be considered in burner design. Stability and storability of the mixtures are also important issues. Several companies have or soon will have pilot plant facilities for production of 25–120 tons/day of CWM slurry. Since the slurries must be carefully formulated to achieve acceptable storing and pumping qualities, CWM will likely be produced commercially and delivered to boiler sites. Often, small percentages of additives are used to improve stability and lower viscosity.

Physical properties of interest in CWM are summarized in Table 2.8. Rheological properties for CWM with three typical coals are shown in Figures 2.6–2.8. These measurements were made by observing pressure drop (ΔP) and volumetric flow rate (Q) in a cylindrical tube of specified length (L). Flow curves are plotted on shear stress (τ_w) vs. strain rate ($\dot{\gamma}$) coordinates, where τ_w and $\dot{\gamma}$ are obtained from the pipe flow data as follows:

$$\tau_w = D\Delta P/4L \tag{2.1}$$

$$\Gamma = 8v/D = 4Q/\pi R^3 \tag{2.2}$$

and, from the Rabinowitsch–Mooney relation for generalized Newtonian or non-Newtonian fluids (Hanks, 1980):

$$\dot{\gamma} = \Gamma[3 + d(\ln \Gamma)/d(\ln \tau_w)] \tag{2.3}$$

TABLE 2.8.   Physical Properties of Coal–Water Mixtures

Property	Comments
Coal solids loading	Presently around 70%, with 30% moisture. Difficult to use high-moisture coals.
Coal size distribution	Standard grinds being used, some CWM use tailored bimodal particle sizes to increase solids percentage. Ultrafine grinding for beneficiation may lead to higher solids and smaller sizes.
Additives	In the range of 1–4% by weight, based on the coal. Several have been tried and many are commercially developed and proprietary. Modified cornstarch, polycarboxylates, and naphthalene sulfonates have been successfully used. Mixture stability and reduced viscosity result.
Viscosity (rheology)	CWM are non-Newtonian and viscous properties vary with shear rate and sometimes time (e.g., thixotropic). Apparent viscosity increases with high coal loading, lower temperature, and declining additive percentage and changes also with coal type.

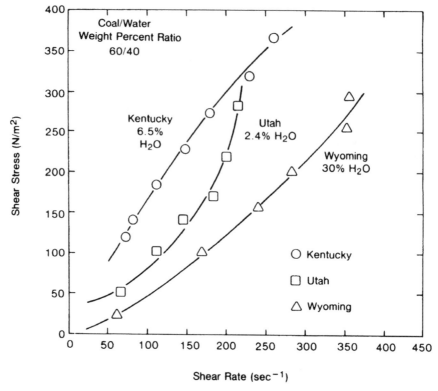

Figure 2.6. Rheogram of Utah, Wyoming, Kentucky coal/water mixtures of 60 wt % coal and 40 wt % water.

Thus, from measurements which provide Γ and τ_w, the strain rate $\dot{\gamma}$ is determined from Eqn. 2.3. The data in Figures 2.6–2.8 correlate the strain rate ($\dot{\gamma}$) vs. wall shear stress (τ_w) for the CWM tested. The dispersant used was a commercially available, high-molecular-weight formaldehyde condensate of naphthalene sulfonate.

Figure 2.6 shows the impact of coal type for three different coals whose properties are summarized in Table 2.9. These data are for mixtures with 60% coal (wet) and 40% added water. However, the Wyoming subbituminous coal already had an inherent moisture level of over 25% (Table 2.9). Effects of added water level are shown in Figure 2.7, while effects of dispersant are shown in Figure 2.8. Increasing water level and dispersant level markedly reduce the apparent viscosity level. Ekmann (1982) and others (Fourth, Fifth Symposia, 1982, 1983) have studied the rheological properties of coal–water mixtures, including those with commercially viable levels of over 70% coal solids.

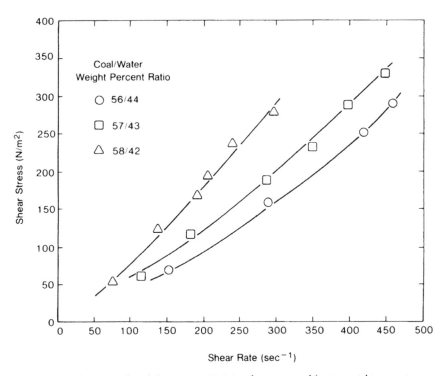

Figure 2.7. Rheogram of Utah bituminous CWM with various coal/water weight percentages.

2.7. PROPERTIES OF OTHER SOLID AND LIQUID FOSSIL FUELS

Table 2.10 summarizes the properties of several other fossil fuels, including tar sands, oil shale, and solvent-refined coal. These fuels can be divided into two classifications:

Primary or naturally occurring fuels	Secondary or processed fuels
Coals	
Oil shale	Solvent-refined coal
Tar sands	Coal char
Peat	Shale oil bitumin (liquid)
Wood	Tar sands carogen (liquid)
Oil	Coal liquid

The table further illustrates the wide variability in physical properties among

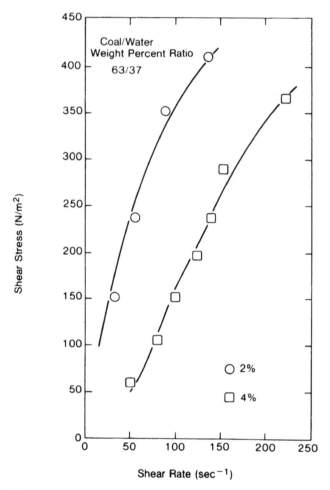

Figure 2.8. Rheogram for Utah bituminous CWM with 2–4 parts per hundred of coal dispersant (pphc).

these fossil fuels and coal fuels. It is beyond the scope of this book to consider the detailed physical or combustion characteristics of all of these liquid or solid fossil fuels. However, Table 2.10 does provide a basis for relating these fuel types to the characteristics of various petroleum feedstocks. One striking difference in coal, as compared to petroleum, bitumin, or carogen, is the highly aromatic nature of the liquid products derived from coal.

TABLE 2.9. Properties of Coals in Coal–Water Mixtures

Property	Utah bituminous	Kentucky bituminous	Wyoming subbituminous
Ultimate analysis (wt. %, dry)			
Ash	8.7	23.3	6.9
H	5.7	4.4	4.4
C	70.2	61.0	67
N	1.4	1.4	1.1
S	0.5	4.3	0.5
O	13.5	5.6	20.2
Proximate analysis (wet basis)			
Moisture	2.4	8.6	27.8
Ash	8.5	21.4	5.0
Volatiles	45.4	32.4	32.9
Fixed carbon	43.7	38.2	34.3
Higher heating value			
kJ/kg	29,534	25,432	20,032
(BTU/lb)	(12,700)	(10,915)	(8,614)

2.8. ILLUSTRATIVE PROBLEMS

1. From the data of this chapter, compare the high heating values of a low-moisture, high-volatile B bituminous coal, a high-moisture lignite, and oil shale on the following bases:
   a. moisture and ash-included basis
   b. moisture-free basis
   c. moisture-free and ash-free basis
What conclusions can be made about the thermal penalties of ash and moisture in these fuels?

2. Based on the data of Table 2.6 and by looking at melting and boiling points of the trace elements and their oxides or other common, naturally occurring compound forms, which of the elements pose a potential health hazard? It may be useful to look at health hazard limits (TLV values (Threshold Limit Values)) and potential emission levels for a typical industrial boiler. Results could be compared with selected observations (see IGT, 1979, Sec. III.2).

3. From the data of Table 2.7, compute the mass-mean and number-mean particle sizes for the fluidized-bed coal and the pulverized coal shown. Then determine if these two distributions are adequately described by a log–normal or Rosin–Rammler distribution functions.

TABLE 2.10. Selected Physical and Chemical Properties of Other Primary and Secondary Fossil Fuels

	Primary fuels					Secondary fuels			
Fuel	Wood	Peat	Oil shale	Wyoming[R1] subbituminous coal	Oil sands	Pyrolysis char	Solvent-refined coal	FMC COED char	Comparison No. 6 oil (low sulfur)
Proximate analysis									
Moisture (%)	40[R6]	90[R2]	1.4[R8]	11.66	4	0.65[R1]	5[R5]	2.5	
Volatile matter (%)		65[(1)]	12.3	40.80	13	11.60	56.1[R1]	8.3	
Fixed carbon (%)		27[(1)]	3.7	42.31	(3)		4 3.7[1]	12.5	
Ash (%)		8	82.6	5.23	83	12.23	0.23[1]	76.7	
Ultimate Analysis									
Carbon (%)	52[R6]	51[R2]	79.2[R4]	58.46	83.0[R8]	79.55[R1]	88.6	78.3	87.46
Hydrogen (%)	6.3	5	10.5	5.55	10.2	2.30	5.6	1.8	11.75
Nitrogen (%)	0.1	3	0.6	1.07	0.5	1.23	1.3	1.7	0.32
Sulfur (%)	0.0	1	1.2	0.58	3.8	4.60	0.13	2.7	0.35
Oxygen (%)	40.5	32	6.5	29.11	2.5	4.09	4.4	2.7	0.06
Ash (%)	1.0	8	(4)	5.23	(4)	12.23	(4)	12.8	0.06

Heating values, (kJ/kg[13])							
As received	$9,300^{R3}$	$11,400^{(2)}$	5350^{R8}	22,260	$5,350^{(5)}$	29,600	35,400
Higher heating value	$20,800^{R6}$	19,800		27,700	$41,400^{(6)}$		44,500
Chemical structure	(7)	$(8)^{R2}$	$(9)^{R4}$	(10)	(11)	(12)	36,700
Liquid products			Shale oil				
Gravity (deg. API)			20.8				
Viscosity [sus at 333 K(140°F)]			73		38.3		18.2
Pour point [K(°F)]			292 (66)	at 300 K	35.5		385
					239 (−30)		289 (60)

(1) Calculated on a water-free basis.
(2) 50% water.
(3) Figures for tar sands reports as bitumen mineral and water, with no separation of the bitumen into volatile and fixed components.
(4) Not included in analysis.
(5) With mineral content.
(6) Bitumen alone.
(7) Cellular.
(8) Decomposed cellular.
(9) Resins and cyclic alkanes.
(10) Highly aromatic
(11) Asphaltenes, resins and heavy oils
(12) Largely elemental carbon.
(13) To convert to BTU/lb, divide by 2.33.

R1 Data from Schweiger (1979).
R2 Data from Bodle et al. (1978).
R3 Data from Berry (1980).
R4 Data from Yen and Chilingarian (1976).
R5 Data from McRanie (1979).
R6 Data from Tillman (1978).
R7 Data from Patel (1979).
R8 Data from Hendrickson (1975).

4.   Compute the pressure drop that would occur in transporting a 65% Utah bituminous coal–35% water mixture with 2% additive at the rate of 30 kg (65 lb) dry coal per hour in an 0.64 cm ($\frac{1}{4}$ in.) horizontal pipe, 1.5 m (5 ft) long from the supply tank to the nozzle tip of a laboratory combustor. How much would this pressure drop be reduced by use of 4% dispersant? How does this pressure drop compare with that for the same mass flow rate of water?

COAL PARTICLE IGNITION
AND DEVOLATILIZATION

3.1. INTRODUCTION AND SCOPE

Coal particle reactions are an essential aspect in all coal processes. Often, these reactions are among the rate-limiting steps that control the nature and size of coal processes. Further, these reactions have a direct impact on the formation of fine particles, nitrogen and sulfur-containing species, and other pollutants. This section provides an overview of these coal particle reactions.

Table 3.1 summarizes several coal processes that were identified in Chapter 2, with particular emphasis on characteristic particle sizes, residence times, and approximate particle reaction times. This table further illustrates the wide range of conditions under which coal is processed. This, coupled with the extensive variety of coals of interest, defines the enormous scope and complexity of this subject area. Chapters 3 and 4 deal with the following aspects of coal particle reactions:

1. The two-component concept of coal reactions: coal devolatilization and char oxidation.
2. Coal particle heat up and ignition.
3. Coal devolatilization, including illustrative experimental data for rates and products, volatiles combustion, and common methods for modeling.
4. Coal oxidation, with typical test data, factors influencing rates, and methods for modeling.
5. Intrinsic reactivity of char.
6. Other aspects, including brief mention of minerals matter, slag formation, and pollutant formation.

TABLE 3.1. Typical Coal Particle Reaction Rates

Process	Particle size	Pressure (atm)	Flame temperature (K)	Typical residence time
Direct combustion				
Pulverized	10–100 μm[a]	1	1750[b]	1 s[a]
Fluidized bed	1–5 mm[a]		1150[a]	100–500 s[a]
Stoker	1–5 cm[a]		1750[a]	3000–5000 s[a]
MHD	10–100 μm			
Coal/oil mixtures	Slurry			
Gasification				
Fixed Bed	0.3–5.0 cm[c]	1–27[b]	1420[c]	>1 hr
Fluidized bed	<8 mm[c]	1–68[b]	1200[b]	0.5–1 hr[c]
Entrained flow	0.01–0.025 mm	1–82[b]	1640–1920[b]	5 s[c]
In situ	5–30 cm			
Liquefaction				
Pyrolysis	0.01 mm–2 cm	1–35[d]	590–1260[d]	0.1–10 s[d]
Direct liquefaction	15–100 μm			
Solvent extraction		100–270[d]	700[d]	40–100 m[d]
Catalytic		136[d]	700[d]	30–60 m[d]
Indirect liquefaction	(see gasification			
Coking	0.15–0.6 cm[c]	1	1200–1300[c]	16.4–18 hr[c]

[a]Data from Essenhigh *et al.* (1979).
[b]Data from Bodle and Schora (1979).
[c]Data from EPRI (1979).
[d]Data from Nowacki (1979).
[c]Data from Lowry (1945).

7. Illustrative problems dealing with coal particle reactions, together with solutions.

3.2. ELEMENTS OF COAL REACTIONS

Development of a complete description of coal processes will require incorporation of a "coal particle reaction" model. One single model framework will not likely be adequate for all coal varieties and sizes. A schematic diagram of a reacting coal particle is illustrated in Figure 3.1. The "model" suggests that the particle, at any time in the reaction process, is composed of:

1. Moisture
2. Raw coal
3. Char
4. Ash (mineral matter)

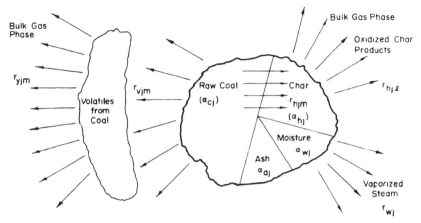

Figure 3.1. Schematic of coal particle, illustrating constituents and reaction processes. (Figure used with permission from Smoot and Pratt, 1979.)

Also, the particle may be surrounded by volatilized matter. A general description of experimental observations for pulverized-coal particles, and assumptions used in developing this model concept, are outlined in detail by Smoot and Smith (1979). An adequate description of coal reaction processes requires the modeling of each of these particle elements.

Coal reactions are generally divided into two distinct components (Gray *et al.*, 1976):

1. *Devolatilization of the Raw Coal.* This part of the reaction cycle occurs as the raw coal is heated in an inert or oxidizing environment; the particle may soften (become plastic) and undergo internal transformation. Moisture present in the coal will evolve early as the temperature rises. As the temperature continues to increase, gases and heavy tarry substances are emitted. The extent of this "pyrolysis" can vary from a few percent up to 70–80% of the total particle weight and can take place in a few milliseconds or several minutes depending on coal size and type, and on temperature conditions. The residual mass, enriched in carbon and depleted in oxygen and hydrogen, and still containing some nitrogen, sulfur, and most of the mineral matter, is referred to as "char." The char particle is often spherical (especially for small particles), has many cracks or holes made by escaping gases, may have swelled to a larger size, and can be very porous internally. The nature of the char is dependent on the original coal type and size and also on the conditions of pyrolysis. Figure 3.2 shows microphotographs of a raw bituminous pulverized coal and a char from that coal.

2. *Oxidation of the Residual Char.* The residual char particles can be oxidized or burned away by direct contact with oxygen at sufficiently high

a

b

Figure 3.2. Selected scanning electron microscope pictures (33-μm Pittsburgh coal dust, 0.050 kg m^{-3}, 0.21 m s^{-1} flame). (a) Feed coal ($\times$200); (b) char from 10 mm behind the flame front ($\times$1000). (Figure used with permission from Smoot and Pratt, 1979.)

temperature. This reaction of the char and the oxygen is thought to be heterogeneous, with gaseous oxygen diffusing to and into the particle, adsorbing and reacting on the particles surface. This heterogeneous process is often much slower than the devolatilization process, requiring seconds for small particles to several minutes or more for larger particles. These rates vary with coal type, temperature, pressure, char characteristics (size, surface area, etc.), and oxidizer concentration. Other reactants, including steam, CO_2, and H_2, will also react with char, but rates with these reactants are considerably slower than for oxygen.

These two processes (i.e., devolatilization and char oxidation) may take place simultaneously, especially at very high heating rates. If devolatilization takes place in an oxidizing environment (e.g., air), then the fuel-rich gaseous and tarry products react further in the gas phase to produce high temperatures in the vicinity of the coal particles. The discussion that follows in this chapter and Chapter 4 attempts to characterize the nature and rates of these coal reaction processes. In this chapter coal particle heatup, particle ignition, devolatilization, and combustion of volatiles are considered. In Chapter 4 heterogeneous reaction of coal is treated, with presentation of rate data for reaction in oxygen and CO_2 and with consideration of intrinsic reactivity.

3.3. IGNITION

3.3a. Background

Ignition of coal is a particularly complex issue, and is not yet adequately described or correlated. The literature contains much information relating to ignition, under such topics as induction periods, quenching distances, flame propagation, flammability limits, flashback, blowoff, and extinction. Williams (1965), Kanury (1975), and Glassman (1977) provide general treatments of ignition in gaseous systems. Field et al. (1967) gave an earlier treatment of ignition in pulverized coal, while Singer (1981) and Elliott (1981) provide some practical information on coal ignition systems. Essenhigh (1981) also includes a recent review of ignition phenomena in coal.

Ignition characteristics of coal are strongly dependent upon they way the coal particles are arranged or configured. Three arrangements can be identified:

1. Single coal particles
2. Coal piles or layers
3. Coal clouds

For single particles, no interaction among particles occurs. This arrangement provides basic data and may provide practical insight in very dilute coal flames. Coal piles or layers occur in moving beds, fixed beds, coal storage piles, and dust layers. Coal clouds exist in pulverized-coal combustors, furnaces, entrained gasifiers, coal dust mine explosions, and in fluidized-bed systems (more dense in the latter case). Coal ignition characteristics in these various configurations differ substantially.

No single definition of ignition for coal seems appropriate. In general terms, ignition can be described as a process of achieving a continuing reaction of fuel and oxidizer. Ignition is most often identified by a visible flame. However, reactions can proceed very slowly at low temperatures, without a visible flame. Ignition is sometimes said to occur when the rate of heat generation of a volume of combustibles exceeds the rate of heat loss. Ignition is more complex in condensed phases and particularly in heterogeneous solids such as coal. In coal, as noted above, particles can react slowly by oxygen attack on the coal surface, leading to spontaneous ignition of a coal pile, as evidenced by a supporting gas flame.

In either inert or reactive hot gases, the coal will react internally, softening and devolatilizing, as the particle increases in temperature. In the process, gases and tars are released. While no flame is visible, and this process is not called "ignition," the onset of this thermal decomposition is an ignition-like process. These off-gases and tars can also be ignited in a surrounding oxidizer. This ignition process could involve only gases or also the tars. Further, the remaining char can be ignited in oxygen by surface reaction processes. This reactant could also be CO_2, H_2O, or H_2, and a visible flame may not result, yet coal reactions continue.

Ignition is often characterized by the time required to achieve a certain temperature or a visible flame or a certain consumption of fuel, for a specified set of conditions. However, no unique ignition time exists either. Potentially important variables that influence coal ignition temperature and time include the following:

Coal type	System pressure
Volatiles content	Gas composition
Particle size	Coal moisture content
Size distribution	Residence time
Gas temperature	Coal concentration or quantity
Surface temperature	Gas velocity
Mineral matter percentage	Coal aging since grinding

Variables that dominate the ignition process depend strongly on the configuration of the coal particles, as will the range of ignition temperatures and times.

3.3b. Theory of Ignition

Williams (1965) and Remenyi (1980), among others, treat the general theory of ignition. It is beyond the scope of this section to provide a comprehensive treatment of theories of ignition. However, some comments will be useful in interpreting experimental observations. Sources of ignition energy can be convective (e.g., hot gas), radiative (e.g., hot particle cloud), conductive (e.g., hot wall), chemical (e.g., fuel surface oxidation), electrical (e.g., spark), etc. Ignition is somtimes said to occur when generation or addition of energy exceeds the rate of energy loss. Thus, ignition theories have often been based on an energy balance for a volume of reactive mixture. Early theories of ignition for a uniform mixture based on energy balance included Semenov (1935) and Frank-Kamenetski (1955). These simple theories showed the relationships between ignition and flammability limits (see Kanury, 1975) and identified some key variables for uniform fluids.

More recently, Annamalai and Durbetaki (1977) and Annamalai (1979) have specifically applied the Semenov theory, extended to consider heterogeneous reaction, to coal particle ignition. More general theories would be required to rigorously account for ignition-related phenomena. Such theories must include reaction, species diffusion, gas–solid heat transfer, and wall effects.

Williams (1965) noted the relationship between propagation of a flame in a uniform gas mixture and the ignition process. He noted that: "Ignition will occur only if enough energy is added to the gas to heat a slab about as thick as a steadily propagating adiabatic laminar flame to the adiabatic flame temperature." Gaseous laminar flames (with initial mixture at STP) are about 0.1 cm thick, while laminar air coal dust flames about an order of magnitude thicker. Thus, it might be expected to be much more difficult to ignite a coal dust/air mixture.

A generalized theory of flame propagation in laminar coal dust flames was published by Smoot et al. (1976). The theory extended earlier theories based on heat transfer. The results of this theory give some insight into ignition in coal dust systems. It was shown theoretically that the following variables influenced the ignition and propagation process:

1. Coal type
2. Coal size
3. Coal size distribution
4. Oxygen percentage
5. Auxiliary gaseous fuels
6. Dust concentration
7. Initial temperature
8. Initial pressure
9. Rate of energy addition

The ignition process was influenced predominantly by the following physical and chemical processes: rate of coal heatup from gas and hot particles, rate of coal pyrolysis, rate of diffusion of gaseous fuel and oxidizer species, and rate of gas-phase reaction of pyrolysis products and oxygen. The theory indicated that reaction of oxygen on the coal surface was not a major factor in this ignition process. However, others have identified surface oxidation as important in some coal dust ignition processes (e.g., Annamalai and Durbetaki, 1977; Essenhigh, 1981).

3.3c. Ignition Data

It is apparent that a wide range of ignition data will result for coal, depending on configuration of the coal particles (i.e., single particles, layers, clouds), the enclosure, the heat source, the coal type and size, etc. Representative ignition temperatures are included here for ignition of coals under various conditions.

Coal Piles. Investigators (Hertzberg *et al.*, 1977; Kim, 1977; Elder *et al.*, 1977; Stott, 1980) seem to agree that self-heating of coal layers is a heterogeneous surface reaction with O_2. Thus, the self-heating rates are dependent on concentration of O_2, surface area (which is greater for more porous coals), coal rank (with lower-rank coals self-heating more rapidly), freshness of coal crushing, heat-transfer effects, coal temperature, air humidity [below 370 K (100°C), a dry coal exposed to a humid air will self-heat more rapidly, especially a lower-rank coal], and coal moisture. If coal is moist, self-heating must evaporate the water and heat the coal; therefore moisture in coal usually inhibits self-heating after the initial heat of wetting is liberated. However, the heat of wetting can exceed the heat of oxidation below 370 K in a dry, lower-rank coal. Thus, a dry coal will self-heat rapidly in the presence of moist or humid air.

The literature reports that coal temperatures [i.e., the adiabatic, self-heating temperatures (ASHT)] above which self-heating will occur vary from 291 K (18°C) for lignites to 378 K (105°C) for bituminous coals in an adiabatic calorimeter with low air flow. Dry coals at 303 K that would not self-heat in dry air will autoignite in seven hours if exposed to air that is moist (Hertzberg *et al.*, 1977).

Autoignition of a coal is not as precisely defined as in gases and vapors (Zabetakis, 1965). Some consider it to be runaway self-heating while others consider it as the point where fire or flame can be visually detected. Results for either definition are not significantly separated in time (usually only a matter of minutes). Autoignition temperatures (AIT) reported in the literature are as low as 393 K (120°C) for bituminous coals and could occur in as short a time as eight hours for an adiabatic calorimeter test (Hertzberg *et al.*, 1977). Autoignition temperatures are somewhat particle-size dependent, ranging from 442 K (169°C)

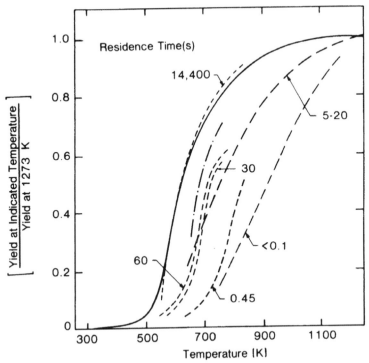

Figure 3.3. Effect of temperature on coal pyrolysis weight loss (yield) at different residence times. [Data from three investigators; reference to original work given by Howard (1981), from which this figure was taken with permission.]

for 2–7 μm size Pittsburgh coal to 500 K (227°C) for 147–200 μm Pittsburgh coal in isothermal calorimeter tests with oven temperature ranging from 436–513 K (Hertzberg *et al.*, 1977).

Once autoignition has occurred and the particle surface temperature reaches the pyrolysis stage [in the 573 K (300°C) range depending on the rank of the coal (Solomon and Colket, 1979)], fuel-rich pyrolysis gases are emitted which can support homogeneous combustion. Results from Howard (1981) shown in Figure 3.3 based on coal-weight-loss data for various coals and residence times show that significant pyrolysis weight loss starts in the range of 570–670 K (300–400°C), depending on residence time and coal heating rate. Above this temperature range, coals will soften and become more porous and further increase the surface area for heterogeneous reaction.

Dust Layers and Clouds. Ignition temperatures of dust clouds and dust layers have been reported for various solid fuels as a function of volatiles

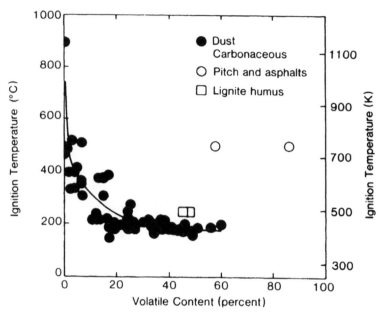

Figure 3.4. Variation in ignition temperature of dust layer with volatile content of carbonaceous dusts. (Figure used with permission from Nagy *et al.*, 1965, courtesy of Bureau of Mines, U.S. Department of Interior.)

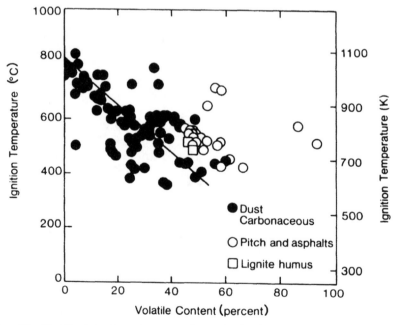

Figure 3.5. Variation in ignition temperature of dust cloud with volatile content of carbonaceous dusts. (Figure used with permission from Nagy *et al.*, 1965, courtesy of Bureau of Mines, U.S. Department of Interior.)

content, as shown in Figures 3.4 and 3.5. For a Pittsburgh coal, the cloud ignition temperature is given as 890 K (617°C) and for a dust layer of the same coal, the temperature is given as 495 K (222°C). Since pyrolysis does not begin for most coals until about 570 K (300°C), dust layer ignition does not likely involve homogeneous mechanisms.

Results of various typical ignition or "onset" temperatures are summarized in Table 3.2. Coal onset temperatures vary from that of coal pile self-heating, which occurs at near-ambient temperatures, to single particle ignition temperatures as high as 1200 K (900°C). All of these onset or ignition temperatures are functions of coal type, size, and condition. Values also depend on experimental method and environment. These representative values emphasize the wide variability in ignition or onset temperatures. Chemical reactions proceed at some rate over a wide range of temperatures. It takes several hours at low temperature for a coal pile to ignite while a coal dust cloud can be ignited in milliseconds at high temperature.

3.4. PARTICLE HEATUP AND TEMPERATURE

3.4a. Heating Processes

Coal reaction processes are very dependent on the rate at which the particles heat and on the maximum particle temperature. Heating of the particle is complicated by coal reaction processes, with devolatilization initiated at about 600 K (300°C) (see Figure 3.3). As the particles continue to heat, convective heating is retarded by the flow of volatile matter from the particle. Reaction of the volatiles in the gas phase changes the surrounding temperature and heatup rate of the particle; exothermic surface oxidation of the char with oxidizer releases energy to the particle and to the gas in the immediate vicinity of the surface. Further oxidation of the surface products (e.g., $CO + O_2 \rightarrow CO_2$) causes additional heat release in the vicinity of the particle. There is no chronological sequence for discussion of these various events, since particle heatup can occur simultaneously with devolatilization, gas-phase reaction, or surface oxidation. Essenhigh (1977) outlines behavior of coal ignition and flame propagation under different heating rates, varying from 10 to over 10^5 K/s. Heating rates of over 10^6 K/s are known to exist in coal processes.

3.4b. Measurements

Recently, optical pyrometers have been successfully used to measure particle and gas temperature histories in reacting coal systems. These measurements provide specific data on coal particle temperature and sometimes gas temperatures in fixed beds, laboratory flames, drop-tube furnaces, and coal

TABLE 3.2. Typical Ignition (or Onset) Temperatures for Various Coal Reaction Processes

Coal process	Reference	Time	Typical ignition (onset) temperature range		Experimental methods
			K	°C	
Coal pile self-heating (ambient conditions)	Kuchta et al. (1980)	hours	303–378	30–105	Adiabatic calorimeter (air flow, 50 cm³/min)
Coal pile autoignition[a]	Hertzberg et al. (1977)	minutes	443–500	170–225	Samples in isothermal electrically preheated cylindrical (8.0-cm-ID) oven
Coal pile thermal ignition[b]	Nagy et al. (1965)	minutes	473–773	200–500	Coal layer on isothermal hot plate
Coal devolatilization	Howard (1981); Solomon and Colket (1979)	milliseconds	573–773	300–500	Several techniques used with various heating rates and residence times
Pulverized-coal cloud ignition	Nagy et al. (1965)	seconds	673–1073	400–800	Coal dust injected into isothermal oven which is open at bottom
Single particle ignition	Essenhigh (1981)	seconds	1073–1173	800–900	"Few" particles on tip of platinum wire in isothermal laboratory furnace

[a]Coal pile autoignition occurs when the coal sample is placed in an environment at some temperature below the ignition temperature, and the coal self-heats to the ignition temperature.

[b]Coal pile thermal ignition occurs when the coal is placed in an environment at some temperature above the ignition temperature and the coal is externally heated to the ignition temperature.

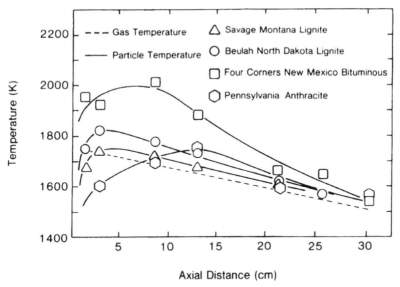

Figure 3.6. Effect of coal type on gas and particle temperatures in a lean CH_4/air laboratory flat flame containing a few coal particles. (Figure used with permission from Seeker *et al.*, 1981.)

combustors. Table 3.3 summarizes results from nine recent studies. Results commonly show peak particle surface temperatures in the range of 2100–2200 K in air or similar O_2–gas mixture. The peak particle temperature can significantly exceed the surrounding gas temperature, partly because of heat release during exothermic surface reaction. As particle burnout continues, the gas temperature approaches the particle temperature.

Figure 3.6 shows measured gas temperature and particle temperatures for several coals as a function of position in a methane–air flat laboratory flame in which a few coal particles were entrained. Particle temperatures were up to 300 K above the gas temperature in the flame, while the differences in particle and gas temperatures depended significantly on coal type. Ultimately, gas and particle temperatures tended to equilibrate. In this flame, particle heatup times were about 20 ms while heatup rates were about 10^5 K/s. The coal dust concentration in this flame is highly dilute when compared to a practical pulverized-coal flame. In the latter case, radiation among particles could be an additional major factor, while coal reaction can also have a significant effect on gas temperature. In systems highly loaded with coal, the gas temperature may exceed the particle temperature as indicated by the results of Cashdollar and Hertzberg (1980) and Mackowski *et al.* (1982). (See Table 3.3.)

Several variables influence the particle temperature. Coals with high volatile content tend to exhibit highest particle temperatues. Coal types with highest

TABLE 3.3. Measured Temperatures of Reacting Coal Particles[a]

Reference	Reactor type	Chemical system		Coal type	Particle size (μm)	Particle temperatures (K)	Estimated heatup rate (K/s)	Gas temperature (K)
		Stoichiometric ratio	Gas system					
Ayling and Smith (1972)	Small, heated entrained flow	Dilute	10–20% O_2	Semi-anthracite	6, 22, 49, 78	1400–2200	—	Entering gases preheated to 1400–1800
Cashdollar and Hertzberg (1980)	(1) 8-L explosion bomb (2) Experimental mine	Close to 1.0	Air	Pittsburgh Seam bituminous	7 (surface weighted mean)	(1) 1400–2100 (2) 700–1300	—	Gas temperature nearly always in excess of particle temperature
Mackowski et al (1982)	Premixed, laminar, flat flame burner	Close to 1.0	23% O_2; 77% N_2	Eastern Kentucky bituminous	64 (number based mean)	1100–2100	10^5–10^6	Gas temperature nearly always in excess of particle temperatures
Mitchell and McLean (1982)	Premixed, transparent, flat flame burner	Dilute	0–25% O_2; 75–100% CH_4–N_2–H_2	(1) Millmerran char (2) Petroleum coke	90 (mass mean)	(1) 1400–1900 (2) 1400–1900	10^5	1300–1800

Nettleton (1965)	Drop-tube furnace	Dilute		Gedling char	400–925	Up to 2100	10^4	1200
Starley et al. (1982)	Fixed-bed furnace	0.7–1.2	Air	UT bituminous	12,500–25,000	1400–2100	—	—
Timothy et al. (1982)	Heated muffle tube	Dilute	O_2–He–Ar mixtures	(1) ND lignite (2) MT lignite (3) AL Rosa (4) UT bituminous (5) IL No. 6 bit	38–45, 90–105	2300 (21% O_2) Up to 3100 with high oxygen	10^5	1250, 1700
Seeker et al. (1981)	Flat flame burner	1.35 (for CH_4–air mixture)	CH_4–air	(1) ND lignite (2) MT lignite (3) AL bituminous (4) NM bituminous (5) UT bituminous (6) PA anthracite	40, 80	1500–2200	10^5–10^6	1750
LaFollette (1982)	Large-scale entrained flow	0.9–1.1	Air	Gillete, Wyoming subbituminous	49 (mass mean)	Up to 2250	10^5	—

^{a}Data from LaFollette (1983)

temperature are, in descending order, high-volatile bituminous, medium-volatile bituminous, lignites, and anthracite. Particles with high mineral matter content exhibit lower temperatures. Smaller particles exhibit higher peak temperatures and heat up more quickly than larger particles. Increased oxygen concentration raises particle temperature significantly. Values as high as 3100 K were observed by Timothy *et al.* (1982) for systems with high oxygen percentage.

3.4c. Predictions

Predictions of gas and particle temperatures during reaction of pulverized coal provide a useful comparison with observations. Predicted gas and particle temperature results for combustion and for gasification of pulverized coal are shown in Figure 3.7. These predictions were made with a generalized multi-dimensional model (Smith *et al.*, 1980) with parameters summarized in Table 3.4. Predictions were made for several different particle sizes along the centerline

TABLE 3.4. Parameters Used in Predictions of Particle and Gas Temperature Histories during Combustion and Gasification of Pulverized Coal

Reactor	Coal combustion	Coal gasification
Primary diameter (cm)	1.6	1.31
Secondary diameter (cm)	5.4	2.87
Chamber diameter (cm)	20.3	20.0
Chamber length (m)	1.52	1.19

Feed streams	Coal combustion	Coal gasification
Primary gas flow rate (kg/hr)	20.2 (air)	26.3 (O_2)
Secondary gas flow rate (kg/hr)	130 (air)	6.48 (H_2O)
Primary coal flow rate, daf (kg/hr)	13.7	23.8
Primary velocity (m/s)	33	42
Secondary velocity (m/s)	34	6
Primary temperature (K)	367	367
Secondary temperature (K)	589	450
Mass-mean particle diameter (μm)	49	42

Major assumptions

Heat of reaction from the heterogeneous char reactions is given to the gas phase in the region of the particle.

Elemental composition of off-gas from the coal particle is constant (includes both devolatilization and char oxidation).

Heat loss is estimated using an overall heat-loss factor; 0% heat loss represents an adiabatic calculation. Heat-loss factor for predictions were 0.20 for combustion and 0.50 for gasification. Predicted results matched measured profiles of gas composition well with these assumptions.

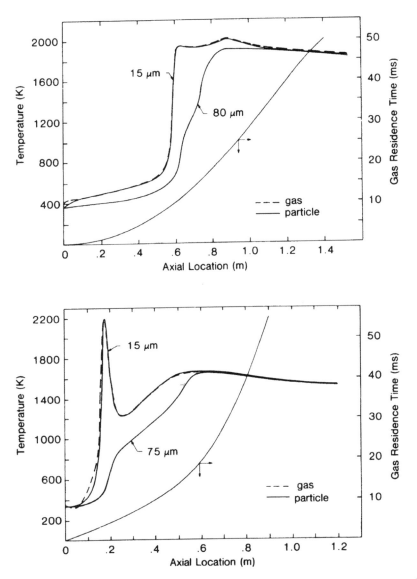

Figure 3.7. Predicted gas and particle temperature histories for combustion and gasification of pulverized coal (parameters are shown in Table 3.4): (a) combustion in air; (b) gasification in steam and oxygen.

of each reactor, but only results for the smallest and largest sizes are illustrated in the figure. Small particles heat up more rapidly than large ones. Over most of the range, the predicted local gas temperatures exceed the particle temperature, sometimes by several hundred degrees, particularly for the larger particle sizes. The residence time also differs for each particle size and for the gas, since velocities of all are different.

In forward regions of the reactor, the gas temperature increases through rapid reaction of the coal volatiles with the oxidizer. Subsequently, the particles are heated principally by convection from the gas to the maximum temperatures. Calculated maximum heating rates for the smallest particles were approximately 2×10^5 K/s for combustion and $1-2 \times 10^6$ K/s for gasification. Peak temperatures on the reactor centerline are about 2000 K for combustion and over 2200 K for gasification. Higher predicted temperatures were noted in regions off the centerline. In these computations, exothermic heat release from the heterogeneous char–oxidizer reaction was assigned to the bulk gas phase near the particle. If all of this heat were assigned to the particles, it is likely that in aft regions of the reactors, predicted particle temperatures would exceed gas temperatures.

3.5.   DEVOLATILIZATION

3.5a.   Experimental Studies

Anthony and Howard (1976) and Howard (1981), and to a less general extent, Horton (1979), Wendt (1980), and Solomon (1980), provide recent reviews of coal devolatilization. Anthony and Howard (1976) also note several reviews of devolatilization over the period from 1963 to 1972. Most of the recent work has emphasized finely pulverized-coal particles at high temperatures. Significant devolatilization does not start until temperatues of 625–675 K (see Figure 3.3). The heating causes thermal rupture of bonds, and volatile fragments escape from the coal. Table 3.5, taken principally from Anthony and Howard (1976) with more recent entries added, provides a detailed summary of much of the experimental methods and test conditions of previous studies of devolatilization. Variables that influence devolatilization rates include temperature, residence time, pressure, particle size, and coal type. Table 3.5 provides a general summary of the observed effects of these key variables.

Final temperature is possibly the most important variable, as illustrated in Figure 3.8 for a lignite coal and a bituminous coal. This figure illustrates several important observations relating to these data:

1.   The extent to which the coal devolatilizes varies greatly for both coals (less than 5% to over 60%) as a function of final temperature. This

TABLE 3.5. Experimental Techniques and Conditions for Coal Devolatilization

Investigator[b]	Technique	Residence time (seconds)	Temperature (K)	Heating rate (K/s)	Pressure (atm)[y]	Ambient gas	Particle size (μm)
Captive sample techniques							
Standard proximate analysis (ASTM, 1974)	Crucible	420	1220	15–20	1.0	Air (lid on)	≤250
Wiser et al. (1967)	Crucible	300–72,000	670–770	15–20	1.0	N_2	246–417
Portal and Tan (1974)	Crucible and basket	15–1200	820–1420	0.5–250	1.0	N_2	<45–88
Gray et al. (1974)	Crucible	420	1220–1470	0.3–20	1.0	N_2	≤200
Hiteshue et al. (1962a, b; 1964)	Hot rod	20–900	750–1470	10	18–400	H_2	250–600
Feldkirchner and Linden (1963); Feldkirchner and Huebler (1965)	Semiflow	10–480	980–1200	100–300 (seems high)	34–168	H_2, H_2O	841–1000
Moseley and Paterson (1965a)	Railway heater	15–165	1090–1220	25	18–95	H_2	150–300
Feldkirchner and Johnson (1968); Johnson (1971)	Thermobalance	Several to 7200	≤1200	≤100	≤100	H_2, N_2, H_2O, etc.	425–850
Gardner et al. (1974)	Thermobalance	25–3000	1120–1220	—	35–69	H_2	500–1000
Loison and Chauvin (1964)	Elect. grid	0.7	≤1320	1500	Vacuum	—	50–80
Rau and Robertson (1966)	Elect. grid	1–1.5	1170–1470	600	1.0	—	250–425
Juntgen and Van Heek (1968)	Elect. grid	0.7	≤1270	1500	Vacuum	—	50–60
Koch et al. (1969)	Elect. grid	7	≤1770	167	Vacuum	—	75–630
Mentser et al. (1970, 1974)	Elect. grid	0.05–0.15	670–1470	8250	Vacuum	—	44–53
Cheong et al. (1975)	Elect. grid	<1–1800	570–1270	≤1000	Vacuum–3.4	—	90–355
Anthony et al. (1974, 1975, 1976)	Elect. grid	0.1–20	670–1370	100–12,000	0.011–100	H_2, He, N_2	53–1000
Graff et al. (1975); Squires et al. (1975)	Ring of coal	1–6 (gas), 10–30 (solid)	870–1270	650	69	H_2	≤44
Suuberg et al. (1978)	Elect. grid		570–1370	$<3(10^4)$	10^2–10^4	He	74–1000
Blair et al. (1976)	Heated ribbon		1070–1970	2–$8(10^4)$	1	He, Ar	500–600
Solomon et al. (1978)	Elect. grid		570–1570	$<10^4$	~0.1	Vacuum	>100

TABLE 3.5. (*Cont.*)

Investigator[b]	Technique	Residence time (seconds)	Temperature (K)	Heating rate (K/s)	Pressure (atm)[c]	Ambient gas	Particle size (μm)
Coal flow techniques							
Stone et al. (1954)	Fluidized bed	10–2500	670–970	—	1.0	N_2	200–600
Peters et al. (1960, 1965)	Fluidized bed	1–15	870–1370	300	1.0	N_2	1000–3000
Pitt (1962)	Fluidized bed	10–6000	670–970	—	1.0	—	200–600
Jones et al. (1964)	Fluidized bed	~2400	700–1370	1000+	1.0	N_2, H_2O	≤1000
Friedman (1975)	Fluidized bed	1800–3600	570–920	—	1.0	H_2O	250–710
Zielke and Gorin (1955)	Fluidized bed	—	1090–1200	—	1–30	H_2, H_2O, etc.	150–212
Birch et al. (1960, 1969)	Fluidized bed	500–9500	770–1220	—	21–42	H_2, etc.	150–710
Eddinger et al. (1966)	Entrained flow	0.008–0.04	900–1270	2500+	1.0	He	6150
Howard and Essenhigh (1967)	Entrained flow	0–0.8	470–1820	22,000	1.0	Air	80% ≤74
Badzioch et al. (1967, 1970)	Entrained flow	0.03–0.11	670–1270	25,000–50,000	1.0	N_2	20
Kimber and Gray (1967a, b)	Entrained flow	0.012–0.34	1050–2270	150,000–400,000	1.0	Ar	22–50
Belt et al. (1971, 1972)	Entrained flow	≤1	1090–1310	—	1–28.2	H_2, N_2	70% ≤74
Sass (1972)	Entrained flow	A few	810–920	10,000	1.0	—	25–80
Coates et al. (1974)	Entrained flow	0.012–0.34	920–1640	—	1.0	H_2, H_2O, etc.	≤74
Moseley and Paterson (1965b)	Entrained flow	0.17–2.5	1060–1270	1000+	50–520	H_2	100–150
Glenn et al. (1967)	Entrained flow	2.4–10.4	1190–1240	—	70–84	H_2, CO, etc.	≤44
Johnson (1975)	Entrained flow	5–14	760–1120	28	18–52	H_2, He	75–90
Shapatina et al. (1960)	Free fall	0.45–14,400	≤820	—	45–490	N_2	150–200
Moseley and Paterson (1967)	Free fall	A few	1110–1270	—	45–490	H_2	100–150
Feldmann et al. (1970)	Free fall	A few	920–1170	—	35–205	H_2	150–300

[a] Data taken mostly from Anthony and Howard (1976).
[b] References in this table are identified in detail in Anthony and Howard (1976) or Smoot (1980).
[c] To convert to MPA, divide by 10.

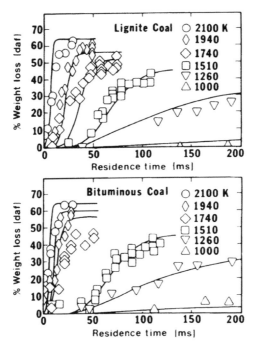

Figure 3.8. Comparison of calculated weight losses with experimental results for devolatilization of lignite and bituminous coals. For Eqns. (3.2) and (3.6), $Y_1 = 0.3$, $E_1 = 104.6$ mJ/kmol (25 kcal/mole), $A_1 = 2 \times 10^5$ s^{-1}, $Y_2 = 1.0$, $E_2 = 167.4$ mJ/kmol (40 kcal/mole), and $A_2 = 1.3 \times 10^7$ s^{-1}. (Figure used with permission from Kobayashi *et al.*, 1977.)

extent of devolatilization can obviously differ significantly from the "proximate" level.

2. The residence time required to devolatilize the pulverized coal is very short (10–200 ms) for these small particles and high temperatures.

3. These two coals, with significantly different chemical characteristics, have similar devolatilization rate characteristics.

These data have emphasized the rate of weight loss during devolatilization. The nature of the pyrolysis products is illustrated in Figure 3.9, again for a bituminous coal and a lignite coal. This figure correlates the results from different investigations. Results show:

1. Significantly different behavior between the two coal types with much more liquid/tar products for the bituminous coal and much more gas products (including H$_2$O) from the lignite.

2. The proportions of various products change with changes in pyrolysis temperature.

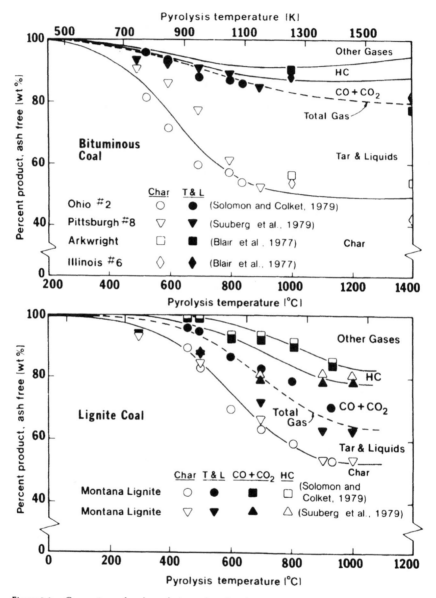

Figure 3.9. Comparison of coal pyrolysis product distributions at various temperatures from three independent investigators. d_p = coal diameter, h = heating rate, HC = hydrocarbons, T and L = tar and liquids. Data not shown for gaseous components for bituminous coal. For Suuberg *et al.* (1979), d_p = 74–1000 μm, in helium, 10^2–10^4 atm, $h < 10^4$ K/s; for Blair *et al.* (1979), d_p = 500–600 μm, in helium/argon, 1 atm, $h = 2$–8×10^4 K/s; for Solomon and Colket (1979) $d_p > 100$ μm, vacuum, 1 KPa (0.01 atm), $h < 10^3$ K/s.

These data are for rapid heating rates and for particles ranging from 74–1000 μm, and for a wide range of pressures. For the range of small particles tested, neither the product distribution nor the rate of devolatilization seem to vary significantly with the particle diameter. Several studies noted by Anthony and Howard (1976) also provided observations that weight loss was not size dependent for particles up to 400 μm in diameter. However, very large particles behave differently from finely pulverized coal. Larger particles will not heat rapidly or uniformly, so that a single temperature cannot be used to characterize the entire particle. The internal char surface provides a site where secondary reactions occur. Pyrolysis products generated near the center of a particle must migrate to the outside to escape. During this migration, they may crack, condense, or polymerize, with some carbon deposition taking place. The larger the particle, the greater the amount of deposition, and hence the smaller the volatiles yield.

Devolatilization processes during gasification are thought to be quite similar to that during combustion. Major differences are elevated pressure and a more fuel-rich environment that often occur in gasification. Anthony et al. (1975) reported that a lower ambient pressure favors the liberation of a greater mass of volatiles. Weight loss of a bituminous coal had declined from 50 to 55% [dry ash-free] (daf) for pressures under 10 kPa (0.1 atm) to below 40% at 10 mPa (100 atm) (Howard, 1981). This appears to be consistent with other results, since an increase in pressure also increases the transit time of volatiles within the particle. An increase in pressure thus has an effect analogous to an increase in particle size. Howard (1981) more recently reviewed the data available on effects of elevated pressure. Howard notes results of approximately 20 devolatilization studies at elevated pressures up to 50 mPa.

3.5b. Devolatilization Models

Badzioch and Hawksley (1970) postulated that the devolatilization was a first-order reaction process, with the reaction rate being proportional to the amount of volatile matter still remaining in the coal:

$$dv/dt = k(v_\infty - v) \tag{3.1}$$

with

$$k = A \exp(-E/RT) \tag{3.2}$$

This treatment required a method for relating the amount of "total" volatile matter v_∞ to that obtained from proximate analyses. The correlation used was

$$v_\infty = Q(1 - v_c)v_p \tag{3.3}$$

The parameters Q and v_c were empirically determined, although a value of 0.15 for v_c was suitable for all of the nonswelling coals tested. This is a satisfactory single-step reaction to describe pyrolysis. However, it lacks the flexibility required to describe much of the experimental data available and may be inadequate to describe nonisothermal pyrolysis. The fact that the parameters v_p, Q, v_c, A, and E may depend upon the specific coal dust also limits the generality of this model.

Kobayashi *et al.* (1977) suggested that pyrolysis could be modeled with the following pair of parallel, first-order, irreversible reactions:

$$C \xrightarrow{\ k_1\ } (1-Y_1)S_1 + Y_1V_1$$

$$C \xrightarrow{\ k_2\ } (1-Y_2)S_2 + Y_2V_2$$

$$(3.4)$$

with the rate equations

$$dc/dt = -(k_1 + k_2)c \qquad (3.5)$$

and

$$\frac{dv}{dt} = \frac{dv_1 + dv_2}{dt} = (Y_1k_1 + Y_2k_2)c \qquad (3.6)$$

Again, k_1 and k_2 are Arrhenius-type rate coefficients. An important feature of the model is that $E_1 < E_2$. This approach satisfactorily correlates the data of Badzioch *et al.* (1970) and Kimber and Gray (1976), and the more recent data of Kobayashi *et al.* (1977) obtained under conditions of transient temperature as illustrated in Figure 3.8. In fact, the agreement for weight loss is very impressive for both coals and for a range of temperatures. Computations were made with a single set of parameters shown in Figure 3.8. The model is conceptually sound in that the variation in volatiles yield with temperature is explained by a second reaction rather than by a correlating parameter like that of Eqn. (3.3). As before, the general utility may be limited because the parameters Y_1, Y_2, A_1, A_2, E_1, and E_2 will depend on the specific coal dust. The stoichiometric parameters Y_1 and Y_2 can be estimated as the fraction of coal devolatilized during proximate analysis (Y_1) and the fraction that can be devolatilized at high temperatures (Y_2, often near unity).

In another treatment, Anthony *et al.* (1975) postulated that pyrolysis occurs through an infinite series of parallel reactions. A continuous Gaussian distribution of activation energies is assumed, along with a common value for the frequency factor so that

$$\frac{v_\infty - v}{v_\infty} = [\sigma(2\pi)^{1/2}]^{-1}\left\{\int_0^\infty \exp\left[-\left(\int_0^t k\,dt\right) f(E)\,dE\right]\right\} \tag{3.7}$$

with

$$f(E) = [\sigma(2\pi)^{1/2}]^{-1} \exp\left(\frac{-(E-E_0)^2}{2\sigma^2}\right) \tag{3.8}$$

This approach also provided very good correlation of the data from Anthony et al. (1975) as well as the more recent experimental results of Suuberg et al. (1977). However, minimal comparison has been made with the data of other investigators. The model is attractive because it requires only four correlating constants. However, the utility of this model too, may be restricted by the need to determine the parameters v_∞, k, E_0, and σ for the specific coal dust of interest.

Sprouse and Schuman (1981) have compared the lignite data of Figure 3.8 (1000–2100 K) with predictions from Eqns. (3.7) and (3.8) as shown in Figure 3.10a. Parameters shown in Figure 3.10 provided near optimum agreement with the data. The authors note the excellent agreement with four parameters (σ, E_0, k, v_∞), compared to five for comparisons of Figure 3.8. A further comparison for lignite data of Anthony et al. (1975) for lower temperatures (973–1273 K) but for variable heating rates (10^2–10^4 K/s) is shown in Figure 3.10b. Again, agreement is very good with the same parameters. It is also interesting to note the absence of significant effect of heating rate on the maximum extent of devolatilization.

Both methods (i.e., Eqn (3.6) and Eqns. (3.7) and, (3.8)) give very good results. Differences in the number of coefficients (4 vs. 5) is considered secondary, since it is likely that a priori information may be available for coefficients for Eqn. (3.6). For example, Y_1 could be taken as the proximate volatiles level while Y_2 may be near unity on a daf basis. More complex reaction mechanisms were reviewed by Anthony and Howard (1976), and often involved several coal reaction steps.

In addition, comprehensive codes require a devolatilization model that will produce the composition of the volatile gaseous products and the residual char, as well as the rate of volatiles evolution. The two-step model of Eqn. (3.4) produces these compositions. Knowing the elemental composition of the raw coal (daf coal) and specifying the char composition (mostly carbon), the volatiles composition V_1 and V_2 can be calculated by mass balance. The total volatile matter evolved from the competing two-step process produces different amounts and composition of volatile matter from different temperature histories. The infinite series of parallel reactions of Eqns. (3.7) and (3.8) does not lend itself to a description of the composition of the resulting products.

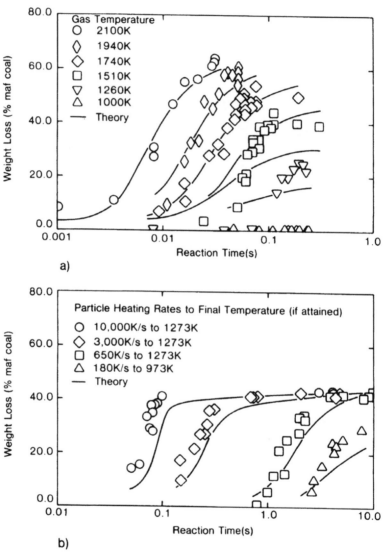

Figure 3.10. Comparison of measured lignite-weight-loss data (0.1 mPa) and predictions from Eqns. 3.7 and 3.8. (a) effects of temperature, laminar flow reactor; (b) effects of temperature and heating rate from electric screen (data from Anthony *et al.*, 1975). $\sigma = 1 \times 10^5$ J/mole, $v_\infty = 63.5$ wt %, $E_0 = 3.15 \times 10^5$ J/mole, $k = 1.67 \times 10^{13}$ s^{-1}. Parameters were not quite optimum but were common in both figures. (Figure used with permission from Sprouse and Schuman, 1981.)

A more recent method for predicting both devolatilization rates and product distribution has been developed by Solomon and co-workers (Solomon and Colket, 1979; Solomon, 1980). Prediction of time and temperature variation in these two properties is based on coal structural group, using general kinetic parameters for each of the coal constituents. Key observations upon which this technique was based are:

1. While the overall rates of devolatilization of coal vary with coal type, the rates of individual functional groups in the coal (i.e., ether groups, hydroxyl groups, tar, etc.) are *independent* of coal type.
2. The chemical composition of the tars is essentially that of the raw organic coal.

Thus, predictions can be made from quantitative measurements of coal composition and the set of parameters for each of the functional groups which are taken to be valid for all coals.

Figure 2.4 showed a model of the raw coal organic structure, while Figure 3.11 illustrates a model for thermal decomposition of this same raw coal. The raw organic coal is thus identified as being composed of a series of chemical constituents as illustrated in Figure 3.12. These include carboxyl (producing CO_2), hydroxyl (producing H_2O), ether (producing CO), aromatic hydrogen, aliphatic hydrogen, nitrogen, and nonvolatile carbon. It has been observed experimentally that the tars produced during the devolatilization process have a very similar structure to the raw organic coal. This close similarity is illustrated in the infrared spectra of Figure 3.13 for four different coals and their associated tars. Agreement for the lignite is not as good as for the bituminous coal. However, for many coals, the general composition of these tars can be identified from the parent coal.

Thermal decomposition is postulated to follow separate activated chemical reaction processes for each of these functional groups as also illustrated in Figure 3.12, including the tar. A given chemical species such as CO, OH, or CH_4 then is evolved from the coal by two independent, first-order processes—one for the species directly, and the second for the fraction of that chemical species in the tar as the tar is evolved. The rates of reaction for thermal decomposition for each species including the tar are assumed to be first order:

$$W_i = W_i^0[1 - \exp(k_i t)] \tag{3.9}$$

where W_i is the weight fraction of the functional group evolved during devolatilization and W_i^0 is the weight fraction of the functional group in the organic part of the raw coal; $k_i = A_i \exp(-E_i/RT)$, and A_i and E_i are determined from time-dependent functional-group composition data during devolatilization. Additional details and constraints involved in this model scheme are discussed

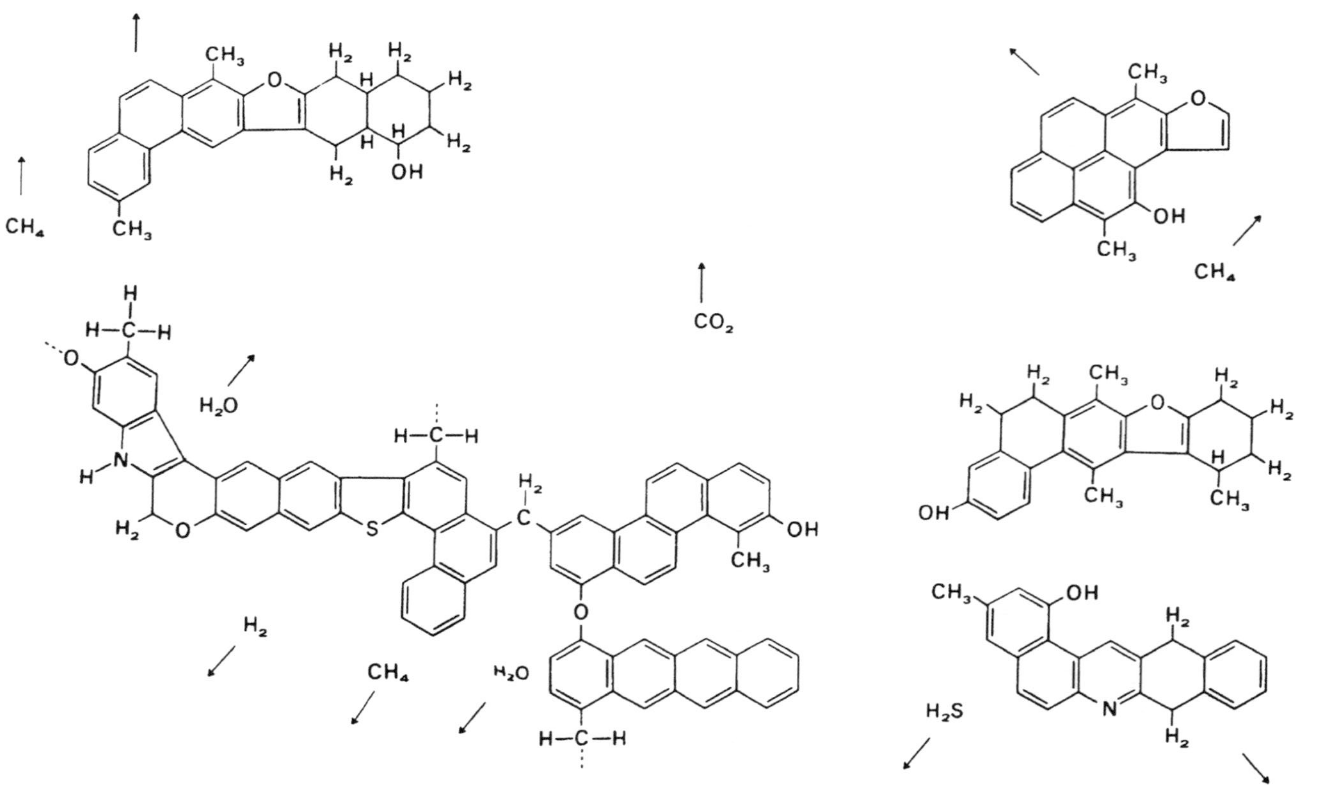

Figure 3.11. Cracking of hypothetical coal molecule during thermal decomposition. (Refer to Figure 2.4 for the postulated original structure before thermal decomposition). (Figure used with permission from Solomon, 1980.)

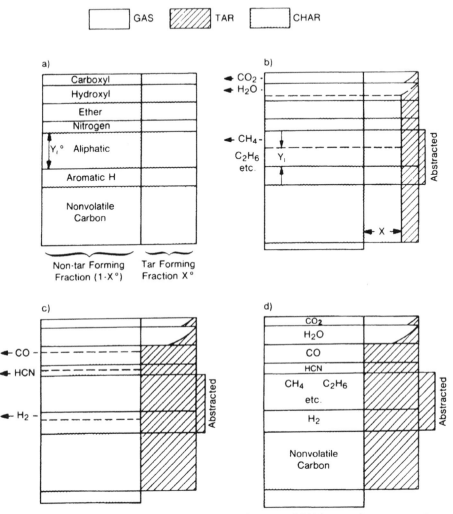

Figure 3.12. Progress of thermal decomposition according to model (a) functional-group composition of coal, (b) initial state of decomposition, (c) later stage of decomposition, (d) completion of decomposition. (Figure used with permission from Solomon, 1980.)

by Solomon (1980), and Howard (1981) where values of kinetic parameters are also presented.

Figure 3.14 shows comparisons of measured and predicted levels of evolved products H_2, tar, and aliphatics, together with residual C, H, and O in the char for a bituminous coal and a lignite for various devolatilization temperatures. Other comparisons for several other coals and other functional groups are

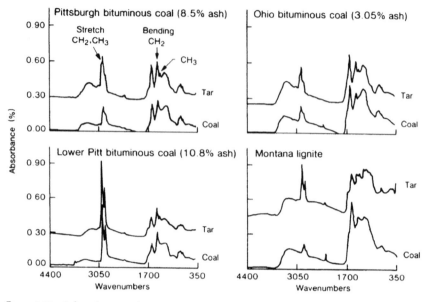

Figure 3.13. Infrared spectra showing comparison of four coals (daf) and the associated tars. (Figure used with permission from Solomon, 1980.)

shown by Solomon (1980). Results are promising; however, the technique still requires further development. For example, kinetic rates for functional groups derived from vacuum thermal decomposition differ from those derived from experiments in other reactors at atmospheric pressure.

3.5c. Effects of High Moisture Content

Specific studies for measurement of devolatilization of coals as a function of moisture level are in progress (Solomon, 1983). This study includes coal–water mixtures and highly cleaned coals. Plans include testing of wet and dry lignite and subbituminous and bituminous coals. It is anticipated that the presence of high moisture levels in coal could have several effects on the devolatilization process, as summarized in Table 3.6. Some insight into the effects of moisture on the devolatilization process could be obtained from a comparison of the bituminous (2.2% H_2O) and lignite coals* (13.6% H_2O) shown in Figure 3.8. However, it is not clear if the raw coals were dried prior to testing. Devolatilization rates for the two very different coals under similar heating conditions were remarkably similar. From this very limited information, if the thermal

---

*The lignite coal had 28.2% moisture at the time of mining, but apparently only 13.6% H_2O at the time of testing.

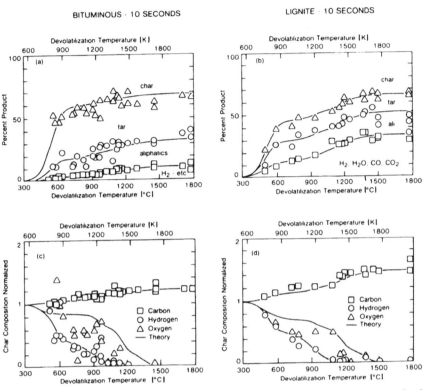

Figure 3.14. Pyrolysis product yields—experiment and theory. The char composition normalized to the composition of the parent coal. (Figure used with permission from Solomon, 1980.)

environment is held near steady, devolatilization rates may not strongly depend on initial moisture content.

3.6. VOLATILES COMBUSTION

3.6a. Background

It was noted above (see Figure 3.9) that a variety of products are produced during devolatilization of coal, including tars and hydrocarbon liquids, hydrocarbon gases, CO_2, CO, H_2, H_2O, HCN, etc. Several hundred hydrocarbon compounds have been identified in the tars produced during devolatilization, mostly aromatic (McNeil, 1981). These products react with oxygen in the vicinity of the char particles, increasing temperature and depleting the oxidizer (e.g., oxygen). This complex reaction process is very important to the control of nitrogen oxides, formation of soot, stability of coal flames, and ignition of char.

TABLE 3.6. Some Possible Effects of High Moisture Content on Coal Devolatilization Rates

Effect	Comments
Lower peak temperature	Presence of 30% water in CWM may reduce peak temperature about 100 K, leading to lower total devolatilization.
Change in internal char structure	Loss of water from rapidly heated particle may change internal pore structure.
Gas-phase composition	Presence of high percentages of H_2O in gas will reduce oxygen concentration and change gaseous reaction rates with volatiles products.
Temperature profile in particle	Residual moisture in coal may reduce heating rate of coal particle.

Some aspects of this process include the following:

1. Release of volatiles from the coal/char.
2. Condensation and repolymerization of tars in char pores.
3. Evolution of hydrocarbon cloud through small pores from the moving particle.
4. Cracking of the hydrocarbons to smaller hydrocarbon fragments, with local production of soot.
5. Condensation of gaseous hydrocarbons and agglomeration of sooty particles.
6. Macromixing of the devolatilizing coal particles with the oxidizer (e.g., air).
7. Micromixing of the volatiles cloud and the oxygen.
8. Oxidation of the gaseous species to combustion products.
9. Production of nitrogen oxides and sulfur oxides by reaction of devolatilized products and oxygen.
10. Heat transfer from the reacting fluids to the char particles.

At very high combustion temperatures, up to 70% or more of the reactive coal mass is consumed through this process (see Figure 3.8).

In an earlier treatment, Field *et al.* (1967) discussed combustion of coal volatiles, while Thurgood and Smoot (1979) also treated this topic more recently. Also, Essenhigh (1981) mentions this subject briefly, noting the significant lack of specific kinetic information.

3.6b. Experimental Data

Recently, Seeker *et al.* (1981) photographically observed the clouds and jets of volatiles being emitted from devolatilizing coal particles shown in Fig. 3.15.

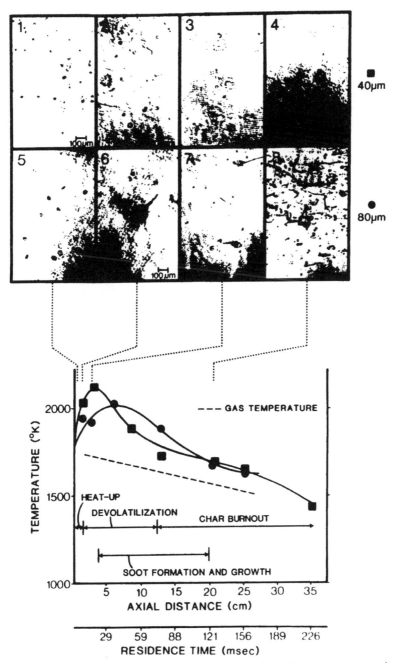

Figure 3.15. Holographic photographs of volatiles evaluation and soot formation in an atmospheric laboratory CH_4/air flame (35% excess air, 1750 K maximum gas temperature). Data are shown at four different locations in the flame for both 40 and 80 μm (mass mean). High-volatile B bituminous coal (from Four Corners, New Mexico) with 41% proximate volatiles, 41% fixed carbon, 19% ash (daf), and 4% moisture. Figure used with permission from Seeker *et al.*, 1981.)

Their tests were done with five pulverized coals of various sizes in a methane-heated laboratory air flame at 1750 K with 35% excess air. They observed, with holographic methods, jetlike volatiles evolution forming a cloud of volatiles from larger (80 μm) bituminous coal particles, with subsequent soot particle (ca. 3 μm) formation and agglomeration. However, smaller bituminous coal particles (40 μm) and anthracite or lignite coals did not produce the soot trails. Figure 3.15 shows these volatile clouds with soot formation for bituminous coal. Each of these little clouds can form a diffusion flame with the oxidizing gas, if the gas temperature and residence time are adequate.

Thurgood and Smoot (1979) tabulated basic reaction rate coefficients for oxidation of hydrogen, methane, acetylene, ethylene, and ethane. In the methane oxidation mechanism, 41 basic radial reactions were included. Engleman (1976) has provided a more extensive review of basic gaseous reaction mechanism, which includes 322 reactions for methane oxidation. It is apparent that a rigorous treatment of the volatiles reaction process is not a realistic expectation in a heterogeneous turbulent flow involving hundreds of chemical species and many more reactions. However, this volatiles reaction process is rapid at high temperatures and for some purposes such as overall coal burnout rates, volatiles reaction rates may not be required. Yet, for other purposes, such as nitrogen oxide formation or soot formation, these volatiles reaction rates may control.

3.6c. Computational Approaches

Two useful approaches for treatment of volatiles reaction are considered here: local equilibrium and global reaction rates. For local equilibrium, it is assumed that the volatiles products and the oxidizing gas are in thermodynamic equilibrium locally if the gas temperature is sufficiently high. Thus, as the volatiles leave the coal, they are assumed to immediately equilibrate with the gas in that local region. From this assumption, the gas temperature and gas composition can be estimated without a knowledge of the evolved chemical species. All that is required is the elemental composition of the volatiles and the heat of devolatilization (thought to be near zero). Data on elements emitted during the devolatilization process can be deduced from measurements (e.g.,

TABLE 3.7. Parameters for Global Hydrocarbon Reaction[a]
(Eqn. 3.12)

Hydrocarbon	A	E/R
Long chain	59.8	12.20×10^3
Cyclic	2.07×10^4	9.65×10^3

[a]Data from Siminski et al. 1972).

see Figures 3.9 and 3.14). This information is often adequate for examining many aspects of coal flames such as exit or local gas concentration and char consumption times, even in turbulent systems.

Sometimes, volatiles and oxidizing gases are not in equilibrium, such as in formation of NO_x from HCN or in formation of soot in fuel-rich regions. Use of global reaction rates provides some estimate of volatiles combustion rates. The approach is to correlate chemical reaction rates of various fuels (CH_4, H_2, C_nH_m, etc.) for the overall reactions such as

$$C_nH_m + (\tfrac{1}{2}n + \tfrac{1}{4}m)O_2 \xrightarrow{k} nCO + (\tfrac{1}{2}m)H_2O \qquad (3.10)$$

where

$$k = A \exp(-E/RT) \qquad (3.11)$$

Edelman and Fortune (1969) and Siminski et al. (1972) proposed global reaction for consumption of hydrocarbons where reaction products are CO and H_2 rather than CO_2 and H_2O. Siminski et al. (1972) have correlated kinetic rates for heavy hydrocarbon combustion for the global reaction:

$$C_nH_m + \tfrac{1}{2}nO_2 \rightarrow \tfrac{1}{2}mH_2 + nCO \qquad (3.12)$$

Different finite rates were specified for long-chained hydrocarbons and for cyclic hydrocarbons. The rate is given as

$$dC_H/dt = -ATp^{0.3}(C_H)^{0.5}(C_O) \exp(-E/RT) \qquad (3.13)$$

where T is the temperature in K, P is the pressure in pascals, C_H and C_O are molar concentrations of hydrocarbon and oxygen in kmol m^{-3}, t is the time in seconds, E is the activation energy in kcal/kmol, and the constants are given in Table 3.7.

Once a global scheme is selected, a determination of the carbon-to-hydrogen ratio to be used for the volatiles pseudomolecule C_nH_m in Eqn. (3.12) must be made. One possible approach is based upon ultimate analysis data of the char. All coal components other than char can be grouped together, and a simple material balance can be made to determine the carbon-to-hydrogen ratio of the group. The group ratio can then be used as a rough estimate of stoichiometric coefficients m and n for the global reaction. When tars represent a major fraction of the volatiles products, this estimate should give reasonable results.

TABLE 3.8. Summary of Global Reaction Rates for Various Major Species in Gaseous Hydrocarbon Flames (gmoles/cm³ s)[a]

$$C_nH_{2n+2} \rightarrow (n/2)C_2H_4 + H_2; \quad \frac{d[C_nH_{2n+2}]}{dt} = -10^{17.32} \exp(-49{,}600/RT)[C_nH_{2n+2}]^{0.50}[O_2]^{1.07}[C_2H_4]^{0.40}$$

$$C_2H_4 + O_2 \rightarrow 2CO + 2H_2; \quad \frac{d[C_2H_4]}{dt} = -10^{14.70} \exp(-50{,}000/RT)[C_2H_4]^{0.90}[O_2]^{1.18}[C_nH_{2n+2}]^{-0.37}$$

$$CO + 0.5O_2 \rightarrow CO_2; \quad \frac{d[CO]}{dt} = \{-10^{14.6} \exp(-40{,}000/RT)[CO]^{1.0}[O_2]^{0.25}[H_2O]^{0.50}\}7.93 \exp(-2.48\phi)$$

$$H_2 + 0.5O_2 \rightarrow H_2O; \quad \frac{d[H_2]}{dt} = -10^{13.52} \exp(-41{,}000/RT)[H_2]^{0.85}[O_2]^{1.42}[C_2H_4]^{-0.56}$$

[a]Data from Hautman et al. (1981).

More recently, Hautman et al. (1981) provided overall rates for H_2, CO, C_2H_4, and for alkanes (primarily from propane data):

$$C_nH_{2n+2} \rightarrow \tfrac{1}{2}nC_2H_4 + H_2 \tag{3.14}$$

$$C_2H_4 + O_2 \rightarrow 2CO + 2H_2 \tag{3.15}$$

$$CO + \tfrac{1}{2}O_2 \rightarrow CO_2 \tag{3.16}$$

$$H_2 + \tfrac{1}{2}O_2 \rightarrow H_2O \tag{3.17}$$

Rate expressions for these four reactions are summarized in Table 3.8 where coefficients for these rates are also shown. This sequence of reactions has been shown to be quite reliable for aliphatic hydrocarbons for a range of SR values (0.12–2), pressures (0.1–0.9 mPa or 1–9 atm) and temperatures (960–1540 K). If Eqns. (3.14) and (4.15) are added, an overall rate for consumption of the aliphatic hydrocarbon C_2H_{2n+2} results:

$$C_nH_{2n+2} + \tfrac{1}{2}nO_2 \rightarrow nCO + nH_2 \tag{3.18}$$

Here, no knowledge of the specific intermediate hydrocarbons (i.e., C_2H_4) is required.

3.6d. Volatiles Reaction Rates

Table 3.9 provides an estimate of times required to consume 50% of several gaseous reactant species, C_nH_m, C_nH_{2n+2}, CO, H_2, and HCN, at 1500 and 2000 K. All estimates were for the fuel and air at $SR = 1.0$. Other assumptions are

TABLE 3.9. Summary of Initial Reaction Rates and Consumption Times of Major Gas Species in Air[a] ($SR = 1$, $P = 0.1$ mPa)

Species	Fuel vol %	Eqn.	Initial rate (kmol/m³ s)		50% Consumption time (ms)	
			1500 K	2000 K	1500 K	2000 K
C_nH_{2n}[b]	1.65	3.18	-4.6	-159	0.23	0.004
(C_nH_m)[c] chain	1.65	3.13	0.016	0.106	21.4	2.4
(C_nH_m)[d] cyclic	2.72	3.13	39	169	0.008	0.002
CO[e]	29.6	3.16	12.3	-213	0.15	0.006
H_2	29.6	3.17	-26.7	-510	0.10	0.004
HCN[f]	10.71	3.19	$-2.3(10^{-9})$	$-4.5(10^{-7})$	4920	17.8

[a] Isothermal conditions.
[b] Assumes octane: rate shown is for decomposition to C_2H_4; consumption times are for decomposition to $CO + H_2$.
[c] Assumes octane.
[d] Assumes benzene.
[e] Assumes 2 mol-% H_2O in gas from coal moisture.
[f] Combustion to CO.

noted in the table. All of the oxidation reactions except for HCN are very fast (<0.2 ms except for aliphatic hydrocarbon at 1500 K by Eqn. (3.13)). With these very fast rates, treatment of volatiles reaction may not generally be required. For cyclic hydrocarbons, results are generally consistent with observations noted by Field et al. (1967) that the slowest, and therefore rate-controlling step, in the high-temperature combustion of hydrocarbons is the oxidation of CO. Aliphatic decomposition rates from the more recent work of Hautman et al. (1981) are much faster than those suggested by Siminski et al. (1972).

DeSoete (1975) have reported global oxidation rates for HCN by oxygen:

$$HCN + O_2 \xrightarrow{k} NO + \cdots \qquad (3.19)$$

where

$$k = 10^{10} \exp(-67{,}000/RT) \qquad (3.20)$$

From this expression, for HCN in air at 0.1 mPa (1 atm) and $SR = 1.0$, 50% consumption times are very long (5 s at 1500 K, 20 ms at 2000 K), compared to oxidation times for CO, H_2, and hydrocarbons. Thus, formation of NO from fuel nitrogen will often depend on the rate of HCN reaction.

Even with some estimate of volatiles reaction rates, the influence of gas turbulence on the rates makes kinetics calculations of volatiles consumption in a practical combustor a more difficult task with more uncertain results (see Chapters 11 and 12).

3.7. ILLUSTRATIVE PROBLEMS

1. Kobayashi *et al.* (1977) have proposed the following sequence of reactions of rapid pyrolysis of a high-volatile bituminous coal:

$$\text{Coal} \rightarrow 0.30\,V + 0.7\,C$$

$$\text{Coal} \rightarrow 1.0\,V$$

where

$$k_1 = 2 \times 10^5 \exp(-25{,}000/RT)\ (\text{s}^{-1})$$

$$k_2 = 1.3 \times 10^7 \exp(-40{,}000/RT)\ (\text{s}^{-1})$$

with $R = 1.987$ kcal/kmol K and T in K. If raw, moisture- and ash-free bituminous coal at 70-μm diameter is exposed to hot inert atmospheric gases at 1940 K for 0.05 s, what percent weight loss will the coal sustain? What if the total time is 0.2 s at 1000 K? How about 10 s at 1000 K? Compare your predictions with Figure 3.8 and explain any differences. Is the devolatilization process complete under these three sets of conditions?

2. For Problem 1, with the data of Figure 3.9, estimate the composition of the volatile matter at 1273 K (1000°C) and compare it with that at 1073 K (800°C).

3. A power-generating company is operating a 1000-MW plant with a high-volatile bituminous coal. They are considering pyrolyzing the coal first to obtain a liquid product for mobile fuel. The coal is 50-μm (mean) diameter. What pyrolysis temperature do you recommend? How much liquid product will be obtained? How much gaseous product? What combustion problems may develop in the power-generating unit?

4. For a high-volatile bituminous coal being devolatilized at 1273 K (1000°C) for 10 s, use the method of Solomon to predict both the extent of devolatilization and the final product composition (gas, char, and tar). Information on the composition of the raw coal together with the required kinetic parameters are given in Solomon (1980). Compare results with Problems 1 and 2.

5. Compare the times required to consume 50% of gaseous toluene and
n-butane at 0.1 mPa (1 atm) and 1273 K (1000°C) and with a stoichiometric ratio
of 0.4 (fuel rich). Consider these results in light of reaction rates at higher SR
and T values and speculate on the resulting potential for HC emissions or on
soot formation.

HETEROGENEOUS CHAR REACTION PROCESSES

Chapter 3 outlined general coal reaction processes and considered coal particle heatup, ignition, devolatilization, and volatiles combustion in detail. This chapter considers heterogeneous char reactions in detail.

4.1. REACTION VARIABLES

The time required for consumption of a char particle, which is often large compared to the time scale of turbulent gaseous fluctuations, is a significant part of the coal reaction process, and can range from 30 ms to over an hour. Modeling of the heterogeneous reaction process is complicated by:

1. Coal structural variations
2. Diffusion of reactants
3. Reaction by various reactants (O_2, H_2O, CO_2, H_2)
4. Particle size effects
5. Pore diffusion
6. Char mineral content
7. Changes in surface area
8. Fracturing of the char
9. Variations with temperature and pressure
10. Moisture content of the raw coal

Because of these uncertainties, modeling of this process has been highly dependent on laboratory rate data for the specific coal and test conditions used. Recent reviews of this important topic include Essenhigh (1977, 1981), Laurendeau (1978), Smith (1979), and Skinner and Smoot (1979).

The rate-limiting step in the oxidation of char can be chemical (adsorption of the reactant, reaction, desorption of products) or gaseous diffusion (bulk phase or pore diffusion of reactants or products). Several investigators, such as Walker *et al.* (1959) and Gray *et al.* (1976) have postulated the existence of different temperature zones or regimes which determine which resistance is controlling. In zone I, which occurs at low temperatures, or for very large particles, chemical reaction is the rate-determining step. Zone II is characterized by control due to both chemical reaction and pore diffusion. Zone III, which occurs at high temperatures, is characterized by bulk mass-transfer limitations. Figure 4.1 illustrates these zones graphically and shows the theoretical dependence of the reaction rate on particle diameter and oxidizer concentration. Kinetic data obtained by any investigator must be interpreted in light of the conditions under which the data were obtained.

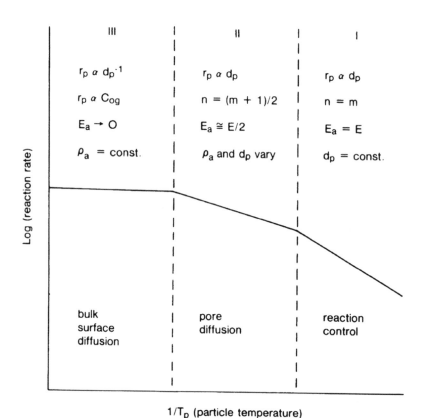

Figure 4.1. Rate-controlling regimes for heterogeneous char oxidation. (Figure used with permission from Smith, 1979.)

In zone I, the experimentally observed activation energy will be the true activation energy, and the reaction order will be the true order, since chemical reaction is the rate-determining step. In zone II, the observed activation energy will be roughly one-half the true value, while the observed or apparent reaction order n will be related to the true order m as follows (Smith, 1979):

$$n = \tfrac{1}{2}(m + 1) \tag{4.1}$$

In zone III, where bulk phase mass transfer is the limiting resistance, the apparent activation energy will be small, generally in the neighborhood of $20,000 \text{ kJ kmol}^{-1}$ (Gray et al., 1976). These results are also summarized in Figure 4.1.

In general, most experimenters have attempted to obtain data on the intrinsic chemical reaction rates of carbon or char with various gases. To do this, a variety of schemes have been tried to minimize the effects of diffusion. This includes efforts to conduct tests at low pressure or low temperature, by microsampling of gases very close to the solid surface, by use of high-velocity gas streams to decrease the boundary layer thickness, and by use of small particles (Field, 1964; Field et al., 1967). Field et al. (1967) have dismissed data obtained by the technique of reduced pressure because of difficulties in extrapolating the results thus obtained to atmospheric pressure and greater than atmospheric pressure.

The rates of carbon oxidation by steam and carbon dioxide are of the same order of magnitude, and are generally much lower than the reaction of carbon with oxygen, while the hydrogenation reaction (i.e., H_2 + char) is several orders of magnitude slower than the steam–char and CO_2–char reactions. For example, Walker et al. (1959) estimated the relative rates of the reactions at 1073 K and 10 kPa pressure as being 3×10^{-3} for H_2, 1 for CO_2, 3 for H_2O, and 10^5 for O_2. At higher temperatures, the differences in rates will not be so extreme, but rates for H_2, CO_2, and H_2O will still remain relatively unimportant as long as comparable concentrations of oxygen are present. Often, only the heterogeneous carbon oxidation reaction with oxygen needs to be considered for combustors (Field, 1964). In coal gasifiers, however, the reactions with steam and CO_2 can be quite important, especially after the rapid depletion of oxygen. Batchelder et al. (1953) calculated that the oxygen is essentially consumed in about 10 ms at the gasification temperatures encountered in a pulverized-coal reactor.

The H_2–char reaction will, under some conditions, be as important as the carbon–steam reaction (Walker et al., 1959). von Fredersdorff and Elliot (1963) cite experimental evidence that under the conditions of 3–20 mPa H_2 pressure and 750–1200 K, 200–400 μm char particles are completely reacted in about $\tfrac{1}{2}$ hr. Anthony and Howard (1976) have found that nearly complete conversion of raw coal can be achieved in a matter of a few seconds using high H_2 pressure; however, no particle sizes were mentioned.

In examining rate data from various sources, it is apparent that temperature and reactant concentration are not the only variables which influence reaction rates. The composition of the feed material, its pretreatment, and its thermal history are also important. Some experimenters have used pure forms of carbon, while others have used char. In general, it has been found that the more pure forms of carbon are less reactive than the chars. It has also been found that chars from the lower-rank coals are more reactive than those from higher-rank coals. Differences in reactivity at different heating rates and temperatures have been noted in chars formed from the same parent coal, but some of the differences in reactivity noted here may be due to additional volatiles reactions from the low-temperature chars, which were not totally pyrolyzed.

Other important variables are the internal and external surface areas of the char. Walker *et al.* (1953) have performed experiments with CO_2 and carbon which show that higher internal area leads to a higher reaction rate. Thus, under some conditions, reactant gases also diffuse into the particles before reacting. The observed increase in reaction rate with increasing burnoff, followed by a subsequent decline in rate, has been explained by some investigators in terms of expanding of the pores to expose more and more internal area (Thomas, 1977). The surface area continues to increase up to a certain point; then, due to coalescence of pores and thermal annealing, the area and reaction rate decreases. Wen (1972), on the other hand, explains the observed decline in terms of different macerals (microscopic organic constituents in the char) having differing activities. He postulates that the initial high activity is due to the reaction with aliphatic hydrocarbon sites and oxygenated functional groups. Once these are depleted, the low activity is due to the residual carbonaceous coke.

Another factor which may affect the oxidation rates is the presence of catalytic impurities in the char. Batchelder *et al.* (1953) cited studies which indicated that oxides and carbonate salts act as catalysts for all four of the reactions being considered: $C-CO_2$, $C-H_2$, $C-H_2O$, and $C-O_2$. Tomita *et al.* (1977) recently conducted experiments in which various minerals were added to high-purity carbon to test their effect on the rates of the various oxidation reactions. They found no catalytic effect for the minerals tested on the $C-O_2$ reaction, inhibition of the $C-CO_2$ reaction with included minerals such as pyrite, gypsum, kaolinite, and calcite, and both positive and negative catalytic effects on the hydrogasification reactions. As might be expected, catalytic effects, whether positive or negative, tend to be more important at low temperatures. At high combustion temperatures, their effects may be small.

4.2. GLOBAL REACTION RATES

It has been common practice (Smith, 1982) to relate measured char consumption rates to the external char surface area, even when pore reaction may occur.

The resulting rate is termed "global." The basic approach and selected data are discussed below.

4.2a.   Char Oxidation Data

Global laboratory data from several investigators for oxidation (with O_2) are shown in Figures 4.2–4.4 for bituminous coal chars, anthracites and semi-anthracites, and coke and pyrolysis char, respectively, for foreign coals. For

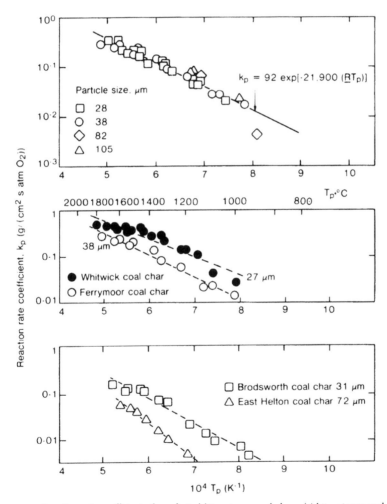

Figure 4.2.   Reaction rate coefficients for selected bituminous coal chars: (a) bituminous coal char (Field, 1969); (b) non-swelling bituminous coal (Field, 1970); (c) swelling bituminous coal (Field, 1970). (Figure used with permission from Smith, 1971b.)

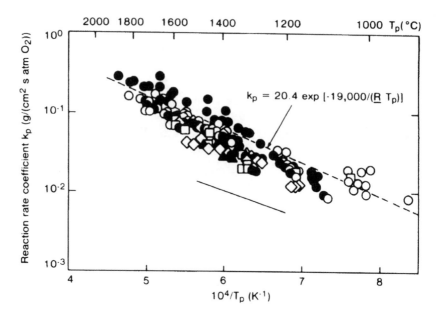

Combustion rate data for anthracites and semi-anthracites

Author	Smith (1971a)	Smith (1971b)	Field (1970)	Field & Roberts (1967)
Symbol	●	○	◇	▲
Size (μm)	78,49,22	72,42	28,25	31,24,23

Author	Marsden (1964) (single value)	Hein (1970)	Beér et al. (1964)
Symbol	△	□	—
Size	polydisperse	polydisperse	polydisperse

Figure 4.3. Reaction rate coefficients for anthracites and semianthracites. (Figure used with permission from Smith, I. W., Feb., 1979.)

these data, the specific coal reaction rate (R_{po}, (g oxidizer consumed) $m^{-2} s^{-1}$, based on external surface area), expressed as a first-order, heterogeneous reaction is

$$R_{po} = \rho_{og}/[(1/k_m) + (1/k_r)] \tag{4.2}$$

where ρ_{og} is the oxidizer concentration (g/m^3) in the bulk gas, k_m is the oxidizer

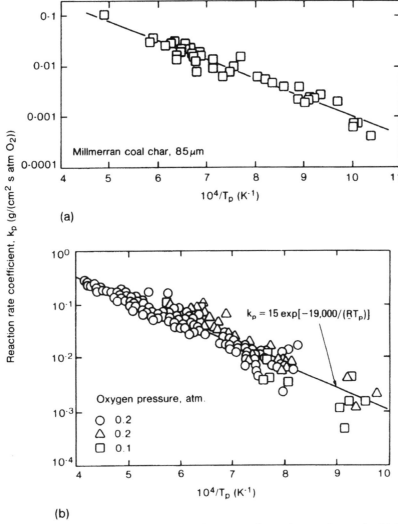

Figure 4.4. Reaction rate coefficients for a flash pyrolysis char and a petroleum coke: (a) flash pyrolysis char (Young, private communication); (b) petroleum coke (Mulcahy and Smith, 1971; Smith 1971a; Young (private communication)). (Figure used with permission from Smith, I. W., Feb., 1979.)

mass-transfer coefficient to the particle surface, and k_r is the reaction rate coefficient for O_2 with the char. The specific coal reaction rate R_p is computed from weight of the particle, the mean initial external surface area, and the reaction time. The consumption rate of the char is related directly to that of the oxidizer R_{po} once the reactants and products are specified (see Eqn. (4.6)).

The particle temperature T_p is back-calculated from measured gas temperature. Then, k_m, the mass-transfer coefficient for oxidizer transfer from the bulk gas to the particle surface, is computed. With R_p, ρ_{og}, and k_m, the specific reaction rate coefficient k_r is computed for each experiment. k_r is assumed to have an Arrhenius form of temperature dependence:

$$k_r = A \exp(-E/RT_p) \tag{4.3}$$

For a family of experiments at various T_p values, reaction rate values are plotted as $\ln k$ vs. $1/T_p$, as illustrated in Figures 4.2–4.4 with one additional variation. Smith (1979) elected to plot the parameter k_p, which is the reaction rate coefficient in the expression $r_{po} = k_p A_p P_{O_2}$. Thus, k_p is based on pressure and has units of $g/cm^{-2} s^{-1} (atm \ O_2)^{-1}$. By comparison, k_r has units of $g/(cm^2 \ s\text{–}g/cm^3)$ or cm/s. Thus, from k_p values of Figures 4.2–4.4, k_r values can be computed from $k_r = k_p T_g R/M_{O_2}$.

Observations from these data include the following:

1. For a given char or coke, the variation with temperature is reasonably well correlated in Arrhenius form, supporting an activated reaction process.
2. At a given temperature, reaction rate coefficient values differ greatly among the chars. For example, at about 1500 K, k_p values for the several cases illustrated vary from about 0.01 to 0.1, or by an order of magnitude.
3. For these small dust particles (20–100 μm), k_p is not a function of particle size, suggesting that the external surface area adequately correlates the char oxidation rate, even though reactions may be occurring internally.
4. From the slopes of the data correlations in Figures 4.2–4.4, the activation energy E can be determined. For these data, E values range from about 75 to 142 kJ/g-mole (18–34 kcal/mole).
5. These data are correlated with the assumption that the surface oxidation rate is first order in oxygen concentration, as specified by Eqn. (4.2) and indicated by the units on k_p. There is still considerable uncertainty on reaction order and activation energy as illustrated by the data of Table 4.1. From this summary of char–oxygen data from several investigators, reaction orders from zero to unity have been identified, while E/R values (K) vary from 3000 to over 30,000.

Recently, Goetz et al., (1982) reported char reaction rate parameters for four typical U.S. coals. Results were reported for reaction with O_2 and CO_2. Measurements were made in a drop-tube furnace. Properties of the parent coals are shown in Table 4.2. Chars were prepared from 200 to 400 mesh parent

TABLE 4.1. Rate Parameters for Char Reaction with Oxygen for Rate Equation $= r = AT^N \exp(-E/RT)P_{O_2}^{a}$

Reference	A (kg m^{-2} s^{-1} kPa^{-n} K^{-N})	E/R (K)	N	Order n	Particle type	Size graded	Sizes (μm)	Temperature range (K)
Smith and Tyler (1974)	1.32×10^{-1}	16,400	0	0	Brown coal char	Yes	22, 49, 89	630–1812
Field et al. (1967)	8.6×10^{2}	18,000	0	1	Various fuels	Varied	Varied	950–1650
Essenhigh et al. (1965)	—	20,100	0	0, 1[b]	Carbon	Yes	2.54×10^{4}	—
Howard and Essenhigh (1967)	—	3000–6000 15,000–32,700	0	1[c] 0	Bituminous char	No	0–200	—
Nettleton (1967)	—	6500–25,000[d]	1.75–3.5	1	Various coals	Yes	420–1000	1100–1500
Hamor et al. (1973)	$9.18\text{-}10^{-1}$	8200	0	0.5	Brown coal char	Yes	22, 49, 89	630–2200
Smith (1971b)	2.013	9600	0	1	Semianthracite	Yes	5, 22, 49, 78	1400–2200
Smith and Tyler (1971)	5.428	20,100	0	1	Semianthracite	Yes	6, 22, 49, 78	1400–2200
Sargeant and Smith (1973)	2.903[e]	10,300[e]	0	1	Bituminous char	Yes	18, 35, 70	800–1700

[a] Data from Skinner and Smoot (1979).
[b] 0 for $T < 1000$ K; 1 for $T > 1000$ K.
[c] 1 indicates adsorption control before flame front; 0 indicates desorption control in tail of flame.
[d] Most values between 11,600–14,600 K.
[e] Calculated from plot in Sergeant and Smith (1973).

TABLE 4.2. Analyses of Parent U.S. Coals[a]

Analysis	Texas (Monticello) lignite A		Wyoming (Jacobs Ranch Range) Subbituminous C		Illinois No. 6 (Freeman) Hv bituminous Cb		Pittsburgh No. 8 (Delton) Hv bituminous Ab	
	As received	Dry ash-free	As received	Dry ash-free	As received	Dry ash-free	As received	Dry ash-free
Proximate (wt %)								
Moisture (total)	29.3	—	28.0	—	19.4	—	0.8	—
Volatile matter	30.0	54.0	32.2	48.6	31.3	43.6	38.1	42.1
Fixed carbon	25.5	46.0	34.2	51.4	40.4	56.4	52.6	57.9
Ash	15.2	—	5.6	—	8.9	—	8.5	—
Total	100.0	100.0	100.0	100.0	100.0	100.0	100.0	100.0
Ultimate (wt %)								
Moisture (total)	29.3	—	28.0	—	29.4	—	0.8	—
Hydrogen	3.2	5.7	3.5	5.2	3.6	5.1	5.1	5.6
Carbon	39.4	71.0	49.3	74.3	53.0	73.9	75.9	83.7
Sulfur	1.0	1.9	0.4	0.6	3.0	4.2	2.5	2.8
Nitrogen	0.7	1.3	0.8	1.1	1.0	1.4	1.4	1.5
Oxygen (diff.)	11.2	20.1	12.4	18.8	11.1	15.4	5.8	6.4
Ash	15.2	—	5.6	—	8.9	—	8.5	—
Total	100.0	100.0	100.0	100.0	100.0	100.0	100.0	100.0
Higher heating value								
kJ/kg	16,092	28,964	19,871	29,871	21,813	30,406	31,755	35,010
(Btu/lb)	(6920)	(12,455)	(8525)	(12,845)	(9380)	(13,075)	(13,655)	(15,055)

[a]Data from Goetz et al., 1982.

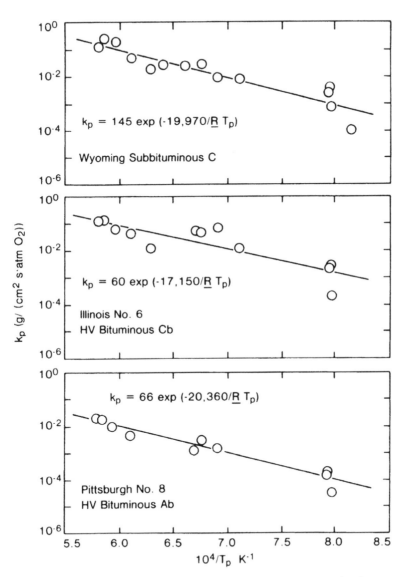

Figure 4.5. Char reaction rate coefficients in oxygen for three U.S. coals. (Figure used with permission from Goetz *et al.*, 1982.)

coal in the furnace at 1750 K in a nitrogen atmosphere prior to oxidation tests. Extent of combustion was determined from oxygen depletion, while extent of reaction with CO_2 was determined from CO concentration.

Char combustion was performed at 0.1 mPa (1 atm) with $P_{O_2} = 1-5$ kPa (0.01–0.05 atm), and over a temperature range of 1250–1730 K. Key assumptions in data reduction were first-order reaction, CO as the surface product of combustion, and shrinking core particle based on external surface area. Figure 4.5 shows the reaction rate coefficients (k_p values) for three of the U.S coals of Table 4.2. Additional observations relating to these data include the following:

1. Specific rate coefficients vary by about an order of magnitude for the three different U.S. coals. The decrease was specifically associated with a measured decrease in internal surface area.
2. Char rate coefficient decreased with increasing coal rank.
3. The reaction rate values are of the same order as those of the non-U.S. coals of Figures 4.2–4.4.
4. Small variations in activation energy 71–84 kJ/g-mole (17–20 kcal/g-mole) were observed among the three coals.

Specific reaction rate coefficient data for surface reaction with CO_2 are shown in Figure 4.6 for the four U.S. coals. Values also vary by over three orders of magnitude among the four coals. Again, rates varied directly with coal rank, and more specifically with measured char internal surface area. Reaction rate coefficients for CO_2 are also up to six or seven orders of magnitude lower than values for O_2 at a given temperature. Activation energies for CO_2–char reaction are much greater than for O_2–char reaction, ranging from 146 to 234 kJ/g-mole (35–56 kcal/g-mole). These values provide specific rate data for a family of commonly used U.S. coals.

Recent results for O_2–char reactions are also reported by Dutta and Wen (1977), Mandel (1977), Froberg and Essenhigh (1979), and Young and Smith (1981). Other specific data are also available for reaction of various chars with CO_2, and with steam and H_2. Much of these data are summarized by Skinner and Smoot (1979) and more recently and more exhaustively by Essenhigh (1981). Recent data on coal and char reactions with carbon dioxide (Mandel, 1977; Yang and Steinburg, 1977; Dutta et al., 1977), steam (Linares-Solano et al., 1979; Otto et al., 1979a, 1979b; Feistel et al., 1978), and hydrogen (Chanhon and Longenbach, 1978) have also been reported. Reactions of char with CO_2 and H_2O will be particularly important in gasification applications where oxygen is rapidly depleted before the char is consumed.

4.2b. Effects of Coal Moisture on Char Oxidation

The presence of large quantities of moisture will have some impact on the oxidation of residual chars. This would be particularly true at high temperatures

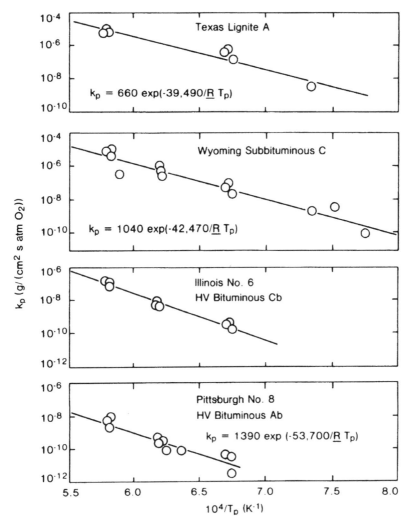

Figure 4.6. Char reaction rate coefficients in CO_2 for four U.S. coals. (Figure used with permission from Goetz *et al.*, 1982.)

and low oxygen concentrations such as occur during gasification. Essenhigh (1981) and Skinner and Smoot (1979) review analytical aspects and experimental data for the C–H_2O reaction. This reaction is about 10^3–10^5 times slower (Walker *et al.*, 1959) than the C–O_2 reaction at combustion temperatures. At higher temperatures, the differences are reduced.

In a review of 166 experimental studies of heterogeneous reaction of carbonaceous substances with an oxidizer (e.g., CO_2, H_2O, O_2, H_2, air, etc.), only one investigator (Ref. 327, 328 in Essenhigh, 1981) reported work with C and O_2 in dry and moist conditions. In this study, the order of the reaction between O_2 and graphite at 0.1 mPa (1 atm) and at low temperature (only 850–925 K) was reported to have increased from about 0.6 to 0.65 with use of moist coal. However, conditions were far removed from those of combustion or gasification. Evidence shown subsequently suggests that variation of moisture level may not be a major issue in the direct combustion of coal.

4.2c.   Effects of Mineral Matter

Properties of mineral matter in coal were considered in Chapter 2, where as many as 35 elements were identified in the ashes of typical coals (Table 2.6). Percentages of these elements in ash range from high values (Si, Al, Fe, Ca) to only trace quantities. Further, ash quantities in coals vary greatly from a few percent to half or more of the total coal. Large quantities of highly variable mineral matter can have significant impact on the combustion and gasification of coal.

Major reviews have been reported on the properties of ash (Bryers, 1978, etc., Gluskoter et al., 1981), on the effects of mineral matter on the combustion of coal particles (Essenhigh, 1981), and on combustion systems (Reid, 1981; Wall et al., 1979; Nettleton, 1979). Yet the behavior and effects of ash remain a major unresolved issue in the utilization of coal. New physical and chemical cleaning methods are also being developed to reduce the percentage of ash in coal prior to combustion (Liu, 1982).

The presence of ash and of the specific minerals in ash can have several potential effects on the combustion of coal, including the following:

1. *Thermal Effects.*   Large quantities of ash change the thermal behavior of particles. Ash consumes energy as it is heated to high temperatures and changes phase.
2. *Radiative Properties.*   Radiative properties of ash differ from those of char or coal (Wall et al., 1979); and the presence of the ash provides a solid medium for radiative heat transfer when the carbon is consumed.
3. *Particle Size.*   Char particles, toward the end of burnout, tend to break into smaller fragments (Sarofim et al., 1977). This breakup process is undoubtedly related to the quantity and nature of mineral matter in the char.
4. *Catalytic Effects.*   Various minerals in the char have been shown to cause increases in char reactivity, particularly at low temperature. For example, Essenhigh (1981) reports a factor-of-30 increase in lignite char reactivity at 923 K as calcium percentage in the char (achieved

through demineralization and ion exchange) was varied from near zero to 13%. However, these catalytic effects are likely far less prominent at high temperature. Walker *et al.* (1968) provides a comprehensive review of the effects of catalysts on carbon consumption and also discusses mechanisms.

5. *Hindrance Effects.* Mineral matter provides a barrier through which the reactant (e.g., oxygen) must pass to reach the char. Particularly toward the end of burnout, it is possible that high quantities of mineral matter will impede combustion. This can be worsened by softening and melting of the mineral matter (Bryers, 1978).

The above effects deal directly with the impact of mineral matter on the consumption of char. Mineral matter also plays a vital role in the operation of practical combustion systems (Wall *et al.*, 1979; Reid, 1981). These effects include fouling and slagging on reactor walls and heat-transfer tubes, sulfur pollutant formation, radiative heat transfer, corrosion, vaporization of trace metals, and formation of fly ash.

4.2d. Global Modeling of Char Oxidation

It is assumed that the oxidizer (e.g., O_2) must diffuse from the bulk gas, adsorb onto the surface of the particle, and react with the carbon in the char. The rate of reaction will relate to the diffusion and chemical reaction rates and to the specific products formed. For example, if CO were the product, the reaction rate may be more rapid than if CO_2 were the product, since less oxygen would be required.

If the particle is porous, then diffusion of the 'product into the porous volume and reaction on the larger internal surface area will influence the overall reaction rate. Because the details of this complex internal structure are not always known, it has been common to relate the char reaction rate to the external surface area. This type of char model is referred to as a global model.

Assuming a heterogeneous surface reaction, the reaction rate (kg/s) for a single spherical char particle is

$$r_k = v_s M_p k_r \xi_p A_p C_{op}^n \tag{4.4}$$

where C_{op} is the molar concentration of the oxidizer in the gas phase at the surface of particle; A_p is the external surface area of the sphere (or equivalent sphere); k_r is the rate constant for the heterogeneous reaction; M_p is the molecular weight of the reactant in the particle (e.g., carbon); v_s is the stoichiometric coefficient to identify the number of moles of product gas per mole of oxidant (i.e., 2 for CO product; 1 for CO_2); ξ_p is the particle area factor to account for internal surface burning (effective burning area of the entire particle/external area of the equivalent spherical particle); and n is the apparent order of the reaction. Smith (1979, 1982) has reviewed the question of the products of char oxidation, and suggests that carbon monoxide is most likely.

By analogy to Newton's law of cooling, the oxidizer diffusion rate is often expressed as

$$r_{do} = k_m M_o A_p (C_{og} - C_{op}) \qquad (4.5a)$$

This analogy is only correct for equimolar counter diffusion in a stagnant film or for very low rates of mass transfer. More generally, the diffusion rate requires a contribution from the bulk flow or motion induced by the mass transfer itself:

$$r_{do} = k_m M_o A_p (C_{og} - C_{op}) + r_d C_{og} / C_g \qquad (4.5b)$$

where r_d is the total diffusion rate and C_g is the total molar gas concentration. The total diffusion rate r_d includes the contribution of the oxidizer r_{do} and thus Eqn. (4.5b) is implicit in the diffusion rate of the oxidizer.

The particle reaction rate, determined from the diffusion of oxidizer, is

$$r_p = v_s r_{do} M_p / M_o \qquad (4.6)$$

When the particle is burning in a quasi-steady manner, the rate of diffusion of the oxidizer must also equal the rate of consumption of the oxidizer (i.e., $r_{do} = r_{ko} = r_p$). Then, if the order of the reaction (n) is unity, Equations (4.4)–(4.6) are often combined straightforwardly to eliminate r_{do} and the unknown quantity C_{op} to give

$$r_p = A_p v_s M_p C_{og} / [(1/k_r \xi_p) + (1/k_m)] \qquad (4.7a)$$

When $\xi_p = $ unity, this result is equivalent to Eqn. (4.2), which was the basis for reporting the data of Figures 4.2–4.6. Although Eqn. (4.7a) is attractive and often used, it neglects the bulk diffusion term as shown in Eqn. 4.5b. The correct expression is implicit in the reaction rate:

$$r_p = A_p v_s M_p C_{og} \left/ \left(\frac{1}{k_r \xi_p} + \frac{1}{k_m} + \frac{r_{pt}}{A_p m_g C_g k_r \xi_p k_m} \right) \right. \qquad (4.7b)$$

where r_{pt} is the total reaction rate of the particle (Smith, 1979). When the reaction order is nonunity, Eqns. (4.2)–(4.6) are still appropriate and can be solved iteratively.

For a single particle, the rate of mass change is thus:

$$\frac{dm_p}{dt} = -r_p = \frac{d(\rho_{ap}V_p)}{dt} = \frac{d[\rho_{ap}(\pi/6)d_p^3]}{dt} \qquad (4.8)$$

Two assumptions are commonly made in tracking particle burnout:

1. The particle is highly porous and burns internally as well as externally with variable density and near-constant diameter; or,
2. The particle is a uniform solid and burns with constant density and shrinking diameter.

In the first case:

$$dm_p/dt = V_p d\rho_{ap}/dt \qquad (4.9)$$

while in second, if the particle is assumed to be spherical:

$$dm_p/dt = \rho_{ap}(\pi/2)d_p^2 d(d_p)/dt \qquad (4.10)$$

Alternatively, Smith (1979, 1982) presents correlations of particle size *vs.* fractional char burnoff that could be used to establish the relationship between d_p and ρ_{ap}.

To compute the time required to consume a char particle, the following is required (for the variable-diameter case):

Parameter	Units	Definition	Source
ρ_{ap}	kg/m^3	Bulk or apparent density of the char particle	Lab measurements (see Table 2.4)
A_p	m^2	External surface area of equivalent sphere	Initial value specified; can vary during burnout
v_s	—	Stoichiometric coefficient	2 for CO 1 for CO_2
M_p	kg/kmol	Molecular weight of the combustible fuel	12 for carbon
k_r	m/s	Reaction rate (for $n = 1$)	$A \exp(-E/RT_r)$
A, E	m/s kcal/kmol	Reaction rate constants	Lab data (Figures. 4.2–4.6)
ξ_p	—	Surface area factor	From lab measurements—usually unity to correspond with basis for reported data
k_m	m/s	Mass-transfer coefficient	$\sim 2\,D_{om}/d_p$ with low Re_p and r_p
D_{om}	m^2/s	Oxidizer diffusion coefficient	Data tables

The internal surface area of the char can be far greater than the external area. However, if the basic char oxidation data have been based on the external area, then this factor (ξ_p) is contained in the reported value of k_r, and will thus be taken as unity.

While the use of this model and the data correlations provide a basis for estimating char oxidation rates, many questions remain. Influences of pore diffusion affect the values of k_r and are not modeled by Eqns. (4.2) and (4.3). Devolatilization influences the nature of the char, and the char also changes characteristics during burnout (Smith, 1979, 1982). Reaction order (n) has been established over small variations in oxygen partial pressure. Much less information is available for CO_2, H_2O, or H_2 (Essenhigh, 1981). Impurities in the coal apparently influence char reactivity. Ash may accumulate on the particle surface, increasing resistance to diffusion of oxidizer and products. The particles can fracture into several smaller pieces toward the end of burnout (Sarofim et al., 1977). High reaction temperatures can lead to slag droplet formation on the surface of the char.

Because of the complexity of these coal particle reaction processes, it is emphasized that a realistic model must be closely related to laboratory data for the same type of coal under similar conditions of pressure, temperature, size, and oxidizer concentration.

4.3. INTRINSIC CHAR REACTION RATES

4.3a. Background

Methods outlined above for describing char reaction are based on the external area of the char particle and do not formally account for any of the following:

1. Continued swelling and pyrolysis during char oxidation.
2. Porosity of the char and changes in the pore structure as the char is consumed.
3. Internal diffusion of oxidizer and products in the porous char.
4. Internal reaction of oxidizer and char surface.
5. Influences of carbon structure and char impurities on consumption.
6. Fracturing of particles toward the end of consumption.

A more realistic treatment of char reaction would be to formally account for the internal char reaction. If the specific internal structure were described (A_i, defined as square meters of total area per kilogram of particle) and the true order of the reaction known (m), the instantaneous rate of consumption of char could be computed from the intrinsic reaction rate k_i (reaction rate based on the area of the pore walls) in the absence of internal diffusion limitations (compare to Eqn. (4.4)):

$$r_{pi} = k_i A_t C_{og}^m \qquad (4.11)$$

Work has been conducted to measure r_{pi} and m for various carbonaceous solids and to provide computational methods for predicting A_t during char consumption and to account for internal diffusion resistance.

4.3b. Intrinsic Rate Data

It has been clearly established that char oxidation can occur in the internal pores of char particles. Smith (1982) notes that internal surface areas vary from about 1 to 10^3 m^2/g for chars and cokes.

Char reactivity (grams of char consumed per minute per gram of char present) with CO_2 was recently measured (Wegener, 1982) for five process-development chars at 1173 K (900°C) in a small fixed bed of char particles. Char samples had been preheated in an inert atmosphere to temperatures above 1173 K (900°C) before performing CO_2 reactivity tests in order to eliminate volatiles products. Table 4.3 summarizes some of the characteristics of the chars, including internal surface area (BET, Brunaver–Emmett–Teller) measured by adsorption of nitrogen gas. Figure 4.7 shows char reactivity as a function of temperature. Reactivities of the various charts differ by nearly two orders of magnitude. Reactivities of the two chars with internal surface areas (BET) of 200–220 m^2/g are an order of magnitude or more above that for the three chars whose internal surfaces areas are less than 20 m^2/g. Internal surface area is not the only factor governing char reactivity. However, from the results of Table 4.3 and Figure 4.7, it is obviously a major factor. These observations were also noted by Goetz et al. (1982) for char reactivities of the parent coals of Table 4.2, as illustrated in Figure 4.8. Reactivity data are shown for both thermogravimetric analysis (TGA) and high-temperature gasification methods, and relate to internal surface area.

Char reactivity in oxygen at low pressure and lower temperatures (550–750 K) have also been measured (Radovic and Walker, 1984) for the same process chars of Table 4.3. Strong variation of reactivity with temperature and with char type shown in Figure 4.7b are similar to those in Figure 4.7a at higher temperature in CO_2. However, the order among char reactivities differs somewhat for the two different sets of tests. These differences are likely due in part to differences in preparation methods (1173 vs. 973 K), and to differences in grinding or classification prior to testing, as well as to the different temperature, pressure, and oxidant.

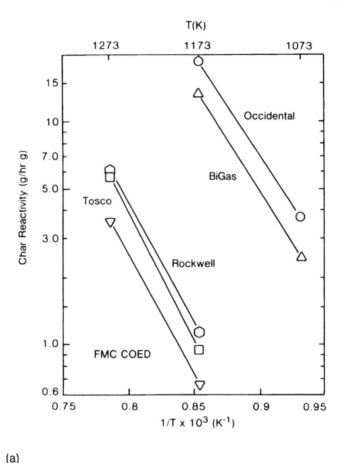

(a)

Figure 4.7. Effect of temperature and char type on reactivity: (a) char reactivity in CO_2 (0.2 mPa), (unpublished data for Wegener, 1982); (b) TGA 0.1 mPa air (data from Radovic and Walker, 1984).

For comparison with the char reactivity data of Figure 4.7 at low and intermediate temperatures, char reactivity has also been measured at high temperature in a drop-tube furnace (Wells *et al.*, 1984). Tests were made in air for a particle temperature range of about 1500–2100 K for the chars of Table 4.3. Results are shown in Figure 4.8 for classified char fractions, where the average mass-mean diameter was about 68 μm. These test data have been evaluated with methods documented by Smith (1982), with external diffusion effects considered. The reaction rate coefficient values are based on per unit of external particle surface area. However, the external areas of each of the char

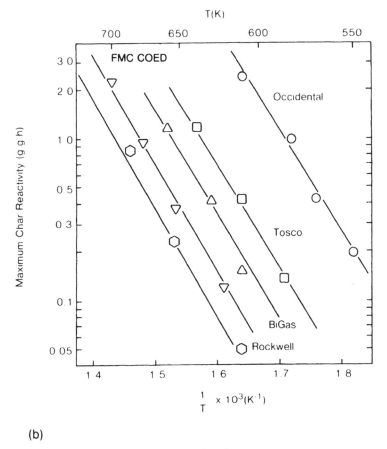

(b)

Figure 4.7 (cont.)

types were comparable, so comparison of reactivities among chars on a mass basis would produce a similar result.

Results of Figure 4.9 show very high reactivities at these higher temperatures. Further, much smaller differences are observed among the chars, compared to the dramatic differences shown in Figure 4.7. Results imply that the reaction occurs far less on the internal surfaces at higher temperatures. Thus, variations in the internal surface area do not cause such marked changes in reaction rate. External diffusional rates have become an important controlling factor in consumption rate at these high temperatures, together with surface reaction.

These results, when compared to those of Figure 4.7, dramatically illustrate the variation in char reactivity among chars over a wide temperature range.

TABLE 4.3. Properties of Process-Development Chars for Reactivity Tests[a]

Process ID	Parent coal type	Char ash[b] (%)	Char d_p (μm)	Char (% volatile matter)	Char preparation temperature[c] (K)	BET (N$_2$) surface area[d] (m²/g)	Char reactivity[e] (g/min per grams of char)
Occidental	Wyodak subbituminous	12	63	15	870–1070	205	0.31
BiGas	Montana Rosebud bituminous	22	260	7	1270	227	0.16
Rockwell	Kentucky bituminous	18	110[f]	10	1170–1270	3	0.022
Toscoal	Utah bituminous	18	90[f]	17	770	8	0.019
FMC COED	Pittsburgh bituminous	14	60[f]	10	670–1170	17	0.013

[a] Reactivity and BET (N$_2$) measurements from Wegener (1982). Characterization data from BYU Combustion Laboratory (1982).
[b] Dry basis.
[c] All chars were initially heated to above 1170 K (900°C) for subsequent char reactivity tests.
[d] After grinding.
[e] At 50% carbon conversion and 1173 K.
[f] Reduced to this size from original process char by grinding.

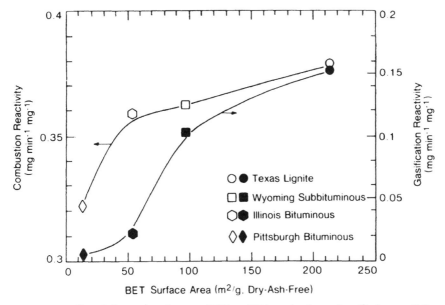

Figure 4.8. Effect of char pore surface area (BET) on TGA combustion and gasification reactivity for 200×400 mesh chars. (Figure used with permission from Goetz *et al.*, 1982).

They also suggest that control of internal structure is not so important to char consumption at elevated temperatures.

Smith (1982) reported intrinsic reaction rates of several porous carbonaceous materials in oxygen including chars, cokes, graphite, and highly purified carbons, as shown in Figure 4.10. Data are in terms of specific particle consumption rate R_{pi} at atmospheric pressure in pure oxygen. The correlation of Figure 4.10 is shown for data over 11 orders of magnitude in char reactivity for temperatures from 700 to 2000 K. Yet, reactivity also varies by up to four orders of magnitude for a given solid fuel at fixed temperature. According to Smith (1982), the wide range of intrinsic reactivities observed for various substances at a single temperature is due to differences in both the carbon structure and to impurities in the solids and gaseous reactants. Intrinsic reactivity values for highly purified carbons show improved correlation with temperature.

4.3c. Conceptual (Macroscopic) Model of Intrinsic Reactivity

Experimental results suggest that knowledge of internal surface area and internal diffusion effects are important factors in determining heterogeneous char consumption. In spite of the difficulty of developing an adequate model with the limited data available, recent articles or reviews (e.g., Laurendeau,

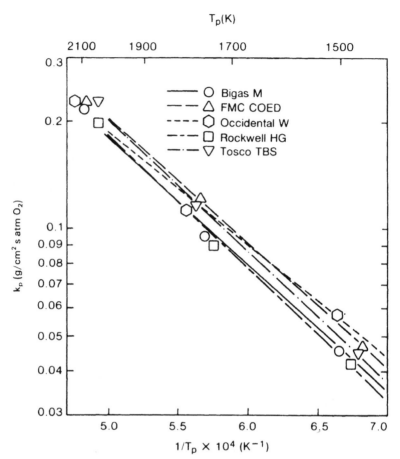

Figure 4.9. Reaction rate coefficients of selected process chars at high temperature in air at near-atmospheric pressure.

1978; Essenhigh, 1981; and Smith, 1982) have stressed the importance of implementing intrinsic rate expressions. Intrinsic treatment allows the greatest potential of differentiating among the various char reactivities on the basis of fundamental kinetic parameters. This necessitates a knowledge of not only the intrinsic kinetics and reaction conditions, but also a knowledge of the initial pore structure and its changes during char consumption. Two different classifications of pore structure models have been defined—macroscopic and microscopic (Laurendeau, 1978). Macroscopic modeling of pore diffusion constitutes the majority of the existing char models. It utilizes an effective diffusivity throughout the particle, and includes the classical unreacted shrinking core model and the progressive conversion model. An intrinsic coal particle consump-

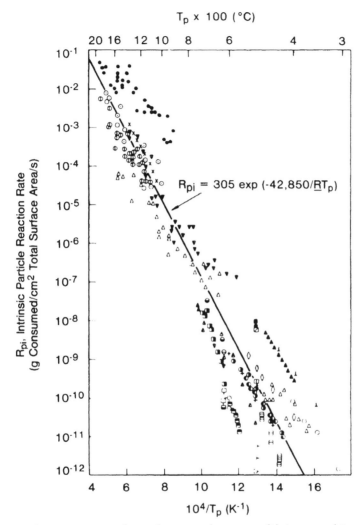

Figure 4.10. Intrinsic reactivity of several porous carbonaceous solids in oxygen (at an oxygen pressure of 0.1 mPa (1 atm)). (Figure used with permission from Smith, 1982.)

tion model which incorporates a macroscopic pore model is considered in this subsection.

Smith and co-workers (1974, 1978, 1982) have utilized a series of simplified theoretical equations for calculating particle consumption rates. These equations relate the intrinsic kinetics for a given porous structure and gaseous environment to char reactivity based on external surface area or weight of unreacted char. The intrinsic reaction rate coefficient k, is defined by Eqn. (4.11) in the absence of any mass transfer or pore diffusional limitation as the rate of particle reaction per unit of total surface area per (unit of oxidizer concentration)^m. The exponent m is the true order of reaction.

Thus, the mass rate of consumption of a char particle with mass transfer or pore diffusional limitations is

$$r_{pi} = \eta k_i A_t C_{os}^m$$

$$= \eta k_i A_t C_{og}(1 - \chi)^m \tag{4.12}$$

where $\chi = r_{pi}/r_{p\infty}$ and η is the effectiveness factor, which is defined as the particle consumption rate over the maximum value in the absence of pore-diffusional limitations. The second equality, shown by Smith (1982), provides an expression for any m in terms of the known, bulk gas/oxidizer concentration C_{og}. Thus, computation of the particle consumption rate from intrinsic reaction rate data requires a value for η together with the total particle surface area and intrinsic rate coefficient.

Based on a formal treatment of internal pore diffusion and internal particle surface reaction, Mehta and Aris (1971) have shown the relationship between the effectiveness factor η, the true order of reaction m, the effective diffusion coefficient of oxidizer in the particle pores, and other parameters already defined. This relationship is shown graphically in Figure 4.11, where a new variable ϕ, the Thiele Modulus, is defined (Smith, 1982):

$$\phi = (V_p/A_p)(A_t\rho_{ap}k_iC_{os}^{m-1}/D_e)^{0.5} \tag{4.13}$$

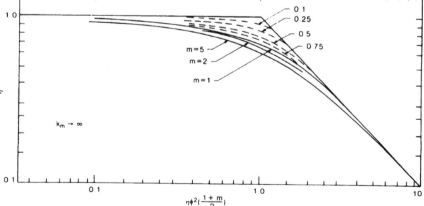

Figure 4.11. Effectiveness factors vs. $\eta\phi^2$ for infinite k_m and various m values. (Figure used with permission from Mehta and Aris, 1971.)

The development of this graphical relationship was for the case where external diffusion from the bulk gas to the particle surface was fast ($k_m \rightarrow \infty$). However, the general use of the relationship on Figure 4.11 is appropriate, since Eqn. (4.12) has accounted for the effects in external diffusion.

The application of the preceding equations is possible only if a value of the effective pore diffusion coefficient D_e is available. D_e is often calculated from the relation (Smith, 1982):

$$D_e = D_k \theta / \tau \tag{4.14}$$

where

$$\theta = [1 - (\rho_a / \rho_s)]$$

In the above equation, θ is the porosity, τ is the tortuosity, ρ_s is the density of the solid material in the particle, and D_k is the Knudsen diffusion coefficient. Tortuosity τ is an empirical factor used to adjust computed D_e values to measured values, and has been observed to reach values as high as 10 (Smith, 1982). Knudsen diffusion substantially dominates other forms of diffusion when the surface area is associated with pores less than 1 μm. The Knudsen diffusion coefficient may be approximated by

$$D_k = 97 r_a (T/M_o)^{0.5} \tag{4.15}$$

where r_a is the average pore radius m given by

$$r_a = 2\phi \tau^{0.5} / A_t \rho_{ap} \tag{4.16}$$

This brief presentation of a macroscopic theory of intrinsic particle reaction provides a basis for defining the characteristics of the various zones of particle reaction outlined in Figure 4.1. In zone I, the kinetically controlled region, $\eta \rightarrow 1$, $C_{os} \rightarrow C_{og}$, and $\chi \rightarrow 0$ in Eqn. (4.12). For zone II of Figure 4.1, pore diffusion controls, and can be expressed analytically as a function of the effective diffusivity, leading to an explicit expression for the particle consumption rate, in terms of the intrinsic rate coefficient k_i. In zone III, bulk diffusion controls, and the particle consumption rate, which is independent of k_i, is computed directly from Eqn. (4.5) with $C_{op} \rightarrow 0$.

Application of Eqn. (4.12) to the computation of r_{pi} requires specific values of internal surface area, porosity, apparent particle density, particle size, tortuosity, and intrinsic surface reaction rate, most of which are generally not known. This problem is further complicated if the time for particle consumption is required, since Figure 4.1 must then be integrated. The reaction rate r_{pi} is time dependent since the internal pore structure varies with time, causing changes in η, A_t, τ, and ρ_a. Thus, either experimental data are required for the time variation of these parameters, or some theory is required to estimate these parameters in time. The next section reviews available theories briefly, based on information taken principally from Sloan (1982).

4.3d. Extended (Microscopic) Intrinsic Reactivity Models

The microscopic model of pore diffusion attempts to describe diffusion through a single pore and then to predict the overall particle consumption rates by an appropriate statistical description of the pore size distribution. Microscopic models treat variation of diffusivity with position and burnoff, intrinsic rate expressions with variable reaction orders, and development of pore intersection and expansion due to reaction. Additional complications may include the effects of blind pores, pore coalescence, and particle fragmentation.

Some recent models, which approach a microscopic treatment, include those of Simons and co-workers (1979a, 1979b), Gavalas (1980, 1981), Srinivas, Amundson, and co-workers (1980, 1981, 1982), Bhatia and Perlmutter (1980, 1981), and Kriegbaum and Laurendeau (1977). The last two models are not considered to be as comprehensive or applicable as the first three. The models of Simons *et al.*, Gavalas, and Amundson and co-workers approach the characterization of char oxidation from three different viewpoints as far as the statistical derivation of pore structure evolution is concerned. These models differ in the approach by which the total particle consumption is found. From Simons, information about the overall consumption rate is collected by monitoring species fluxes at the exterior surface. It may not be appropriate to force model equations to be congruent with local pore structure changes.

The Gavalas and Amundson models may be described as modules which may be used to solve the species continuity equations for local species concentrations and fluxes. The addition of several differential equations increases substantially the complexity of this approach. Table 4.4 compares the three models and lists some of their differences.

The model advanced by Lewis and Simons (1979) integrates a semiempirical form for the pore distribution function with the species continuity equation in each pore to obtain an analytical solution. A statistical derivation and the empirical data suggest an r^{-3} (r is pore radius) distribution for the pore number density. This suggests that each pore that reaches the exterior surface of the char particle is the trunk of a tree with smaller pores branching from larger pores. Each tree trunk of radius r has an associated internal surface area within its branches on the order of r^3. Thus, the pore tree will not parallel the behavior of a simple cylindrical pore since the trunk of the tree is responsible for feeding

TABLE 4.4. Comparison of Microscopic Pore Char Models

Characteristics	Simons and co-workers (1979a, 1979b)	Gavalas (1980, 1981)	Srinivas, Amundson, and co-workers (1980, 1982)
1. Initial pore shape	Tree or river stream	Cylindrical	Cylindrical micropores, spherical macropores (optional)
2. Consideration of Knudsen and molecular diffusion	Yes	Yes	Yes
3. Statistical pore distribution function (pdf)	Provided for the user and is proportional to (pore radius)$^{-3}$	Based on pore intersections with a surface and the initial porosity distribution	Absorbed in terms of moments
4. Change of the pore distribution function (pdf) with conversion	Yes	No	Not directly applicable
5. Utilization of Langmuir-type rate expressions	One format only	Yes	Yes
6. Computational computer time	Small	Large	Large
7. Nonisobaric, nonisothermal	No/no	Yes/optional	Yes/optional
8. Capability to monitor local species concentration and pore structure parameters	No	Yes	Yes

an extremely large surface area in the branches. This implies that small trees will be kinetically limited, while the larger trees will be diffusion limited.

Simons proposed a simplified form of a Langmuir expression for the char–oxidizer reaction that included the effects of adsorption and desorption. The analytical solutions from the kinetically limited pores, the diffusion-limited pores and adsorption-limited pores, etc., are then linked together to find the critical radii which separate the regimes. Once the results for the char reaction rate of a single tree are obtained, the char consumption rate is integrated over the pore distribution function to obtain an analytic expression for the total rate. The change of the pore structure with time is determined by numerically integrating a set of time-dependent, ordinary differential equations.

The random capillary model of Gavalas (1980, 1981) considers the porous structure of char as a set of straight cylindrical capillaries which intersect each other and partially overlap. The probability density function which characterizes the porous structure is defined as the density of intersections of the axes of any pore with a surface element and is assumed to remain constant with carbon conversion. The Gavalas density function is only weakly related to the pore size distributions of the Simons and Amundson models. Because the density function remains constant (i.e., the locations of the pore axes remain fixed), the surface porosity reaches a critical value during burnoff where the pore walls merge. This causes the exterior surface to recede at a rate commensurate with the coalescence of the surface pores.

The species continuity equation is solved with an average pore size, surface area, and porosity which is representative of the local pore size distribution. A concentration distribution is calculated within the pore and keeps track of particle consumption at each radial location within the particle. The pores are assumed to be "well communicating," which infers that the concentration distribution calculated for a given radial location is applicable to any pore size in the immediate vicinity of that location. The initial particle porosity is specified as a function of discrete pore radii, which may be an advantage in some cases.

The model of Amundson and co-workers (1981) is an attempted expansion of earlier work performed by Hulburt and Katz (1964) and Hashimoto and Silveston (1973a, 1973b). The approach requires the solution of a differential equation for pore size distribution. Integration in terms of moments has the appealing advantage of solving for porous structure characteristics directly as a function of time. The difficulty lies in estimating model parameters for coalescence, engulfment, enlargement, and initiation of closed pores. The differential equations for moments are solved in conjunction with the species continuity equations along the particle radius. Srinivas and Amundson (1981) have condensed the original model of Zygourakis et al. (1982) by assigning the entire internal surface area to the cylindrical micropores. Various simplifications

TABLE 4.5. Other Important Aspects of Coal Particle Reaction

Item	Description	Key references
1. Particle fracturing	Particles may swell during devolatilization, giving rise to significant increases in porosity; then near the end of particle burnout, char particles may fracture into several smaller fragments, changing their structure, size and shape, and their motion in the gas.	Sarofim et al. (1977); Horton (1979)
2. Larger particles	Much of the information presented herein has focused on finely pulverized-coal dust (150 μm). In fluidized-bed and fixed-bed coal processes, particles are much larger and behave differently. Cracking, internal heat and mass transfer, product condensation, and repolymerization influence the burnout.	Nuttel and Roach (1978); Anthony and Howard (1976)
3. Nitrogen oxide	Most coals have about 1–2% of nitrogen, called fuel nitrogen. Nitrogen pollutants (NO, NO_2, HCN, NH_3) form readily from this nitrogen during combustion. Typically, 30% of the fuel nitrogen can be converted, largely to NO. NO levels can be controlled through control of the combustion process.	Wendt (1980); Malte and Rees (1979); Levy et al. (1978); Sarofim and Flagen (1976); Blair et al. (1979)
4. Sulfur pollutants	Sulfur levels in coal vary considerably (e.g., 0.5–5%) and exist as organic (i.e., in the coal structure) or pyritic (e.g., FeS). Some of the pyritic sulfur can be removed prior to combustion; in some processes (e.g., fluidized beds), the sulfur can be removed with limestone or other additives during combustion. Most of the sulfur forms as H_2S, SO_2, or SO_3 in the exhaust.	Malte and Rees (1979)
5. Trace-metal pollutants	All coals have a variety of trace metals, some of which are potentially harmful (e.g., Be, Pb, As, P, etc.). Some can exist in volatile forms, while others can escape in very small particulates formed during the combustion process.	Mims et al. (1979); Smith (1980); Neville et al. (1980)
6. Other oxidants	Char particles can also be consumed through reaction with steam or H_2. Reaction rates of these reactants are much slower than with oxygen but are important and often controlling in gasifiers, or fuel-rich combusters.	Skinner and Smoot (1979)
7. Ash/Slag formation	Coals contain varying amounts of ash. Some coals have over 50% ash, but 5–25% is common. These minerals form into very small particles or droplets during the combustion process. They also form layers on walls, reducing heat-transfer rates and corroding surfaces. Ash ingredients may also catalyze heterogeneous reactions.	Wall et al. (1979); Nettleton (1979); Taylor and Flagen (1980); Ulrich (1979)
8. Coal particle clouds	Much of the information reported herein on coal reactions is from tests with small quantities of coal. In practical reactors, the particles form clouds and their behavior may differ from that observed for individual particles.	Smoot and Pratt (1979)

render the dimensionless moment equations analytically solvable which decreases the number of necessary computations.

4.4. OTHER ASPECTS OF PARTICLE REACTION

Several important aspects of coal particle reaction have not been addressed herein. Table 4.5 provides a summary of some of these considerations. These issues serve to further illustrate the scope and complexity of coal reaction processes.

4.5. ILLUSTRATIVE PROBLEMS

1. **a.** How long would it take to burn a 50-μm bituminous char particle (assume pure carbon and that the particle burns out at constant diameter) in (1) pure oxygen and (2) pure CO_2 at 1773 K (1500°C) and 0.1 mPa (1 atm) pressure? Does diffusion or surface reaction dominate the process? Assume 50% porosity of the coal char particle initially.

b. Repeat this computation for a 2000-μm particle and compare the results.

2. Which of two particles with the same mass of carbon will burn out more rapidly in air at atmospheric pressure and 2273 K (2000°C): (a) a 50-μm-diam carbon particle with 0% porosity? (b) A 200-μm-diam particle with high porosity?

3. What do you estimate would be the differences in burning pulverized lignite coal particles (with 10% ash and 30% moisture) in a stoichiometric amount of air if the lignite were dried or not dried prior to the combustion process?

4. A dilute stream of Pittsburgh seam bituminous pulverized coal (50-μm diameter) is fed into a cylindrical coal furnace at the top. The gas is 100% oxygen and the inlet weight percent of coal is 10%; the O_2/coal system is completely mixed at the furnace entrance. Estimate the extent of char burnout at the exit of the 1.5-m reactor (20-cm ID). Assume that:

a. Coal contains 50%(wt) volatiles of empirical composition CH_2.
b. There is no moisture or ash.
c. Char is 100% carbon.
d. There is nonswelling coal and constant char diameter.
e. Radiation is neglected.

f. Devolatilization occurs at an infinite rate.

g. Gas-phase reactions occur at an infinite rate.

h. There is no recirculation of products.

i. Inlet velocity into reactor is 3.05 m/s (10 ft/s); the cross-sectional area of the entrance and reactor are the same and constant.

j. The pressure is constant at atmospheric pressure.

k. The initial total coal/air stream is at 300 K.

l. Volatiles heat of formation is that of ethylene.

m. Transpiration effects can be neglected.

n. Char burnout is taken to be controlled by first-order surface oxidation.

o. The gas and particles have the same temperature and velocity.

p. The reactor is adiabatic.

Other reasonable simplifying assumptions can likely be made that will not greatly affect the estimate of the extent of char burnout.

5. For the Pittsburgh bituminous coal and the Montana lignite coal of Table 2.4 and of Figure 3.9 and for a peak particle temperature of 1473 K (1200°C), compute the final gas composition and the final gas–particle temperature in a stoichiometric mixture of coal and air. Assume that all of the hydrogen and oxygen are evolved from the coal during devolatilization, that ash is inert, and that all of the other elements (C, S, N) leave in proportion to the initial weight percentages in the raw coal. Also, the extent of devolatilization could be taken at 1473 K (1200°C), even though the computed gas and particle temperature may differ from this value. Compare differences in computed properties for the two coals as a result of the combustion of the coal volatiles in air with some char remaining.

6. Compare the times required to consume 50% of toluene and n-butane at atmospheric pressure and 1273 K (1000°C) and with a a stoichiometric ratio of 0.8 (fuel rich). Consider these results in light of reaction rates at higher SR and T values and speculate on the resulting potential for HC emissions or on soot formation.

7. Using Figures 4.5, 4.8, and 4.10, which show specific and intrinsic char reaction rates, select a constant temperature (say 1500 K) and deduce the implied ratio of internal to external surface area. What assumptions are inherent in your analysis?

8. Assuming reactivity to be constant at values reported in Table 4.3, compute the time required to consume 70% of a 100-μm Occidental char particle and compare to the time required to consume 70% of a 100-μm FMC COED char particle, both at 1173 K (900°C). Explain the reasons for this significant difference.

9. From the data of Figures 4.5 and 4.6, compute the initial reaction rates of 50-μm Pittsburgh bituminous and Wyoming subbituminous char particles in air at 2000 K. Compare these results with rates of both chars in CO_2 at 2000 K. Discuss the effects of parent coal type and oxidizer type on char reaction rate.

10. How long would it take to gasify a 50-μm Pittsburgh bituminous coal char particle in a gas at atmospheric pressure at 2000 K. Assume the gas is 30 molar percent of CO_2 and neglect reaction with steam. How much would the time be reduced at 2500 K? At 10 times atmospheric pressure and 2000 K?

11. Consider the atmospheric intrinsic oxidation of a 50-μm porous char particle in air to produce carbon monoxide, where $k_i = 5.43 \exp(-40{,}000/\underline{R}T_p)$ [kg C/m^2 s (kPa O$_2$)^m]. Assume first-order isothermal reaction at 1600 K and neglect drag effects. Calculate the consumption rate of the particle, r_p. Justify any assumption regarding a rate-limiting regime. Other data includes $\rho_a = 1230$ kg/m^3; $A_t = 2.5 \times 10^5$ m^2/kg; $\rho_s = 1800$ kg/m^3; $D_{O_2-\text{air}} = 3.5 \times 10^{-4}$ m^2/s.

COMBUSTION OF COAL IN PRACTICAL FLAMES

5.1. INTRODUCTION AND SCOPE

Chapters 3 and 4 dealt with characteristics and reaction rates of small quantities of coal particles under very carefully controlled laboratory conditions. This chapter and Chapter 6 consider the nature and characteristics of practical, coal-containing flames. This chapter treats direct combustion flames, while the next chapter emphasizes practical flames in coal gasification. In principle there is little difference in these two flame types. Direct combustion flames are those where the reactions of coal take place to near completion with CO_2 and H_2O being among the major product species and where a high-temperature gas is the principal purpose.

Direct combustion of coal has been identified in some of the earliest recorded history. According to Elliott and Yoke (1981), the Chinese used coal as early as 1000 B.C., while the Greeks and Romans made use of coal before 200 B.C. By 1215 A.D., trade in coal was started in England, while pioneering uses of coal (e.g., coke, coal tars, gasification) have continued to be advanced since the late sixteenth century. The fixed-bed stoker system was invented in 1822, the firing of pulverized coal occurred in 1831, and fluidized beds were invented in 1931 (Elliott and Yoke, 1981).

Specific purposes for use of the direct combustion process include:

1. Power generation
2. Industrial steam/heat
3. Kilns (cement, brick, etc.)
4. Coking for steel processing
5. Space heating and domestic consumption

Most of the world's coal is consumed in pulverized form for power generation (see Chapter 2). The various forms in which the coal is used also differ substantially. Coal particle diameters vary from micron size through centimeter size. Use is made of virginal coal, char, and coke. Pulverized coal is also slurried with water or other carriers, both for transporting and for direct combustion.

Various processes that are used to directly combust or to gasify coal can be classified in several different ways. This chapter presents a discussion of the general features of these practical flames, together with various classification methods. A brief description of various direct coal combustion processes is also included. This information thus provides a basis for mathematical description (i.e., modeling) of these practical flames, which follows in later chapters. It is beyond the scope of this book to provide an extensive discussion of coal process hardware and operating characteristics. The reader is referred to Elliott (1981) for extended treatments on these topics.

5.2. CLASSIFICATION AND DESCRIPTION OF PRACTICAL FLAMES

5.2a. Flame Classification

Practical flames can be classified in several ways, including the following:

Classification	Examples
1. Flow type	Well stirred Plug flow Recirculating
2. Mathematical model complexity	Zero dimensional One dimensional Multidimensional Transient
3. Process type	Fixed bed Moving bed Fluidized bed Entrained bed (i.e., suspension firing)
4. Flame type	Well stirred Premixed Diffusion
5. Coal particle size	Large Intermediate Small

TABLE 5.1. Classification of Practical Coal Flames

Flow type	Process application	Typical flame type/control	Particle size	Typical mathematical complexity
Perfectly stirred reactor	Fluidized bed	Well mixed, kinetically controlled	Intermediate, 500–1000 μm	Zero dimensional
Plug flow	Moving beds Steady fixed beds Shale retort	Solid reaction/heat-transfer control	Large 1–5 cm	One-dimensional, steady
Plug flow	Pulverized-coal furnace Entrained gasifier	Mixing specified, Solid reaction/heat-transfer control	Small, 10–100 μm	One-dimensional, steady
Plug flow	Coal mine explosions Coal process explosions Flame ignition/stability	Premixed flame, Kinetic and diffusion control	Small, 10–100 μm	One-dimensional, transient
Recirculating flow	Power generators Entrained gasifiers Industrial furnaces	Diffusion flames, complex control (mixing, gas kinetics, solid kinetics)	Small, 10–100 μm	Multidimensional, steady or transient

Table 5.1 summarizes these classification groups and illustrates typical relationships among the various classifications. For example, a fluidized-bed combustor may be approximated as a well-stirred reactor (zero dimensional), and may be assumed to be well mixed and kinetically controlled. An entrained gasifier may be considered as a plug-flow (one dimensional), with solid kinetics and heat transfer controlling the reaction rate. A coal dust explosion may be treated as a transient, premixed, one-dimensional flame with kinetic and convective control. Field *et al.* (1967) and Essenhigh (1977, 1981) also discuss classification of flames.

5.2b. Classic Flame Types

Figures 5.1–5.3 illustrate various features of coal flame types. The following classic types of flames are shown:

1.   *Well-Stirred Reactor* (Figure 5.1).   An ideal fluidized-bed combustor is an example of a well-stirred reactor. This kind of idealized reactor, with instantaneous, complete mixing of solids and plug flow of gases, is simple to describe, since fluid motion need not be described in detail and properties vary only with time. Heat transfer and solid kinetics are dominant processes. When the bed is not perfectly stirred, as in practical fluidized beds, mass transfer and bubble dynamics are important issues.

2.   *Premixed Flame* (Figure 5.2).   A premixed coal dust flame may exist in a coal mine explosion, a coal pulverizer explosion, or a recirculation zone where a diffusion flame is stabilized. This type of flame can be very small (millimeters to meters in thickness), and is very complex with diffusion, solids reaction, heat transfer, and gas kinetics being important rate-controlling processes. Of particular importance to laminar flames of this type are counterdiffusion of gaseous reactants and products and convection between the hot gas and incoming cooler particles.

3.   *Diffusion Flame* (Figure 5.3).   A diffusion flame is characterized by injection of separate streams of fuel and oxidizer into the reactor. In this flame, the mixing of fuel (gaseous or solid) and oxidizer plays an important role. However, gas kinetics, particle kinetics, and heat transfer are also often important. This type of flame is characterized by very complex fluid mechanics that may include recirculating and swirling flows as illustrated by Figure 5.3. Interactions of the turbulence and the reactions further complicate this kind of flame. This flame type will predominate in pulverized-coal combustors, furnaces, and gasifiers.

These general flame classifications often form the basis for modeling of practical flames.

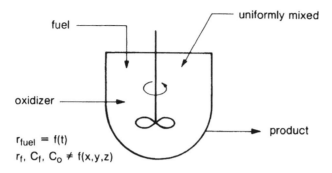

a)

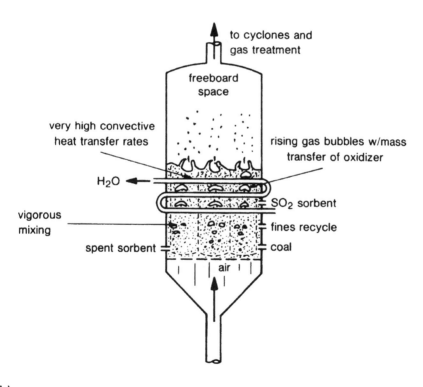

b)

Figure 5.1. Schematic diagrams of well-stirred reactor and fluidized bed: (a) ideal, well-stirred reactor; (b) fluidized bed reactor (from Rosner, 1980).

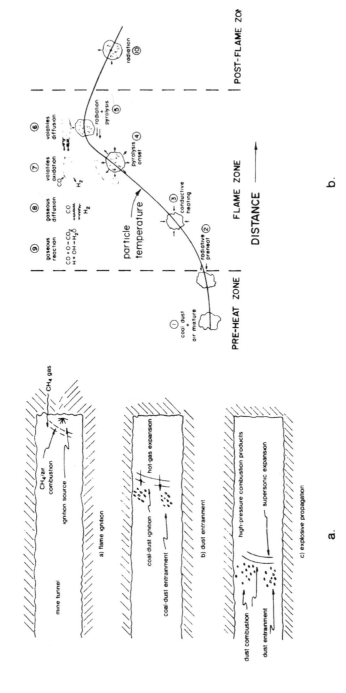

Figure 5.2. Example and characteristics of premixed coal dust flames. a. Propagating flames in coal-mine explosions, illustrating a practical example of a premixed coal dust flame. b. Structure of thin, laminar premixed flames.

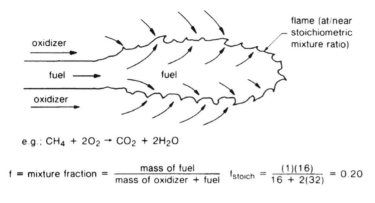

e.g.: $CH_4 + 2O_2 \rightarrow CO_2 + 2H_2O$

$f = \text{mixture fraction} = \dfrac{\text{mass of fuel}}{\text{mass of oxidizer + fuel}}$ $f_{stoich} = \dfrac{(1)(16)}{16 + 2(32)} = 0.20$

a)

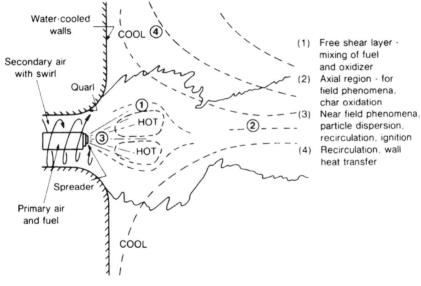

(1) Free shear layer - mixing of fuel and oxidizer
(2) Axial region - for field phenomena, char oxidation
(3) Near field phenomena, particle dispersion, recirculation, ignition
(4) Recirculation, wall heat transfer

b)

Figure 5.3. Schematic of simple and complex diffusion flames: (a) simple diffusion flame; (b) complex diffusion flame (essential features from Wendt, 1980).

5.3. PRACTICAL PROCESS FLAME CHARACTERISTICS

Direct coal combustion processes are most frequently classified according to bed type: fixed or moving beds (large particles), fluidized beds (intermediate-sized particles), and suspended or entrained beds (small particles). Examples are coal stokers, fluidized-bed combustors, and pulverized-coal furnaces, respectively. Table 5.2 provides a more detailed comparison of these three distinct types of direct coal combustion processes. By looking at Table 5.2, it is clear that particle size is a major distinguishing characteristic among these types of processes, with the coal size varying from about 10 μm to 5 cm, depending on process type. Flame temperature also varies with the fluidized bed operating at uniquely lower temperatures (1100–1200 K) compared to fixed- and entrained-bed systems. These predominant differences have a marked impact on residence times of the coal particle in the bed, which vary from less than a second to more than an hour. This observation is well illustrated by measurements and predictions in Figure 5.4 (Essenhigh, 1981). Here particle burning times for particles from 100 to about 5000 μm vary from about 100 ms to 500 sec.

The data of Figure 5.4 were obtained for several chars and other carbonaceous substances in air at pressures from 0.1 to 0.45 mPa. Calculated burning times were for 1500 K. Computation of the burning time in Figure 5.4 is essentially through integration of Eqns. (4.7) and (4.10) with $k_r \to \infty$, with constant temperature, oxygen concentration, and particle density and with negligible convective effects on the oxygen diffusion rate (i.e., very small rates of mass transfer with use of Eqn. (4.7a) in a stagnant film). Thus

$$\frac{dm_p}{dt} = \rho_{ap}\left(\frac{\pi}{2}\right)d_p^2\frac{d(d_p)}{dt} = -r_p \cong \frac{-2D_{om}\pi d_p^2 \nu_s M_p C_{og}}{d_p} \tag{5.1}$$

or

$$d_p\frac{d(d_p)}{dt} \cong \frac{-4\nu_s M_p C_{og} D_{om}}{\rho_{ap}} \tag{5.2}$$

Upon integration to complete burnout (i.e., particle diameter approaches zero):

$$t_b = \left(\frac{\rho_{ap}}{8\nu_s M_p C_{og} D_{om}}\right)d_{p_0}^2 = K_D d_{p_0}^2 \tag{5.3}$$

The experimental data of Figure 5.4 confirm this predicted variation of burn time with the square of particle diameter.

Measured values correlate well with predicted burning times based on the rate of diffusion of oxygen to the particle surface for particles above 100 μm. For smaller particles, internal pore diffusion and chemical reaction effects become important, particularly at lower temperature. This burning of particles was, of course, the key topic of Chapter 4. Based on these marked differences in burning rate with particle size, designs of various process beds differ substantially.

TABLE 5.2. Comparison of Various Characteristics of Practical Direct Coal Combustors

Process type	Fixed/moving bed	Fluidized bed	Suspended bed
Coal size (μm)	10,000–50,000[a]	1500–6000[a]	1–100[a]
Bed porosity (%)	Low	95–99[a]	Very high
Operating temperature (K)	< 2000 K	1000–1400[a]	1900–2000[b]
Residence time (seconds)	500–5000	10–500	< 1 s
Coal feed rate (kg/hr)	Up to 40,000[b]	Up to 40,000[a]	Up to 120,000[b]
Distinguishing features	Countercurrent or cross flow	Particle–particle interaction important; much of heat transfer due to conduction	Very small particles, high rates
Advantages	Established technology, low grinding, simple	Low SO_x and NO_x pollutants; less slagging; less corrosion; low-grade fuel possible[a]	High efficiency, large-scale possibilities, high capacity
Disadvantages	Emissions, especially particulates,[a] less efficient than other methods	New technology, feeding fuel[a]	High NO_x, fly ash, pulverizing expensive
Key operational variables	Fuel feed rate, air flow rate	Bed temperature, pressure, fluidizing velocity	Air/fuel ratio
Commercial operations	Stokers	Industrial boilers	Pulverized-coal furnaces and boilers

[a]Data from Elliot (1981).
[b]Data from Singer (1981).

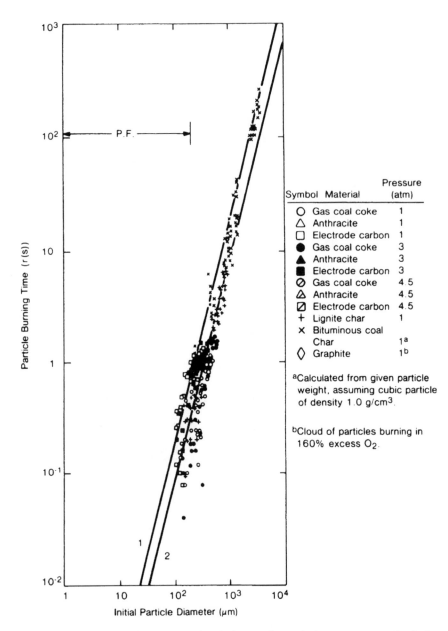

Figure 5.4. Comparison of experimental and theoretical particle burning times. 21% v/v O_2, calculated from Eqn. 5.3 for 1500 K. Curve 1: $\rho_{ap} = 2.0\,g/cm^3$. Curve 2: $\rho_{ap} = 1.0\,g/cm^3$. (Figure used with permission from Essenhigh, 1981.)

5.4.   PHYSICAL AND CHEMICAL PROCESSES IN PRACTICAL FLAMES

5.4a.   Flame Processes

The most distinguishing feature of practical flames is that several physical and chemical processes can occur simultaneously. These processes can include:

Convective and conductive heat transfer
Radiative heat transfer
Turbulent fluid motion
Coal particle devolatilization
Volatiles–oxidizer reaction
Char–oxidizer reaction
Particle dispersion
Ash/slag formation
Soot formation
And others

Which of these processes occur, and more importantly, which control the coal reaction processes differ, depending upon the type of practical flame and upon the operating conditions.

Most of the world's coal production is consumed in finely pulverized form. Practical flames with pulverized coal include premixed flames and, in particular, diffusion flames. In order to provide a foundation for a mathematical description of these flames, the specific features of these two flame types are outlined. Laboratory measurements are used to illustrate the nature of these flames.

5.4b.   Pulverized-Coal Diffusion Flames

The basic features of diffusion flames were illustrated in Figure 5.3. In these flames, fuel and oxidizer, not completely premixed, are delivered to a reactor vessel. The most common practical diffusion flame is that in large-scale utility and industrial furnaces where pulverized coal (usually less than $150\,\mu$m) is pneumatically conveyed with a small percentage of air into a furnace through several ducts. The larger percentage of the total air, which is preheated to higher temperatures, enters through different ducts, placed close to the coal-carrying ducts. The general features of one specific type of large-scale furnace is illustrated in Figure 5.5, together with a specific view of the burners. The specific numbers and combinations of these inlet ducts vary greatly among various furnaces (see Ceeley and Daman, 1981). In order for the coal to burn to completion, the coal dust–air "primary" stream and the hot air "secondary"

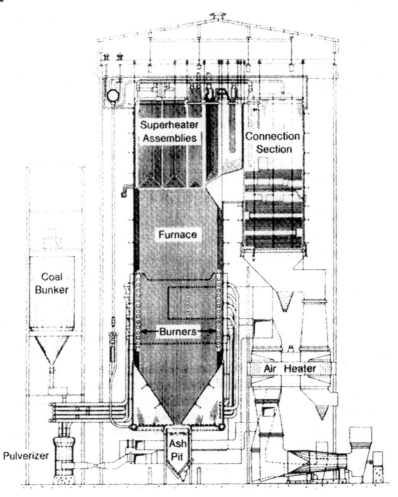

Figure 5.5. Typical large scale pulverized-coal furnace with multiple inlet streams of coal and air. (Figure used with permission from Singer, 1981.)

stream must mix. This turbulent mixing process that occurs in these diffusion flames has a major impact on the burning of the coal.

Flame Variables. Major operational variables that influence the behavior of diffusion flames include:

1. Reactor size and shape
2. Wall materials and temperature
3. Coal type and size distribution

4. Moisture percentage
5. Ash percentage and composition
6. Primary stream temperature, velocity, mass flow, and coal percentage
7. Secondary stream temperature, velocity, mass flow, and swirl level
8. Burner configuration and location (i.e., arrangement of ducts)
9. And others

Flame Measurements. Detailed characteristics of pulverized-coal diffusion flames have been studied experimentally for several years. Work at the International Flame Research Foundation has been in progress for many years, but most of this work is not available in the open literature. Studies that reveal details of these coal diffusion flames most commonly use intrusive probes for measurement of velocity, temperature, and gas and solids composition. Such measurements have been made for various coals. Chapter 8 provides a review of several different sources of these detailed data for practical coal diffusion flames.

Typical results for two very different coals are shown here to illustrate the features of these flames. Data are taken principally from Thurgood *et al.* (1980), Harding *et al.* (1982), and Asay *et al.* (1983). Measurements were made in a laboratory-scale combustor, illustrated in Figure 5.6.

The combustor was oriented vertically to facilitate particle and ash removal and to minimize asymmetric effects due to particle settling and natural convection. The combustor consists of a series of sections, each lined with castable, high-purity aliminum oxide insulation. One of the sections contained a sampling probe for removal of gaseous and particulate material.

The tests were devised in order to determine several key rate parameters summarized in Table 5.3, including gas and particle mixing and coal reaction. Effects of coal type, moisture percentage, swirl numbers of the secondary stream, stoichiometric ratio (SR), coal particle size, secondary stream velocity, and injection angle of the secondary stream are illustrated here. Test conditions of Table 5.4 are similar to operating conditions existing in industrial furnaces. Test results are presented for two coals. The first is a Utah (Deseret Mine), high-volatile bituminous coal with a proximate analysis (weight percent) of 2.4% moisture, 45.4% volatiles, 43.6% fixed carbon, and 8.6% ash. Ultimate analysis (weight percent, moisture-free) was 5.7% H, 70.2% C, 1.4% N, 0.5% S, 8.9% ash, and 13.4% O. The Utah coal was pulverized to 70% through 200 mesh, and the mass-average coal particle diameter of the standard size coal was 49.9 μm. This coal was also obtained by centrifugal classification to produce a smaller coal dust with a mass-average diameter of 20.1 μm. The Wyoming sub-bituminous coal was taken from the Belle Ayre Mine. The as-received coal contained 5.0% ash, 27.8% moisture, 32.9% proximate volatiles, and 34.3% fixed carbon. Elemental composition (daf) was 4.7% H, 72.0% C, 1.2% N, 0.56% S, and

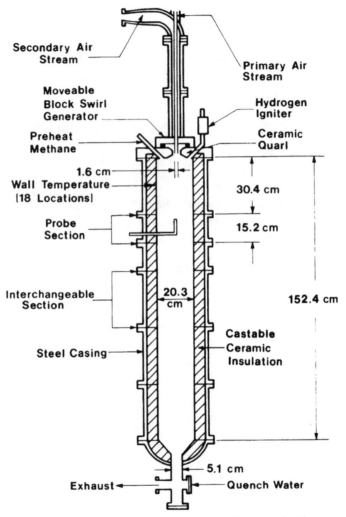

Figure 5.6. Schematic of atmospheric laboratory combustor. (Figure used with permission from Harding *et al.*, 1982.)

21.6% O. The undried coal was pulverized to 80% through 200 mesh, with a 3% moisture loss occurring during grinding. The undried coal contained 25% moisture at the time of testing, while the dried coal contained 4.5% moisture.

Flames without Swirl. Figure 5.7 illustrates radial profiles for the particle mass flux, percent coal burnout, gas mixture fraction, and O_2 concentration for test condition B. For this and also for test conditions A, C, and D, which were

TABLE 5.3. Rate Parameters Determined from Combustor Measurements[a]

Parameter to be determined	Assumptions	Approach	Method of analysis	Results
Extent of gas mixing	Inert N_2 and Ar, negligible N_2 from coal	Ar tracer added in primary gas stream	Gas chromatography	Gas mixture fraction locally inside combustor
Extent of particle mixing	Ash is inert, nonvolatile, and insoluble	Ash tracer in feed particles, with char sample weight, sample time, and probe area	Ultimate analysis of feed coal and product char	Particle mixing parameter
Extent of particle reaction	Ash is inert, nonvolatile, and insoluble	Ash tracer in feed particles with char sample weight	Ultimate analysis of feed coal and product char	Percent particle burnout
Elemental release	Ash is inert, nonvolatile, and insoluble	Comparison of feed and product char composition	Ultimate analysis of feed coal and char product	
Gas composition	CO_2 dissolution in probe quench water is negligible	Direct gas analysis	Gas chromatography; dual carrier system	Direct measurement on water-free basis

[a] Data from Thurgood et al. (1980).

TABLE 5.4. Summary of Experimental Test Conditions for Pulverized-Coal Diffusion Flames in a Laboratory Combustor

Parameter (units—test condition)	A	B	C	D	E	F	G
Test purpose	Base	Injector angle	Secondary velocity	Particle size	Secondary swirl	Coal moisture	Coal type
Coal type	Utah bituminous	Utah bituminous	Utah bituminous	Utah bituminous	Utah bituminous	Wyoming subbituminous	Wyoming subbituminous
Mean coal diameter (μm)	49.9	49.9	49.9	20.1	49.9	<45	<45
Primary stream							
Velocity (m/s)	29.3	29.3	29.3	29.3	30.5	13.5	13.5
Temperature (K)	356	356	356	356	356	300	300
Gas mass flow (kg/hr)	20.2	20.2	20.2	20.2	22.0	18.4	18.4
Coal mass flow (kg/hr)	13.7	13.7	13.7	13.7	13.7	10.2	10.2
Coal moisture (%)	2.4	2.4	2.4	2.4	2.4	25	4.6
Secondary stream							
Velocity (m/s)	34.5	34.5	55.0	34.5	—	3–8	3–8
Temperature (K)	589	589	589	589	589	590	590
Air mass flow (kg/hr)	130	130	130	130	107.6	30–82	30–82
Injection angle (°)	0	30	0	30	0	0	0
Overall							
Stoichiometric ratio (SR)	1.15	1.15	1.15	1.15	1.0	0.57–1.17	0.57–1.17
Secondary swirl number (S)	0	0	0	0	1.4	—	—
Flame type	Lifted	Lifted	Lifted	Lifted	Attached	Attached	Attached

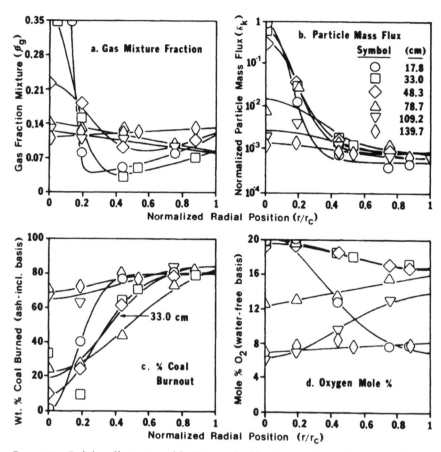

Figure 5.7. Radial profile at six axial locations in the laboratory combustor for test condition B with Utah bituminous coal. (Figure used with permission from Thurgood *et al.*, 1980.)

conducted without swirling secondary flow, the gas mixing was essentially complete in the combustor before particle reaction had begun. This occurrence resulted principally from the relatively slow particle heatup in this combustor in the absence of swirling flow.

Centerline decay of the gas mixture fraction in Figure 5.8 shows that gas mixing is most rapid with a high secondary velocity and 30° secondary injection. Other test variables had little effect on gas mixing rates. Data of Figure 5.9 for centerline decay of particle mass flux show that particle dispersion was not greatly influenced by any of the test variables considered. However, particle dispersion occurred at a much slower rate than gas mixing. Centerline decay data for coal burnout (fraction of unreacted coal) are shown in Figure 5.10.

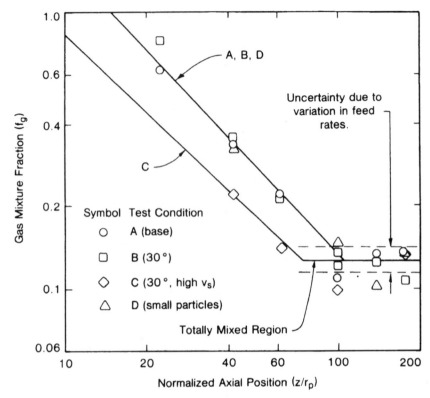

Figure 5.8. Axial gas mixing profiles for Utah bituminous coal in a laboratory combustor. (Figure used with permission from Thurgood *et al.*, 1980.)

Burnout is enhanced by use of smaller particles and reduced by use of high secondary velocity. Changing injection angle had little impact on coal burnout.

Data were also obtained from the ultimate analyses on the relative rates of release of elements (C, H, O, N, S) from the coal at various points in the combustor. Results (Harding *et al.* 1982) of Figure 5.11 indicate that the extent of element release was largely a function of the extent to which the coal was burned and was far less dependent on either test condition or location within the combustor. Results further indicated that hydrogen (see Figure 5.11c) and oxygen were liberated more rapidly than the coal as a whole, while carbon (Figure 5.11a), nitrogen (Figure 5.11b), and sulfur were emitted at about the same rate as total coal mass release.

From these data, rates of particle heatup, gas mixing, particle dispersion, and coal particle reaction were estimated. The characteristic time for gas mixing is defined as the time from initial injection of the gases into the combustor until complete mixing of the primary and secondary streams occurred, as shown by

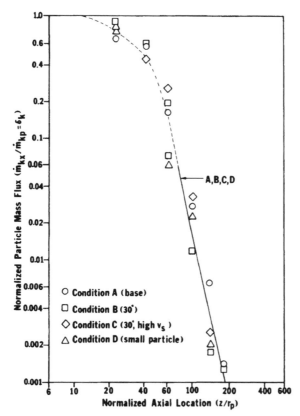

Figure 5.9. Axial particle dispersion profiles for Utah bituminous coal in a laboratory combustor. (Figure used with permission from Thurgood *et al.*, 1980.)

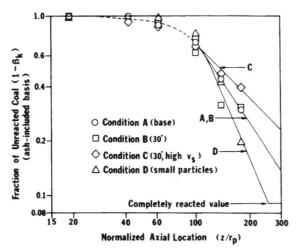

Figure 5.10. Axial profiles of unreacted Utah bituminous coal. (Figure used with permission from Thurgood *et al.*, 1980.)

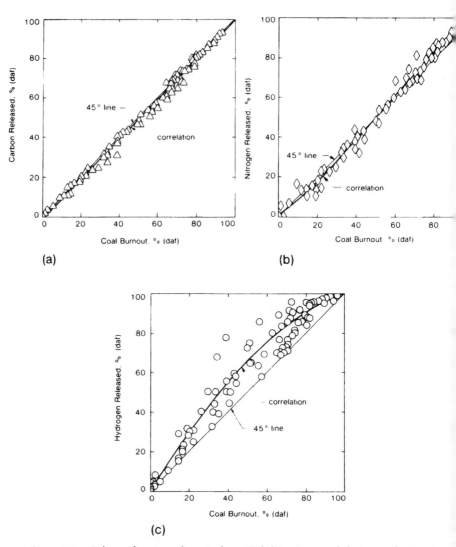

Figure 5.11. Release of various elements from Utah bituminous coal during combustion in a laboratory combustor: (a) carbon, (b) nitrogen, (c) hydrogen. (Figure used with permission from Smoot *et al.*, 1984.)

Figure 5.8. The characteristic time for particle dispersion is defined as the residence time from injection into the combustor until the time at which the particles are completely dispersed in the combustor (see Figure 5.12). The totally dispersed value is the ash flux in the combustor divided by the ash flux in the primary stream. Since the ash is taken to be inert, when the ash is totally

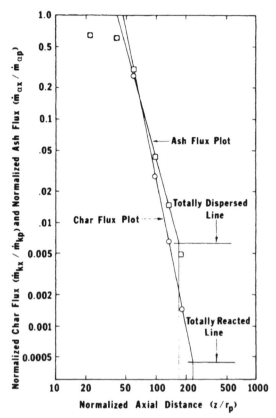

Figure 5.12. Normalized char and ash flux profiles for test condition A of Table 5.4 with Utah bituminous coal in a laboratory combustor. (Figure used with permission from Thurgood *et al.*, 1980.)

dispersed, this ratio is equivalent to the ratio of the primary jet area to the reactor area. The overall reaction time is defined as the residence time from the onset of particle reaction, as determined from ash content of particles, to the completion of particle reaction, as illustrated in Figure 5.13. This time includes both devolatilization and char oxidation times. The totally reacted value is equal to the mass flux of ash through the combustor divided by the mass flux of coal in the primary stream. These characteristic mixing, dispersion, and reaction times are presented in Table 5.5. The characteristic times of gas mixing, particle dispersion, and particle reaction are all of nearly the same magnitude and in the range of 100–400 ms.

The rate processes differ somewhat with experimental conditions. Particularly, the time for particle reaction was decreased through use of the smaller particles. However, percent of burnout for the smaller particles at the combustor

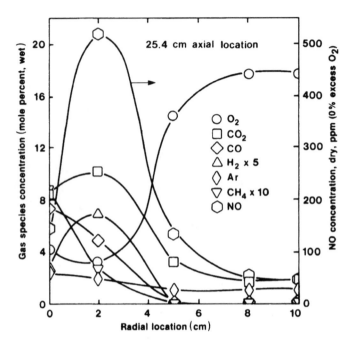

Figure 5.13. Radial species profiles in swirling flow for test condition E with Utah bituminous coal in laboratory combustor, 25.4 cm axial location. (Figure used with permission from Harding *et al.*, 1982.)

TABLE 5.5. Summary of Characteristic Times for Combustion-Related Rate Processes Estimated from Measurements[a]

| Combustion process | Characteristic time (ms)[b] | | | |
	Condition A	Condition B	Condition C	Condition D
Gas mixing	230–260	170–230	150–200	140–160
Particle mixing	310–330	210–250	260–290	170–190
Particle heatup	200–210	150–220	130–190	120–140
Particle reaction	310–340	290–320	300–330	170–220

[a]Data from Thurgood *et al.* (1980).
[b]Based on residence-time values from model velocity predictions and confirmed by limited measurements:

Combustor axial station (cm)	0	20	40	50	60	80	100	120
Residence time (ms) (conditions A, B, and C)	0	16	86	144	184	240	270	296
Residence time (ms) (condition D)	0	20	80	106	120	140	162	190

exit was only five percent greater than for the larger particles, since the higher gas temperature resulted in higher gas velocity, which reduced the combustor residence time for the smaller particles.

For these data, gas mixing was largely complete before coal reaction started. The relative influence of various rate processes will differ considerably among various combustors and furnaces, being a function of combustor scale, operating conditions, coal type, coal size, etc.

Effects of Swirl.   Use of swirling secondary flows changes mixing patterns and causes the onset of combustion at the coal dust–air duct inlet (Harding *et al.* 1982). This behavior is more characteristic of industrial practice. For the Utah bituminous coal, Figure 5.13 shows the effects of secondary stream swirl on the major gas species measured at the axial position of 25 cm in the laboratory combustor of Figure 5.6. Test conditions for this test (E) were shown in Table 5.4. The radial profiles at 25 cm (Figure 5.13) show increasing NO and CO_2 values from the centerline, with peak values about two centimeters off centerline.

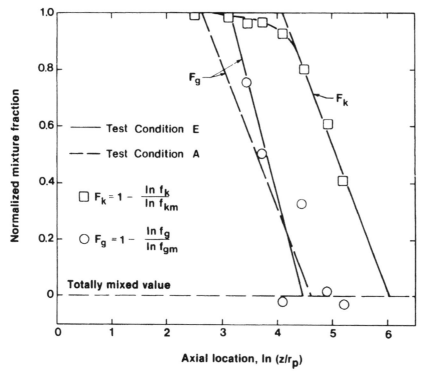

Figure 5.14. Effect of swirl number on extent of gas mixing and particle dispersion for Utah bituminous coal in a laboratory combustor. (Figure used with permission from Harding *et al.*, 1982.)

O_2 concentration shows opposite trends. CO concentrations peak at the center-line and decline to near zero at 5 cm. At 109 cm profiles are quite flat. A fuel-rich core exists in the upper center of the combustor surrounded by a reaction zone about 2–3 cm thick.

Figure 5.14 shows centerline gas and particle mixture fractions with swirling secondary air flow $(S'_s = 1.4)$. These values are compared with those from condition A of Figure 5.8 for tests with parallel injection of primary and secondary streams, i.e., $S'_s = 0$. The mass flow rates of primary and secondary air were about the same. The rate of gas mixing with swirl appears to be comparable to that determined in the absence of swirling secondary flow. The

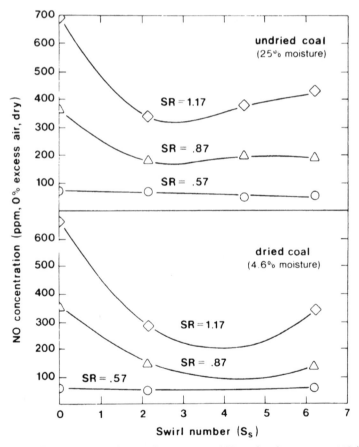

Figure 5.15. Effect of swirl number, stoichiometric ratio (*SR*), and coal moisture on NO level for Wyoming subbituminous coal in a laboratory combustor. (Figure used with permission from Asay *et al.*, 1983.)

data also show that particle dispersion is a much slower process than gas mixing in these swirling flames.

Effects of Coal Moisture. Detailed combustion characteristics of undried (25% moisture) and dried (4.6% moisture) Wyoming subbituminous coal were also determined in the combustor of Figure 5.6. Measurements were made of coal burnout and nitrogen pollutant concentrations for various secondary stream swirl numbers and stoichiometric ratios. The stoichiometric ratio (SR = inlet air/air required for complete coal combustion) was varied by changing the secondary air mass flow rate. Test conditions F and G for these two sets of tests were also shown in Table 5.4.

Effects of coal moisture at various SR and S_s values on nitrogen oxide concentrations are shown in Figure 5.15. The substantial decrease in coal

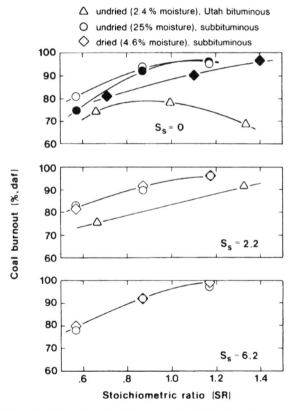

Figure 5.16. Effect of swirl number, stoichiometric ratio (SR), and coal moisture on Wyoming subbituminous coal and Utah bituminous coal burnout. (Closed symbols from carbon balance; open symbols from ash in char). (Figure used with permission from Asay *et al.*, 1983.)

moisture from 25 to 4.6% caused little change in NO level, except at high SR and S_s where a decrease of about 20% in NO concentration was observed. However, a change in the swirl numbers, which causes dramatic changes in the fluid structure of the flame, caused marked changes in NO concentration.

Figure 5.16 shows effects of S_s, SR, and coal moisture on percentage of coal burnout near the reactor exit. Two methods for computing burnout were used. A carbon balance was performed based on carbon-containing species in the gas, and ash was used as an inert particle tracer. Due to ash volatility at flame temperatures as well as moderate ash solubility in sample quench water, these latter values are thought to be somewhat low but were more reproducible. Residence time in the combustor was only 200–300 ms, so coal burnout was not always complete. Measured burnout values are expectedly a strong function of SR, but were influenced little by variation in coal moisture. An increase was observed in burnout with increasing swirl number. Burnout was approximately 80% for $SR = 0.7$ and 90–95% burnout for $SR = 1.2$. Use of swirl thus caused small increases in total burnout, together with a dramatic reduction in nitrogen pollutant level.

Key Variables and Controlling Mechanisms.   These laboratory data for practical but small pulverized-coal flames show effects of several test variables on flame structure, gas composition, pollutant level, and coal burnout:

Variable	Range	Effects
Coal type	Subbituminous, bituminous	Subbituminous more reactive
Particle size	20–50 μm	More rapid ignition for smaller particles
Moisture level	4–25%	Little effect on NO_x or burnout
Stoichiometric ratio	0.6–1.3	Major effect on NO_x, burnout
Secondary swirl number	0–6	Major effect on jet structure, NO_x; leads to rapid flame ignition
Secondary injection angle	0°–30°	Increases jet mixing; little effect on ignition
Secondary velocity	35–55 m/s	Little impact

These variables represent only a small set of possible test conditions in complex practical flames. Data show that marked changes can be made in flame structure through control of test variables. Processes that control these diffusion flames can vary markedly depending on the test conditions. However, turbulent mixing processes and particle ignition processes are clearly important. Providing for a clear understanding of the behavior of these practical flames goes well beyond inspection of experimental data. Subsequent chapters, starting with Chapter 7, deal with the detailed description of phenomena occurring in these diffusion flames.

5.4c. Coal Slurry Diffusion Flames

Applications. The use of coal slurry mixtures to replace coal in oil-fired boilers, for gas turbines, and for ignition of pulverized-coal boilers is of current interest. Coal–water mixtures (CWM) are of particular interest due to economic incentives, and uncertainty in oil availability and cost. Chapters 2–4 outlined characteristics of coal–water mixtures and reviewed briefly information available on devolatilization of coal and heterogeneous reaction of char in the presence of high percentages of moisture. Manfred and Ehrlich (1982) and Bergman and George (1983) recently review the state of the art in coal slurries, including foreign work.

Most of the current CWM interest is in combustion or gasification of pulverized coal. It is not uncommon to combust coals with 30% of moisture without drying. Comparable quantities of steam are commonly added to coal and oxygen during gasification. In these process devices, which are classified as diffusion flames, many physical and chemical phenomena interact in very complex ways, as summarized in Table 5.6. Gas composition, temperature, velocity, etc., vary markedly in the reactor. Thus, it is difficult to interpret basic rates of reaction from such test results. However, data from such experiments do provide information for process development and scaleup. Further, if the data are sufficiently extensive, they provide a basis for evaluation of models. In this subsection, basic combustion results for CWM combustion tests are reviewed briefly.

Combustion of CWM. Direct combustion of coal–water mixtures has only recently been attempted. Manfred and Ehrlich (1982) discuss general aspects of the use of CWM. Test results have been reported by five investigators (Pan *et al.*, 1982; Farthing *et al.*, 1982; McHale *et al.*, 1982; Germane *et al.*, 1983; and Walsh *et al.*, 1984). Table 5.7 provides a summary of the general features of these subscale test programs, while Table 5.8 gives more extensive test results.

Tests have been performed in combustors ranging in size from 100 to 60,000 mJ/hr (heat release rate) and in slurry feed rates from about 25 to 2400 kg/hr. Coal mass percentages in the water slurry ranged from 60 to 70%.

TABLE 5.6. Physical and Chemical Processes That May Occur during Combustion of Coal–Water Mixtures

---

Slurry dispersion
Droplet formation
Droplet dispersion
Coal particle dispersion
Droplet vaporization
Coal moisture loss
Gaseous turbulence
Gas recirculation
Gas–Particle conduction/convection
Gas–droplet conduction/convection
Gas–Particle–Droplet radiative exchange
Droplet–Particle interaction
Particle heating
Fluid-surroundings radiation
Wall-Particle/Droplet collisions
Particle devolatilization
Gas–Volatiles reaction
Gas–Water reaction
Char–Oxygen reaction
Char–Steam interaction
Particle fragmentation

All investigators varied percentages of excess air while coal type, swirl level, slurry feed rate, and secondary air temperature were also varied by some investigators.

Carbon conversion efficiency results are shown in Figure 5.17 from three investigators. Values vary with percent excess air and swirl level and also probably vary with combustor size and slurry formulation. Maximum values of up to 94% were reported. Where a parent coal was tested for comparison, NO_x levels were somehat less, but fouling was worse and carbon conversion was lower. Some problems with these test results include relatively low carbon conversion and potentially inaccurate methods for determining carbon conversion (through carbon in fly ash). Additional test results from ARC (Atlantic Research Corporation) are reported by Scheffee *et al.* (1982). Percent combustion efficiency values of up to 95 to 96% were reported, but details were limited.

TABLE 5.7. Combustor and Slurry Characteristics for CWM Laboratory Combustion Tests

Investigator	Combustion rate mJ/hr (MMBTU/hr)	Slurry composition	Furnace configuration
McHale et al. (1982)	105 (1)	(a) Medium-volatile bituminous (5% ash). (Dickerson, 75% dry coal, 1% additive,* 34% H_2O) (b) High-volatile bituminous (6% ash), (Kentucky, 62% dry coal, 1% additive,* 37% H_2O)	Cylindrical, single nozzle horizontal, 3 m × 0.5 m ID, refractory lined
Pan et al. (1982)	(a) 2000–5000 (2–5) (b) 2–27(10⁴) (20–260)	Pittsburgh bituminous; 60.6% coal (1.7% inherent moisture) 0.5% Lomar—D., 38.9% added water	Rectangular, horizontal water-cooled
Farthing et al. (1982)	4200 (4)	ARC slurry, 66% coal with additive beneficiated, HV eastern bituminous (4–5% ash)	1.5-m ID × 2.4-m length, refractory lined, water-cooled
Germane et al. (1982)	500–700 (0.5–0.7)	ARC slurry, 70% coal, 29% water, 1% additive, medium-volatile bituminous, 6% ash	Cylindrical, vertical 35-cm ID × 2.1-m length, refractory lined, water-cooled
Walsh et al. (1984)	2160–3900 (2–3.6)	Occidental slurry, 69% coal, medium-volatile bituminous, 7–8% ash	1.2 × 1.2 × 10 m section rectangular, refractory lined, and water-cooled

*Proprietary.

TABLE 5.8. CWM Test Characteristics and Results[a,b]

Investigator	Slurry flow (kg/hr)	Burnout Method	Burnout Range	Secondary air temperature (K)	Comments
McHale et al. (1982)	26–576	C in ash	46–92%	450–590	Bimodal coal size (55/45 course/fine) 53-μm mass mean. CH$_4$ assist in some tests; no parent coal comparison.
Pan et al. (1982)	(1) 114–305	C in ash	(1) 81–90	420–525	90% < 200 mesh (46–48 μm mass mean) NO$_x$ 576–712 ppm; also SO$_2$ data.
	(2) 1380		(2) 94		Operated at full and partial loads; no attempts to optimize.
Farthing et al. (1982)	195–230	Gas analysis	89–93	590	360–480 K lower flame temperature; combustor profiles of CO, NO$_x$, O$_2$; used regular oil burner, ARC slurry bimodal size distribution; NO$_x$ 100–200 ppm lower than parent coal; fouling worse than parent coal; no attempts to optimize.
Germane et al. (1982)	30	Gas analysis C in ash	90–96	420	Preliminary test results with swirling flow; no attempt to optimize; stable CWM flame without preheating CWM.
Beér (1983)	99–181	C in ash	99	494–553	Atomizer tests for ignition and stability flow; combustor profiles of CO, NO$_x$, CO$_2$, O$_2$, centerline temperature, and wall radiative heat flux; CO less than 50 ppm, NO$_x$ 520–750 ppm, gas temperature up to 1730 K.

[a]All used air atomized primary slurry nozzle with swirling secondary air flow.
[b]Percent excess air was varied by all investigators.

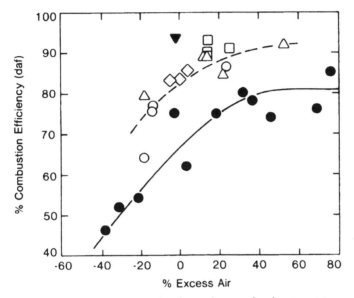

□ Farthing (1982) 4000 MJ/hr
▼ Pan et al. (1982) 27,000 MJ/hr
◇ Pan et al. (1982) 5000 MJ/hr
△○ McHale et al. (1982) 1000 MJ/hr (w/swirl)
● McHale et al. (1982) 1000 MJ/hr (no swirl)

Figure 5.17. Carbon conversion data for combustion of coal–water mixtures.

While these first combustion results were encouraging, some of the unresolved key issues include:

1. Carbon conversion to acceptably high levels (>99% in large-scale systems).
2. Optimization of burner design.
3. Optimization of slurry composition.
4. Demonstration in full-scale utility boilers, including flame stability, slurry stability, carbon conversion, ash/slag control, boiler derating, NO_x/SO_x emissions, and nozzle erosion.

Work is also continuing with detailed laboratory test programs under way at other laboratories. Walsh *et al.* (1984) have reported CWM burnout values in excess of 99% while also providing details of the CWM combustion process. Also, large-scale boiler experiments supported by the Electric Power Research Institute and the U.S. Department of Energy are being planned.

5.4d. Premixed Flames

A secondary major flame class is the premixed flame. When the reactants are premixed, the nature of the flame changes dramatically from that of a diffusion flame.

Background. The premixed flame is as vital to coal-containing flames as it is to the study of gaseous flames. Some of the general characteristics of this type of flame were illustrated in Figure 5.2. This type of flame is important for several reasons:

1. The flame can be stabilized in fixed position without supplementary gaseous fuels over a wide range of coal concentrations ($125-3000$ g/m^3 and diameters $<10\mu$m to about 50 μm).
2. The flame is about $1-4$ cm in thickness at atmospheric pressure and can be probed more readily than thinner gaseous flames.
3. Devolatilization processes govern coal reaction with heterogeneous reaction being less important.
4. Information on flame propagation rates and lean and rich flammability limits are obtained from simple measurements.
5. This coal-containing flame has been described theoretically (Smoot and Pratt, 1979) and thus ability to interpret test results is significantly enhanced.

Much of the interest in this flame type results from critical needs to find ways to control dust explosions in coal mines and in process equipment. It is thought that dust explosions are generated from unwanted ignition of premixed gas flames and subsequent development of premixed coal dust air mixtures. While later stages of a dust explosion are highly turbulent, initial stages may be laminar. Interest in this flame type also stems from flame stability requirements. In many pulverized-coal processes, it is imagined that pockets (eddies) of initially premixed coal dust and air must be ignited and then must subsequently propagate at a sufficient rate to maintain stable combustion.

Experimental Data. Horton *et al.* (1977), Milne and Beachey (1977), Strehlow *et al.* (1974), and Altenkirk *et al.* (1979) are among recent investigators to study premixed coal dust flames in the laminar mode. Various reactor configurations were used in these studies. Little work has been conducted for controlled, turbulent coal dust–air premixed flames. Figure 5.18 shows effects of coal concentration and particle size on flame velocity for a Pittsburgh bituminous coal. Characteristics of this coal are shown in Table 2.5. Data for mixtures of smaller and larger sizes illustrate the dominance of small particles in the flame propagation process. Results also imply that the lean flammability limit decreases

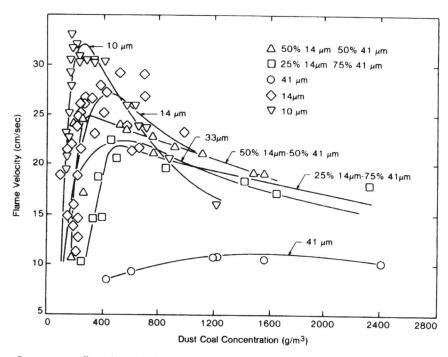

Figure 5.18. Effect of Pittsburgh coal concentration and particle size on laminar flame velocity. (Figure used with permission from Horton *et al.*, 1977.)

with decreasing particle size. While the laminar flame propagation data do not provide precise values for flammability limits, results do generally agree with more recent and more specific measurements from the U.S. Bureau of Mines (Hertzberg *et al.*, 1982) in the modified Hartmann apparatus, shown in Figure 5.19 for the same coal. The more precise Bureau of Mines results show an independence of the lean-flammability-limit dust concentration on particle size, which differs somewhat from the implications of Figure 5.18. However, the magnitude of the lean limit in air (135 g/m³), the absence of a practical rich limit, and maximum particle size for which a coal dust air flame can be stabilized (about 50 μm) are similar in the two sets of test results. Deguingand and Galant (1980) recently reported an upper flammability limit (from 8-liter chamber) for a high-volatile (35% proximate) bituminous coal of about 3900 g of coal per cubic meters of air (3.9 kg of coal per kilogram of air) with a weak dependence on particle size. This value is about 30 times the overall stoichiometric value for the coal and confirms the absence of any practical upper flammability limit. Flammability limits and flame propagation velocities are also influenced by addition of gaseous fuel, as illustrated in Figure 5.20. These data are for flames

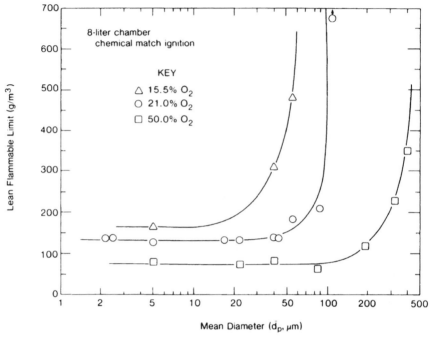

Figure 5.19. Lean flammability limits for Pittsburgh bituminous coal from modified Hartmann apparatus. (Figure used with permission from Hertzberg *et al.*, 1982.)

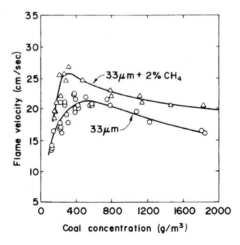

Figure 5.20. Effect of methane addition on Pittsburgh bituminous coal dust/air flame velocity. (Figure used with permission from Smoot *et al.*, 1983.)

with and without 2% (volume) of CH_4 in air for Pittsburgh bituminous coal dust. The flame velocity is increased by about 20%, while the flammability limits may also be extended somewhat.

This unique feature of devolatilizing solids, compared to gases, greatly increases the potential hazard in handling highly loaded air–coal dust mixtures. As the dust concentration increases, the temperature declines and the extent of devolatilization of the coal is lowered, thus self-regulating the local gas fuel/air ratio to maintain a flammable mixture. Figure 5.21 shows the effect of coal dust concentration on measured peak flame temperature in these flames. These data show the decline in peak temperature with increasing dust concentration. Figure 3.8 showed the effect of peak temperature on the extent of devolatilization for the Pittsburgh bituminous coal. The extent of devolatilization

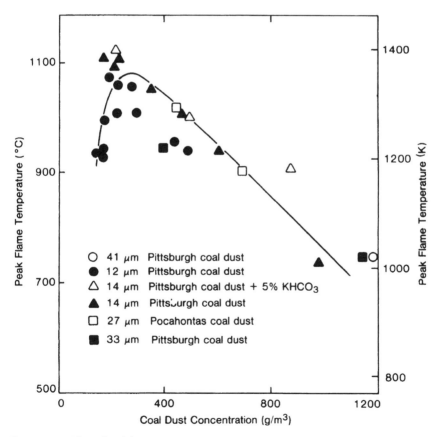

Figure 5.21. Effect of coal dust concentration on maximum measured thermocouple temperature in coal dust/air flames. (Figure used with permission from Smoot *et al.*, 1976).

is shown to drop from about 35% (daf) at 1400 K to about 10% at 1000 K (after 220 ms). At the same time, the dust concentration (Figure 5.21) corresponding to this temperature change would have increased from about 300 g/m^3 at 1400 K to about 1200 g/m^3 at 1000 K. Thus, a factor of about 4 increase in dust concentration is offset by a factor of about 4 decrease in extent of devolatilization. These compensating changes produce a near-constant gaseous air/fuel ratio over a wide range of dust concentrations. Further, it can be readily shown from these results that the resultant gaseous air/fuel ratio is very close to the stoichiometric value for all coal concentrations. Hence, propagation in coal dust is a self-regulating process with no practical rich flammability limit.

Figure 5.22 shows the effects of coal type on flame velocity. The more volatile coals (see Table 2.5) propagate at slightly higher velocity and with a somewhat lower lean flammability limit. However, the influence of coal type is not dramatic, even though proximate volatiles content for these coals varies from near 37% to about 17%. Given the self-regulating behavior of coal in these flames, this result is not surprising.

Figure 5.23 shows the impact of solid suppressants (KHCO$_3$ and rock dust; the latter is used to dust coal mines). Something in excess of 10% of chemically active suppressants in the coal dust such as KHCO$_3$ is sufficient to suppress propagation of the laminar flame, while something over 35% of the chemically inactive rock dust (mostly CaCO$_3$) is required to achieve the same reduction in the flame propagation velocity. These results are generally consistent with flame suppression tests made by the U.S. Bureau of Mines (Burgess et al., 1979) in the Hartmann apparatus, where 18% of the most active suppressant and 57% of rock dust were required for complete inerting. Combined evidences suggest that salt additives suppress flame propagation in CH$_4$–air and coal dust–air flames by interfering with gas-phase propagation processes, with chemical suppression effects being important. Evidences that support this observation include (1) the lack of effectiveness of inert particles (i.e., Al$_2$O$_3$), (2) the effectiveness of potassium- or sodium-containing compounds over rock dust, (3) the lack of correlation of maximum temperature vs. flame speed, and (4) the lack of influence of additives on the overall composition or temperature of the flames.

Premixed Flame Characteristics. The basic nature of these thin laminar flames is quite well identified. The independence of primary variables (dust size, type, concentration, gas fuel additive, suppressant) has been determined and, for the most part, explained. The lack of a practical rich flammability limit is also observed and explained while lean flammability limits have been characterized. However, little or no data are available for well-controlled turbulent coal dust air flames. Also, relationships of thin, laminar flames to thicker, radiatively dominated coal dust flames have not been resolved. Turbulent flames are thicker and radiative processes are more important in more optically thick flames.

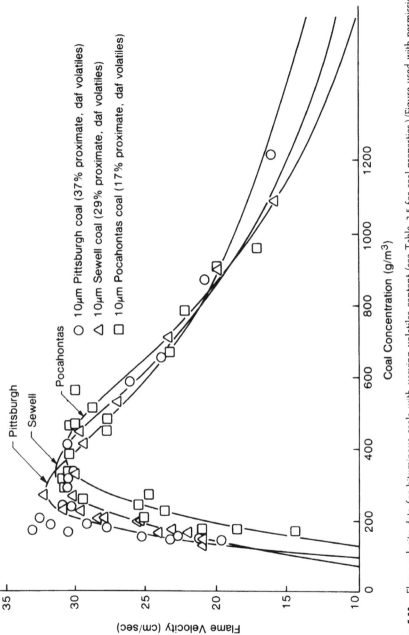

Figure 5.22. Flame velocity data for bituminous coals with varying volatiles content (see Table 2.5 for coal properties.) (Figure used with permission from Smoot et al. 1976.)

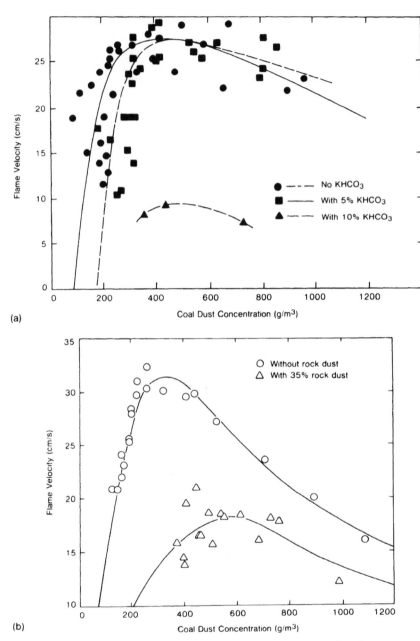

(a)

(b)

Figure 5.23. Effect of solid suppressants on coal dust/air flame velocities for a bituminous coal: (a) 14 μm Pittsburgh coal dust with and without 19 μm KHCO$_2$; (b) 10 μm Sewell coal dust with and without 35% rock dust. (Figure used with permission from Smoot *et al.*, 1976.)

Further, the specific relationships of propagating flames to coal dust explosions has not been clearly identified.

The detailed discussion of some of the characteristics of practical diffusion and premixed combustion flames illustrates the complexity of these flames. Several physical and chemical processes take place in these flames. This and the following discussion of similar flames in coal gasification, establish a common basis for describing these complex flames.

5.5. ILLUSTRATIVE PROBLEMS

1. Given the median particle size and operating temperatures for moving-bed, fluidized-bed, and suspended-bed combustors of Table 5.2, compute the particle consumption times with Eqn. (5.3) and compare results to residence-time values of Table 5.2. Discuss the accuracy of the three separate computations.

2. Derive an analogous equation to Eqn. (5.3) assuming diffusional control for particle consumption, but assuming the particle to be consumed at constant diameter and varying density. Compare the result to Eqn. (5.3) and discuss implications.

3. For the large-scale furnace of Figure 5.5, from the power rating level, estimate coal feed rate, air feed rate (20% excess air), average temperature, and residence time. Is the residence time sufficient to consume coal particles? What is the basis of your answer?

4. By comparing Figures 5.8, 5.9, and 5.14, what can you say about the relative mixing rates of gases and particles? What is the impact of swirling secondary flow on this mixing rate?

5. Why do you suppose that the concentration of NO is lower for dried coal than undried coal (Figure 5.15)? Shouldn't the decrease in temperature caused by the high moisture content of the undried coal reduce NO level? What other factors might influence the NO level?

6. Estimate the reduction that will occur in peak furnace temperature by stoichiometric combustion of a HV bituminous coal/water mixture containing 30% water, compared to direct combustion of the coal with no moisture. Select coal properties from Table 2.5.

7. Compute the estimated time required to burn out a 50-μm Pittsburgh bituminous char particle in air at 2000 K and atmospheric pressure. Compare this value with that required when the air contains water at a concentration that would result from a 30% CWM with stoichiometric air and with the temperature of the air reduced accordingly.

8. See Problem 10, Chapter 8, which is an application of the one-dimensional code to CWM.

9. Use Figures 5.18, 5.21, 3.8, and 3.9 (for bituminous coal), and estimate the temperature, extent of devolatilization, and gas composition for a coal dust/air flame at 400, 800, and 1200 per cubic meters of dust concentration. From these results, explain the nature of the rich flammability limit for dust flames.

GASIFICATION OF COAL IN PRACTICAL FLAMES

6.1. INTRODUCTION AND SCOPE

Gasification of coal is a process which occurs when coal or char is reacted with an oxidizer to produce a fuel-rich product. Principal reactants are coal, oxygen, steam, carbon dioxide, and hydrogen, while desired products are usually carbon monoxide, hydrogen, and methane. The principal purpose for gasifying coal is to produce a gaseous, fuel-rich product. Potential uses of this product are for:

1. Production of substitute natural gas.
2. Use as a synthesis gas for subsequent production of alcohols, gasoline, plastics, etc.
3. Use as a gaseous fuel for generation of electrical power.
4. Use as a gaseous fuel for production of industrial steam or heat.

Gasification of coal has been practiced commercially for nearly 200 years, with the first gas produced from coal in the late 18th century (Perry, 1974). However, the production of gas from coal gradually decreased in the mid-20th century as abundant natural gas resources were widely distributed through general pipeline delivery. A sharply renewed interest in gasification was evident in the mid-1970s with concern over dwindling reserves of oil and gas.

Several commercial gasification processes are in use in the world, particularly in Europe and South Africa (Hebden and Stroud, 1981). In the United States, major gasification demonstration projects are under way in North Dakota and Southern California. It is not the purpose of this chapter to give a detailed accounting of the basic aspects of gasification or a comprehensive description of various gasification processes. Johnson (1981) provides a detailed treatment of the fundamentals of coal gasification, while Hebden and Stroud (1981) give an extensive review of various commercial and developmental gasification processes. Bissett (1978) provides an extensive review of entrained coal gasification research from the Morgantown Energy Technology Center and its predecessor, the U.S. Bureau of Mines. This work dates from 1947. A summary of other U.S. and European entrained gasification research is also included.

In this chapter, general features of coal gasification are identified, and the relationship to coal combustion is discussed. From the standpoint of theoretical description, coal gasification processes are considered to be essentially that of fuel-rich coal combustion. From this concept, similarities and differences in combustion and gasification of coal are identified.

6.2. BASIC PROCESS FEATURES

6.2a. Basic Reactions

A principal overall reaction during the gasification of coal and other carbonaceous materials is

$$C(s) + O_2 \rightarrow CO; \qquad \Delta H_R^{0*} = -110.5 \text{ mJ/kmol} \qquad (6.1)$$

This reaction is exothermic. Further, the reaction of $C(s)$ does not stop at CO, but any free oxygen rapidly reacts with CO in the gas phase to produce CO_2. Thus, for a fuel-rich system, in order to consume the remaining $C(s)$, the much slower endothermic reaction

$$C(s) + CO_2 \rightarrow 2CO; \qquad \Delta H_R^0 = +172.0 \text{ mJ/kmol} \qquad (6.2)$$

must occur.

In order to control high temperatures resulting from $C(s)-O_2$ reaction, and to increase the heating value of the product gas, through addition of hydrogen, steam is usually added as a reactant:

$$C(s) + H_2O(g) \rightarrow CO + H_2; \qquad \Delta H_R^0 = +131.4 \text{ mJ/kmol} \qquad (6.3)$$

This reaction is also endothermic, and must rely on the heat release from the $C(s)-O_2$ reaction for energy requirements. Further, the rate of reaction of $C(s) + H_2O$ is very slow compared to $C(s) + O_2$. However, the resulting product gas has an increased heating value, as illustrated in Figure 6.1. This figure shows

*All ΔH_R^0 at 298 K and atmospheric pressure (Stull and Prophet, 1971).

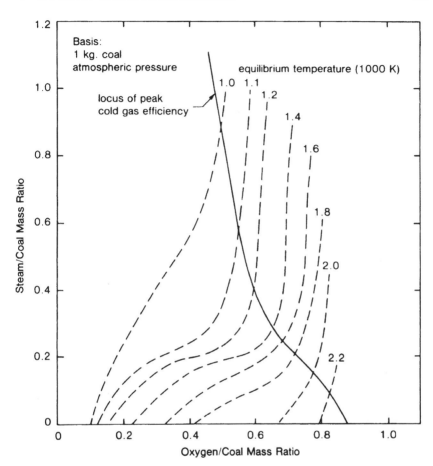

Figure 6.1. Predicted effect of coal/oxygen/steam mixture ratio on equilibrium temperature. (Figure used with permission from Skinner *et al.*, 1980.)

the result of a series of computations for various mixtures of a typical HV bituminous coal with steam and oxygen, assuming thermodynamic equilibrium (Skinner *et al.*, 1980). Temperature is shown to increase with increasing oxygen content and to decrease with increasing steam content. Also shown is the locus of peak cold gas efficiency, which is the ratio of heating value of the product gas at ambient temperature to the heating value of the coal (daf). This value increases with increasing steam and decreases with increasing oxygen. Thus, for development of a given gasification process, a balance between sufficiently high reacting temperature for flame stability and carbon conversion, and acceptably high cold gas efficiency must be achieved.

In a coal/steam/oxygen system, gaseous reactions are also very important, particularly:

$$CO + \tfrac{1}{2}O_2 \rightarrow CO_2; \qquad \Delta H_R^0 = -283.1 \text{ mJ/kmol} \qquad (6.4)$$

$$CO + H_2O(g) \rightarrow CO_2 + H_2; \qquad \Delta H_R^0 = -41.0 \text{ mJ/kmol} \qquad (6.5)$$

The first of these reactions causes rapid consumption of oxygen, increases gas temperature, and forces the requirement for slow, heterogeneous $C(s)$ reaction with CO_2. The second, slightly exothermic water–gas shift reaction, also produces CO_2 from CO and tends to control the final product distribution. In some gasification processes, hydrogen is added as a reactant in order to increase the quantity of methane as a product:

$$C(s) + 2H_2 \rightarrow CH_4 \qquad (6.6)$$

However, with the high cost of hydrogen, this is not common in practical gasification processes. Edmister *et al.* (1952), Batchelder and Sternberg (1950), and particularly Johnson (1981) consider the thermodynamics of coal gasification in greater detail.

6.2b. Gasification Process Types

Chapter 2 outlined key process types for gasification: fixed (or moving) bed, fluidized bed, entrained bed, and other (e.g., molten bath), while Table 2.1 of Chapter 2 compared various features of these processes. Hebden and Stroud (1981) describe the details of several specific gasifiers in each category. Further division is made between commercial and developmental gasifiers, and between those containing nitrogen (i.e., air-blown) in the product gas and those without nitrogen (i.e., oxygen-blown). Air is used when the presence of nitrogen in the product gas is not economically detrimental (e.g., industrial fuel gas), while oxygen is used to produce a nitrogen-free product gas (e.g., substitute natural gas).

A brief summary of the various features of coal gasification process types is given in Table 6.1. Key differences among the various processes relate to coal particle size and operating temperature. Large coal size (6–50 mm) and moderate operating temperatures lead to very long residence times (hours) in fixed or moving beds. Fluidized-bed gasifiers operate with residence times of minutes and with smaller coal size (500–1000 μm) but at lower temperatures, (1100–1200 K) which result from cooling by in-bed tubes. Highly pulverized coals (1–150 μm) are used in entrained systems with very high operating temperatures (up to 2200 K in oxygen-blown gasifiers) that result in very short (<1 s) residence

TABLE 6.1. Comparison of Coal-Gasification-Process Types (Air- and Oxygen-Blown)

	Fixed bed	Fluidized bed	Entrained bed
Residence time	1–3 hr[a]	20–150 min[a]	0.4–12 s
Coal size	6–50 mm	500–2400 μm	10–150 μm
O_2/Coal	0.14–0.81	0.25–0.97	0.28–1.17
Steam/coal	0.28–3.09	0.11–1.93	0.1–1.20
Coal type	Most types no fines	Noncaking coals	All types
Temperature range (K)	1150–1300	600–1470	1150–2500
Pressure range mPa	0.1–2	0.1–10	0.1–30
(atm)	1–20	1–100	1–300
Product gases (mol %)			
$CO + H_2$	39–66	2–80[a]	35.0–0–91
CH_4	2–15	3–68	0.1–17
HHV (BTU/SCF)	250–320	300–800	115–550
Commercial operations	Extensive Lurgi Wellman–Galusha	Slight Winkler	Moderate Koppers-Totzek Texaco
Principal advantages	High turndown ratio Mature technology Low thermal losses	Lower temperature Reduced thermal losses Variety of coal sizes Moderate residence time	Smaller, simple design All coal types Highest capacity/ volume

[a]Information from Perry (1974). All other information from Hebden and Stroud (1981).

times. Rates and the nature of coal reaction differ dramatically among these process types. Other key process variables include operating pressure and reactant composition (i.e., steam/oxygen/coal ratio). High-pressure operation is common in all types of gasifiers and is particularly attractive for combined cycle operation when generating electrical power.

6.2c. Comparison with Combustion

Gasification of coal can be compared to fuel-rich coal combustion. Methods of modeling coal gasification processes are thus essentially identical to those for coal combustion (Smoot and Pratt, 1979). Many similar aspects exist for gasification and combustion, including use of the same types of processes, coal preparation, and grinding, and use of varieties of coal. Table 6.2 summarizes some of the major differences between the direct combustion of coal and its gasification. Even with these differences, methods of computing properties of gasification and combustion processes are essentially identical.

TABLE 6.2. Differences in Direct Coal Combustion and Coal Gasification

	Direct coal combustion	Coal gasification
Operating temperature	Lower	Higher
Operating pressure	Usually atmospheric	Often high pressure
Ash condition	Often dry	Often slagging
Feed gases	Air	Steam, oxygen
Product gases	CO_2, H_2O	CO, H_2, CH_4, CO_2, H_2O
Gas cleanup	Postscrubbing	Intermediate scrubbing
Pollutants	SO_2, NO_x	H_2S, HCN, NH_3
Char reaction rate	Fast (with O_2)	Slow (with CO_2, H_2O)
Oxidizer	In excess	Deficient
Tar production	None	Sometimes
Purpose	High-temperature gas	Fuel-rich gas

6.3. PHYSICAL AND CHEMICAL GASIFICATION PROCESSES

While essentially the same physical and chemical processes occur during gasification and direct combustion, these basic processes interact in different ways with different results. Significant differences among these basic reactions also occur in fixed, fluidized, and entrained gasifiers. The following physical and chemical processes control both combustion and gasification of coal:

1. Turbulent mixing of reactant (except fixed beds)
2. Turbulent dispersion of particles (except fixed beds)
3. Convective coal particle heatup
4. Radiative coal particle heatup
5. Coal devolatilization
6. Gaseous reaction of volatiles products
7. Heterogeneous reaction of char
8. Formation of ash/slag residue

Critical differences in gasification, compared to combustion include the following, particularly in oxygen-blown fluidized and entrained beds:

1. Peak gasification temperatures are often higher in gasification due to absence of the diluent nitrogen. Thus, the extent to which the coal devolatilizes, which is a strong function of peak temperature (see Chapter 3), can be greater during gasification, reaching as high as 80% of the total daf coal (Hedman *et al.*, 1982).
2. These volatiles products quickly mix and react with available oxygen in the gas phase, completely depleting oxygen, and producing the very high temperatures that cause the increased extent of devolatilization.

High concentrations of CO_2 and increased concentrations of H_2O are produced through these gaseous reactions.

3. The residual char is reacted relatively slowly through heterogeneous attack by the CO_2 and steam. This process is faster at higher temperature, but still slow compared to the oxygen–char reaction that occurs in direct combustion.

This series of processes is quite different from direct combustion where oxygen is usually present in excess and dominates the consumption of residual char.

In fixed-bed gasifiers, flow of coal is countercurrent to that of oxygen. Thus, oxygen first encounters and reacts directly with the residual char, producing hot gases that devolatilize the coal in upper gasifier regions. Thus, unlike fluidized and entrained beds, oxygen plays a key role in the last stages of char consumption in the fixed-bed gasifier systems.

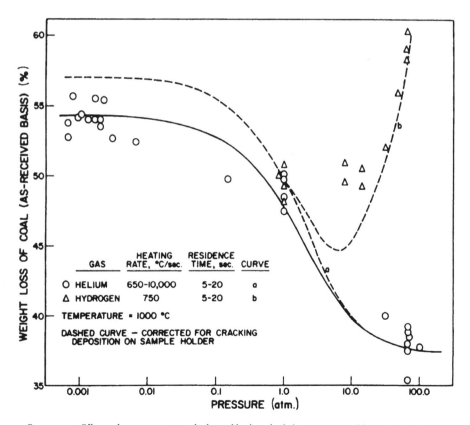

Figure 6.2. Effects of pressure on weight loss of high-volatile bituminous coal heated in hydrogen and helium atmospheres. (Figure used with permission from Anthony *et al.*, 1975.)

Devolatilization rates during gasification are treated as outlined in Chapter 3, with the same treatment used for devolatilization during combustion. Differences during gasification may be due principally to effects of pressure and to the surrounding atmosphere of gases. Figure 6.2 shows measured weight loss of a high-volatile bituminous coal in helium and in hydrogen at 1270 K for pressures up to 100 times atmospheric pressures. The coal weight loss decreases with increasing pressure in inert gases and also initially in hydrogen. At higher pressures, direct hydrogen reaction contributes to the coal weight loss. Howard (1981) treats, to a greater extent, the effects of pressure and of hydrogen on coal devolatilization.

Rates of char reaction differ greatly at a given temperature for different reactants, generally in the order:

$$O_2 \gg H_2O > CO_2 \gg H_2 \qquad (6.7)$$

Smoot and Pratt (1979), Essenhigh (1981), and Johnson (1981) give recent reviews of rate data for char reaction with these reactants. Chapter 4 presented recent data for O_2 and CO_2 reaction of chars from four U.S. coals. Reaction rates for those chars in CO_2 (Figure 4.6) vary by three orders of magnitude among the chars and are six to seven orders of magnitude below O_2 values for the practical temperature range of 1250–1750 K. Thus, consumption of char by CO_2 at temperatures where kinetic rate controls will be dramatically slower than in oxygen and will vary substantially among coals. At higher temperatures, where diffusion of oxygen to the particle surface dominates, effects of coal type and of oxidizer type will be diminished.

It is presumed in the theoretical description of coal conversion processes, that if appropriate laboratory rates for coal devolatilization and char reaction are available, that dramatic differences in the nature of combustion and gasification will be adequately described by the same theory.

6.4. GASIFIER DATA

Data for gasifier design are summarized by Johnson (1981) and by Bissett (1978) for entrained gasifiers. It is beyond the scope of this chapter to present or discuss these data in any detail. Gasification data are of three basic types:

1. Correlation of laboratory data on characteristics of various coals and chars, such as reactivity, surface area, etc. Data of this type were illustrated in Chapters 3 and 4, in such figures as Figures 3.4 and 3.5 (ignition temperatures), Figure 3.8 (coal weight loss during devolatilization), and Figures 4.6 and 4.7 (char reaction rates in O_2 and CO_2).

2.   Correlations of gasifier effluent data for such properties as carbon
conversion, pollutant concentration, product gas composition, and temperature.
These correlations are shown as functions of key gasifier variables such as
reactant gas composition, coal type, pressure, or residence time. Figure 6.3 shows
data of this type where carbon conversion is shown as a function of pressure
and steam–oxygen reactant concentrations from various investigators for
entrained gasification. Table 6.3 summarizes test conditions for these data. Figure

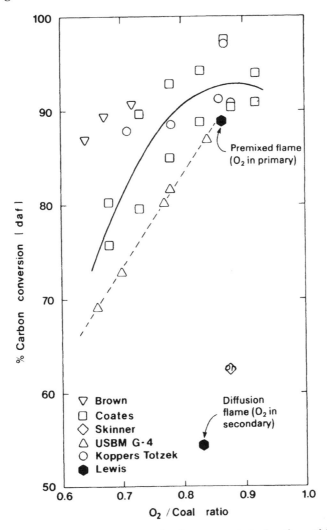

Figure 6.3.   Comparison of carbon conversion values for various entrained gasifiers with bituminous
coal. (See Table 6.3.)

TABLE 6.3. Operating Parameters for Various Gasifiers

Facility	Operating pressure (kPa)	Steam/Coal ratio	Coal feed rate (kg/hr)	Residence time (ms)
Koppers[a]	Atmospheric	0.20–0.54	910–22,050	600–1000
USBM (G-4)[b]	Atmospheric	0.21	220	N/A
McIntosh and Coates (1978)	1052	0.30–0.311	20	130–320
Skinner et al. (1980)	Atmospheric	0.24	25	380
Lewis (1981)	Atmospheric	0.24	25	430
Soelberg (1983)	Atmospheric	0.27	25	400
Brown (1985)	Atmospheric	0.20	27	400
Glassett (1983)	1034	0.27	25	4080

[a] von Fredersdorff and Elliot (1963); Chaurey and Sharma (1979).
[b] Strimbeck et al., 1960.

6.3 also shows that carbon conversion during entrained gasification varies dramatically with oxygen/coal ratio and with operating pressure.

3. The third type of data is also vital to development and evaluation of predictive methods for coal gasification. Here, various properties, such as temperature, residue composition, gas composition, and velocity, are measured throughout a gasifier and from these data, maps or profiles are constructed. These locally measured data provide unique insight into the gasification process. They also can be used for comparison with predictions from gasification theories. The extent of such gasification data is very limited. Skinner et al. (1980) and, more recently, Hedman et al. (1982) reported profiles of gas species concentration for several gases and gas mixture fractions from within a laboratory-scale entrained gasifier. Figure 6.4 shows the data for CO_2, CO, and H_2. Comparisons of theoretical predictions for these data are shown subsequently in Chapter 13. From these data, and associated computations, a more complete description of the gasification process is derived. Data show the very rapid ignition of pulverized coal in an intense reaction region near the inlet of reactants and near the reactor centerline. Up to 80% of the coal has been consumed in this region through devolatilization (according to the theory) with predicted local temperatures as high as 3000 K. Oxygen concentration declines rapidly to zero in this zone. A recirculation zone is shown in the forward region off-centerline. Downstream of the intense reaction zone, residual char is slowly consumed, through (according to the theory) reaction with CO_2 and H_2O, which were present in significant concentrations.

These and other detailed profile data for cold flow and reactive flows are discussed in Chapter 8, while the theory for predicting these complex reactive flows is presented in Chapters 9–14.

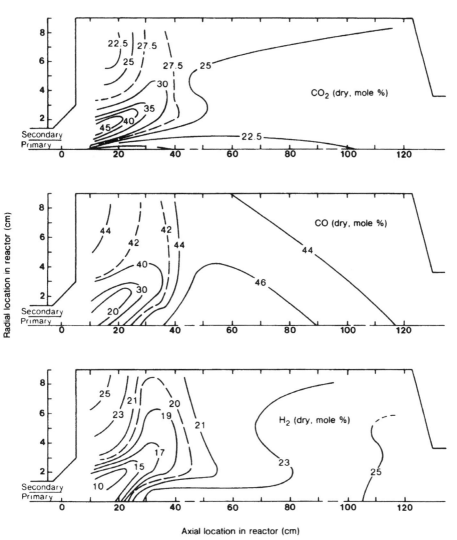

Figure 6.4. Measured species profiles for CO_2, CO, and H_2 from a laboratory-scale entrained gasifier. (Utah bituminous coal, feed rate = 25 kg/hr; mean diameter = 41 μm; oxygen/coal = 0.91; steam/coal = 0.27). (Figure used with permission from Hedman *et al.*, 1982.)

6.5. ILLUSTRATIVE PROBLEMS

  1. For the gasification of Utah bituminous coal (see Table 2.5) in O_2 and H_2O (1 part coal, 1 part O_2, 0.5 parts H_2O, by weight) compute the final product temperature, the final product composition, and the cold gas efficiency. Assume thermodynamic equilibrium, with all reactants initially at 300 K and atmospheric pressure and with gasification also at atmospheric pressure.

2. Given the following conditions for coal gasification processes:

	Char size	Extent of coal devolatilized	Operating temperature (K)	Molar percent CO_2 concentrations in gas
Fixed bed	50 mm	50	1800	9
Fluidized bed	1000 μm	30	1200	22
Entrained bed	50 μm	70	2200	22

Assuming the coal to be Pittsburgh bituminous, the reactant gas to be CO_2, and the rates given by Figure 4.6, compute the time required to consume the char particles of each size. Consider both diffusional and kinetic effects. How much lower would the consumption time be if the oxidant were oxygen?

3. Given the observed dependence of pressure on carbon conversion shown in Figure 6.3, speculate as to the causes of this effect of pressure? What is the estimated relative importance of coal devolatilization, residence time, oxidant diffusion, and char surface reaction?

4. For the gasifier maps of Figure 6.4, and with the test conditions given, compare calculated equilibrium reactor exit gas concentrations with measured values. Carbon conversion (daf) was 70% for this case.

. . . your understanding doth begin to be enlightened,
and your mind doth begin to expand.
Alma (Book of Mormon) 32:34

MODELING OF
COAL PROCESSES

7.1. BACKGROUND AND SCOPE

In the past, comprehensive modeling of coal processes has been restricted by lack of computer speed and capacity and by technical difficulties in describing essential model elements. Lowry (1963) and, subsequently, Elliott (1981) have edited comprehensive reviews of the chemistry of coal utilization which present information on coal origins, characteristics, and reactions. However, comprehensive modeling of coal processes was not addressed. Field *et al.* (1967) discussed many of the elements of modeling of pulverized-coal processes, but emphasized only the more elementary well-stirred and plug-flow models of coal processes. These authors noted that no multidimensional models existed at that time for pulverized-coal systems. Smoot and Pratt (1979) edited a treatment of pulverized-coal models where multidimensional models were considered. More recently, Smoot (1980) reviewed the state of development of models for pulverized-coal flames, both for combustion and gasification.

Until recently, modeling of complex coal furnaces emphasized radiative heat transfer (e.g., Bueters *et al.*, 1974; Lowe *et al.*, 1974; Selcuk *et al.*, 1976), with limited attempts to describe reaction and flow processes. In a recent review of radiative transfer in combustion chambers, Sarofim and Hottel (1978) note that ability to estimate radiative effects in furnaces is limited primarily by inadequate information on the chemical and physical processes governing the concentration and size of particulates. These latter quantities are closely related to the complex flow and reaction processes that take place in coal reactors. Figure 7.1 illustrates several aspects of a complex pulverized-coal flame in an industrial or utility coal boiler. Complete modeling of coal flames must account for all of these aspects. However, for a specific application, not all of these

163

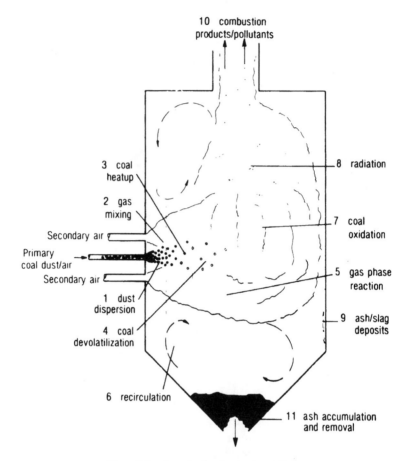

Figure 7.1. Aspects of pulverized-coal flames.

aspects will be important and may not need to be considered to develop an adequate description. The foundations and elements of modeling are presented in this chapter and illustrated through one- and two-dimensional models. Key references to comprehensive modeling are also noted.

7.2. BASIC MODEL ELEMENTS AND PREMISES

In this chapter a "model" is taken to be a comprehensive, computerized code which combines several model components and can be applied to the description of complex processes. The key components used in comprehensive, coal reaction models for application to combustors and gasifiers often include the following:

1. Turbulent fluid mechanics (see Chapter 10)
2. Gaseous, turbulent combustion (see Chapter 11)
3. Particle dispersion (see Chapter 12)
4. Coal devolatilization (See Chapters 3, 12, and 13)
5. Heterogeneous char reaction (see Chapters 4, 12, and 13)
6. Radiation (see Chapter 14)
7. Pollutant formation (see Chapter 15)
8. Ash/slag formation (see Chapter 2)

Several of these elements have been discussed in earlier chapters while others are presented in subsequent chapters, as noted in the above list. It is a major challenge in development of coal models to provide adequate submodels for each of these components.

Key premises that provide the foundation for development of comprehensive models include the following:

1. The behavior of clusters or clouds of particles can be predicted from information on the behavior of individual particles or small groups of particles.
2. Unsteady or quasi-steady behavior can be predicted from basic data obtained from steady-state measurements.
3. Processes as complex as those in coal combustion and gasification are often governed by key controlling steps; obtaining an adequate process description does not require a complete description of all aspects of the combustion process.
4. Use of numerical methods which have been evaluated for simple computational systems can produce acceptably accurate results in more complex systems that defy rigorous evaluation of uniqueness and accuracy.

Since the validity of these premises cannot always be demonstrated directly, the entire foundation of modeling of complex processes relies heavily on comparison with experimental observations (see Chapter 8). Many studies have emphasized fundamental aspects of this problem, such as coal pore diffusion, radiative properties of coals and chars, coal structure and its relationship to reactions, and particle changes during devolatilization. Still, the development of these coal process models requires a large number of specific assumptions. Frequently, these assumptions are not strongly supported by experimental data. Thus, demonstration of validity and accuracy of model predictions is a continuing challenge. The codes are often too extensive to permit separate and complete evaluation of every component. Comparison with gross data, such as outlet temperature or composition, provides little confidence. Comparison with mean values of space-resolved properties is a much stronger test.

A general comprehensive code usually includes input properties, independent and dependent variables, and model parameters. These properties typically include the following:

Independent Variables

| Physical coordinates (x, y, z) | Time (t) |

Dependent Variables

Gas species composition	Particle temperature
Gas temperature	Particle velocity
Gas velocity	Turbulent energy
Pressure	of dissipation
Mean turbulent kinetic energy	Particle size distribution
Mixture fraction (mean and variance)	Elemental composition
	extent of reaction
Bulk density	Radiant heat flux

Input Data for Each Inlet Stream

Gas velocity	Fuel elemental composition
Gas composition	Particle temperature(s)
Gas temperature	Particle size distribution
Gas turbulent intensity	Particle velocity(s)
Gas mass flow rate	Particle mass flow rate(s)
Pressure	Particle bulk density(s)

Reactor Parameters

Specific configuration	Dimensions
Inlet configurations	Wall materials
Inlet locations	Wall thickness

In addition, key physical parameters arising in the model subcomponents must be specified. Parameters required will vary, depending on model assumptions, and are obtained principally from basic laboratory measurements.

Comprehensive models of this type are time consuming to develop, require significant computer time to operate, and the reliability is difficult to determine.

It is therefore appropriate to consider the justification for development of such codes. Reasons frequently cited are to

1. Identify general reactor features
2. Interpret measurements
3. Identify important test variables
4. Identify rate-controlling processes
5. Identify areas requiring additional investigation
6. Assist in scaleup
7. Assist in design and optimization of reactors

Given the high cost of large-scale process development, reliable predictive methods are of significant value. However, the question of model reliability is central to model utility. Are reasonable descriptions of the various processes possible? Are the models financially practical to exercise? Can reliability be demonstrated? Most of these issues are not yet resolved; yet the potential is such that development of these codes is receiving significant emphasis.

A classification of coal process flames was given in Chapter 5, and summarized in Table 5.1. Models of coal processes follow these basic classifications:

Classification	Examples
1. Flow type	Well stirred Plug flow Recirculating
2. Mathematical complexity	One dimensional Multidimensional Transient
3. Process type	Fixed bed Moving bed Fluidized bed Entrained bed
4. Flame type	Premixed Diffusion

In the text that follows, modeling of coal processes is illustrated through discussion of the elements of a one-dimensional riug-flow model and a generalized multidimensional model. These models both apply to pulverized-coal flames. Reference is also made to models for fixed and fluidized beds.

7.3. PLUG-FLOW MODEL

In order to illustrate typical features of coal process models, a one-dimensional model of pulverized-coal combustion and gasification (1-DICOG) is outlined in this section. This code is particularly applicable to suspended or entrained flow systems, especially those that are essentially one dimensional (e.g., long and narrow). However, plug-flow models can also be applied to fixed beds, moving beds, and reverse-flow reactors. This model contains a rigorous treatment of coal particle reactions, but avoids the complexities of multidimensional fluid motion. Such codes are relatively inexpensive to operate and sufficiently general to permit application to a wide range of coal reaction processes. This particular model has been documented in detail in the literature (Smoot and Smith, 1979; Smith and Smoot, 1980) and has been placed at several independent laboratories. Several other one-dimensional codes have also been reported (see Table 7.2). However, this particular one-dimensional code was selected because of its familiarity and availability.

7.3a. Model Assumptions

1-DICOG uses the integrated or macroscopic form of the general conservation equations (Smoot and Pratt, 1979) for a volume element inside the gasifier or combustor as illustrated in Figure 7.2. The following aspects of pulverized-coal combustion and gasification have been included in the model:

1. Mixing of primary and secondary streams (specified as input).
2. Recirculation of reacted products (specified as input).
3. Devolatilization and swelling of the coal.
4. Reaction of the char by oxygen, steam, carbon dioxide, or hydrogen.
5. Conductive and convective heat transfer between the coal or char particles and the gases.
6. Convective losses to the reactor wall.
7. Particle–particle radiation and particle–wall radiation, with transparent gases.
8. Variations in composition of inlet gases and solids.
9. Variation in coal or char particle sizes.
10. Oxidation of the coal devolatilization products.
11. Inclusion of multiple sizes or types of coal particles, each with their own individual properties, composition, and reaction rates.

The following are considered to be the major limitations of 1-DICOG:

1. The model does not predict local fluctuating properties within the pulverized-coal reactor, and predicted mean properties are only a function of axial position.

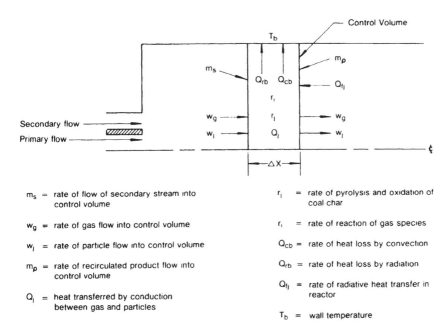

Figure 7.2. Schematic diagram of reactor volume element for 1-DICOG.

m_s = rate of flow of secondary stream into control volume

w_g = rate of gas flow into control volume

w_l = rate of particle flow into control volume

m_p = rate of recirculated product flow into control volume

Q_l = heat transferred by conduction between gas and particles

r_l = rate of pyrolysis and oxidation of coal char

r_l = rate of reaction of gas species

Q_{cb} = rate of heat loss by convection

Q_{rb} = rate of heat loss by radiation

Q_{fl} = rate of radiative heat transfer in reactor

T_b = wall temperature

2. The model does not predict rates of jet mixing or recirculation; rather, these values are required input.

3. The detailed behavior of coal reactions is not yet well understood, which leads to uncertainty in the kinetic description and parameters for the pulverized-coal systems.

4. Some details of the pulverized-coal gasification and combustion processes have been neglected in this version to reduce model complexity and computation time. These include micromixing processes, gas-phase rate-limiting reactions, and gas-phase radiation.

1-DICOG has been developed for an arbitrary number of chemical elements (K in total number). Equations are derived for elemental balances as opposed to species balances, since the gas phase is assumed to be in chemical equilibrium locally. This does not require that neighboring elements are in equilibrium with each other. Use of these elemental balances reduces the number of differential equations and simplifies the required link with the coal reaction model. Volatiles, for example, need not be specified by species composition, which is difficult to identify, but only by elemental composition which can be defined from ultimate analyses of the coal and char. From only the element balances, the calculated energy level and pressure, the complete species composition, gas temperature, and other properties are computed.

The particle is assumed to be composed of specified amounts of raw coal, char, ash, and moisture. The dry, ash-free portion of the coal undergoes devolatilization to volatiles (see Chapters 3 and 12) and char by one or more reactions (M in total number) of the form

$$(\text{raw coal})_j \xrightarrow{k_{jm}} Y_{jm}(\text{volatiles})_{jm} + (1 - Y_{jm})(\text{char}) \tag{7.1}$$

The volatiles react further in the gas phase (see Chapters 3 and 12). The char reacts heterogeneously (see Chapters 4 and 12) after diffusion of the reactant (i.e., O_2, CO_2, H_2O, H_2) to the particle surface by one or more reactions (L in total number) of the form

$$\phi_l(\text{char}) + (\text{oxidizer})_l \xrightarrow{k_l} (\text{gaseous products})_l \tag{7.2}$$

7.3b.   Differential Equations

The differential equations for 1-DICOG are first-order, nonlinear, highly coupled, ordinary differential equations derived for the volume element of Figure 7.2. The mass rate of change of each chemical element k, is

$$d(w_g \omega_k)/dx = A \sum_j r_{jk} + m_{sgk} + m_{\rho gk} \tag{7.3}$$

One equation is solved for each of the K total elements considered. The sources or sinks of elemental mass addition or depletion for the gas phase are only three in number. Mass addition or depletion may take place through reaction with the particles (term 1, right-hand side). Other sources or sinks are due to secondary mixing (term 2) and recirculation (term 3). These last two terms must be specified by the user, since they represent multidimensional effects that might be important in the early regions of the reactor.

The mass rate of change of each particle phase is

$$dw_j/dx = -Ar_j + m_{sj} + m_{\rho j} \tag{7.4}$$

One of these equations is solved for each particle-phase classification. The latter two terms of this equation are not included in 1-DICOG. The only source or sink is the overall particle reaction. The gas-phase energy balance is

$$\frac{d(w_g h_g)}{dx} = h_{sg} m_{sg} + h_{\rho g} m_{\rho g} + A\left(\sum_j Q_j - Q_{cb} + \sum_j (r_j h_{jg} + \xi r_j \Delta h_j)\right) \tag{7.5}$$

The first two terms on the right-hand side depict the energy exchange due to secondary mixing and recirculation, respectively. The final terms enclosed between large parentheses represent all other gas-phase heat-transfer mechanisms. Q_j accounts for the conductive and convective heat transfer from the jth particle type to the gas. Q_{cb} is the convective heat transfer from the gas to the boundary or reactor walls. The last term is the energy carried to the gas by the mass addition from the particle phase. The summation over each of the particle terms accounts for all particle classifications.

The particle-phase energy equation is

$$\frac{d(w_j h_j)}{dx} = m_{s\,j} h_{s\,j} + m_j h_j + A(Q_{f\,j} + Q_{rb\,j} - Q_j - r_j h_j - \xi r_j \Delta h_j) \tag{7.6}$$

Energy exchange due to secondary mixing or recirculation of particles is accounted for in the first two terms. The particle energy equation is similar to the gas-phase energy equation with the exception of two radiation terms and no convective wall term. The radiation terms include particle-phase radiation, but with the assumption of no gas-phase radiation. $Q_{f\,j}$ accounts for particle–particle radiation within the reactor. $Q_{rb\,j}$ is the radiative-heat-transfer rate from the jth particle cloud to the reactor boundaries.

Both Eqns. (7.5) and (7.6) include the term

$$\xi r_j \Delta h_j \tag{7.7}$$

This term allows for the heat of reaction for the heterogeneous reactions to be partitioned between the solid and the gas phases. Since these reactions take place at the boundary between the two phases, it is not immediately apparent which phase should be credited with the energy generated by exothermic reactions or consumed by endothermic reactions. In the approach incorporated in 1-DICOG, this partitioning is specified by the user by setting the value of ξ. If $\xi = 0.0$, all of the energy is given to the gas phase; if $\xi = 1.0$, all of the energy is given to the particulate phase. Any value between 0.0 and 1.0 is also acceptable.

The rate of change of the gas-phase mass flow rate is

$$\frac{d(w_g)}{dx} = A \sum_j r_j + m_{sg} + m_{pg} \tag{7.8}$$

This equation results from the sum of the gas element balances (Eqn. 7.3) over all of the elements.

From the coal particle model (see Figure 3.1), the change of each of the constituents of each of the particle classifications considered is

$$d(\alpha_{c\,j})/dx = r_{c\,j}/n_j v \tag{7.9}$$

$$d(\alpha_{h\,j})/dx = r_{h\,j}/n_j v \tag{7.10}$$

$$d(\alpha_{w_i})/dx = r_{w_i}/n_i v \qquad (7.11)$$

Initially, the reacting particles are composed of specified amounts of raw coal $(\alpha_{c_{io}})$, char $(\alpha_{h_{io}})$, moisture $(\alpha_{w_{io}})$, and ash $(\alpha_{a_{io}})$. Ash is defined as that portion of the initial particle which is inert and is thus constant. This set of differential equations is sufficient to describe the changes occurring in the volume element.

7.3c. Auxiliary Equations

In addition to the required differential conservation equations, a large number of algebraic auxiliary equations is required. It is beyond the scope of this section to present these equations in detail (see Smith and Smoot, 1980). However, they occur in the following categories:

1. *Enthalpy–Temperature Relations.* Nine equations are used to express the enthalpies of the coal particle components in terms of heat capacity and temperature, for each particle type.
2. *Physical Properties.* $5i + 3j + 4$ equations are used to compute the physical properties C_p, k, μ, and D_{im} as functions of composition and temperature, where i is the number of gaseous chemical species in the calculation and j is the number of particle types.
3. *Radiative Heat Transfer.* A radiatively transparent gas phase and radiatively gray particles are assumed with negligible particle scattering. The only radiative energy transfer is thus among the particles within the reactor (Q_{ii}) and between the particle cloud and the reactor walls (Q_{rb_i}). Three different options have been coded for the radiation model (i.e., zone model, diffusion model, and flux model). Each model requires different numbers of equations. Several equations are used to describe the dependence of these models to terms on temperature, dust concentration and size, and radiative properties.
4. *Convective Heat Transfer.* Two equations were formulated to describe convective heat-transfer rates between each particle type and the gas and between the fluid and the walls. Reduction of convective heat transfer by gas evolution from particles is included.
5. *Particle Reactions.* The devolatilization rate was treated according to Eqns. (3.2)–(3.4) with parameters shown in Table 7.1. Char reaction with O_2, CO_2, or H_2O was treated according to Eqn. (4.7) with parameters also shown in Table 7.1. In addition, the coal particle diameter was assumed to vary with the extent of devolatilization, with the swelling coefficient γ specified:

$$d_i = d_{io}\left(1 + \frac{\gamma(\alpha_{c_{io}} - \alpha_{c_i})}{\alpha_{c_{io}}}\right) \qquad (7.12)$$

TABLE 7.1. Selected Kinetic Parameters for Particle Model in 1-DICOG

Process	Reference	Reaction	Rate constant	Parameters
Devolatilization	Ubhayakar et al. (1976)	Raw coal $\xrightarrow{k_1} Y_1$ (volatiles)$_1$ $+(1-Y_1)$(char) simultaneously with reaction 2	$k_1 = A_1 \exp(-E_1/RT_i)$	$Y_1 = 0.39$ $A_1 = 3.7 \times 10^5\,\text{s}^{-1}$ $E_1 = 17.6\,\text{kcal mol}^{-1}$
		Raw coal $\xrightarrow{k_2} Y_2$ (volatiles)$_2$ $+(1-Y_2)$(char) simultaneously with reaction 1	$k_2 = A_2 \exp(-E_2/RT_i)$	$Y_2 = 0.8$ $A_2 = 1.46 \times 10^{13}\,\text{s}^{-1}$ $E_2 = 60.0\,\text{kcal mol}^{-1}$
Char oxidation with O_2	Field[a] (1969)	$2C + O_2 \xrightarrow{k}$ CO	$k = T_i(A + BT_i)$	$A = -1.68 \times 10^{-2}\,\text{ms}^{-1}\,\text{K}^{-1}$ $B = 1.32 \times 10^{-6}\,\text{ms}^{-1}\,\text{K}^{-2}$
Char oxidation with CO_2	Mayers[a] (1934a)	$C + CO_2 \xrightarrow{k}$ 2CO	$k = AT_i^n \exp(-E/RT_i)$	$n = 1.0$ $A = 4.40\,\text{ms}^{-1}\,\text{K}^{-1}$ $E = 38.7\,\text{kcal mol}^{-1}$
Char oxidation with H_2O	Mayers (1934b)	$C + H_2O \xrightarrow{k}$ CO + H_2	$k = AT_i^n \exp(-E/RT_i)$	$n = 1.0$ $A = 1.33\,\text{ms}^{-1}\,\text{K}^{-1}$ $E = 35.1\,\text{kcal mol}^{-1}$

[a]More recent data of Goetz et al. (1982) for four U.S. coals should be considered for char reaction with O_2 and CO_2.

After devolatilization, the particle could be envisioned as a porous sphere (see Figure 3.2), where reaction takes place within the pores. As char combustion proceeds, the particle diameter can be taken to remain relatively constant, but with variable density until breakup or complete burnout of the fuel. Alternatively, the particle may be perceived to react mostly on the surface and thus burn out at a near constant density but as a shrinking particle. These two options were treated in Chapter 4 (see Eqns. (4.9) and (4.10)). In this latter case,

$$d_j = (6\alpha_j / \pi\rho_{aj})^{1/3} \tag{7.13}$$

where ρ_{aj} is the solid density of the jth particle type after complete devolatilization. Both options have been coded in 1-DICOG.

7.3d. Model Predictions and Comparisons

Computations show that the rate of initial particle heatup is an important step in the overall reaction process. The coal or char particles receive or lose

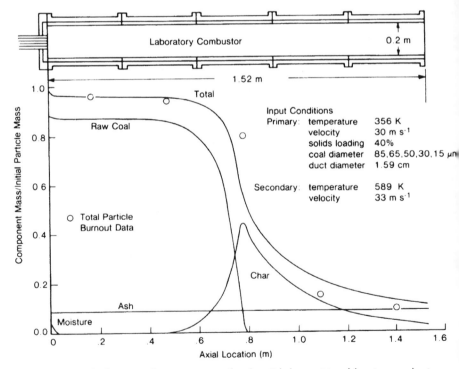

Figure 7.3. Predictions and measurements of coal particle burnout in a laboratory combustor.

energy by radiation from downstream particles and from the vessel walls; they also exchange energy by conduction and convection to the gases which surround them. The rate of energy exchange with the incoming particles determines where in the reactor the particle ignition will occur.

Selected computations to investigate devolatilization were also conducted. As soon as devolatilization begins, the process proceeds rapidly to completion. The devolatilization rate is affected only slightly by particle size, whereas the other particle rate processes are strongly influenced by particle diameter. After devolatilization is initiated, the gaseous products react in the gas phase and the gas temperature rises rapidly. The particle temperature subsequently rises by transfer of heat from the hot gaseous surroundings.

Parametric predictions were made by selectively altering the diffusion rate to determine whether the surface reaction rate or the oxygen diffusion rate was the controlling mechanism during char burnout. Pore diffusion and reaction are accounted for in this formulation only indirectly through the magnitude of the experimental rate constants, which are based on external spherical surface areas for the coal char. It was found for small pulverized-coal particles that surface reaction was rate controlling over oxidizer diffusion.

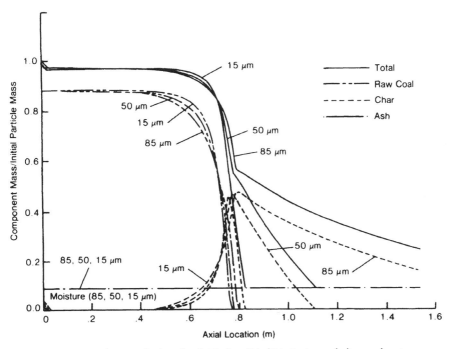

Figure 7.4. Predictions of selected individual particle histories in a polydispersed system.

Thurgood *et al.* (1980) have made local measurements of both gas and solid phases in a laboratory combustor. The reactor was a 20-cm-ID axisymmetric refractory-lined combustor, 1.5 m in length. Figure 7.3 includes a reactor schematic, typical test conditions, and model predictions compared with laboratory combustor measurements of coal particle burnout as a function of reactor length. These calculations and measurements were performed for a high-volatile B bituminous coal. Five size classifications were used to approximate the measured distribution. Figure 7.4 shows the predicted particle histories for three of the five classifications. Figure 7.5 shows the measured and predicted gas-phase mole fraction history for the same case.

The reason for the good agreement between laboratory measurements and predictions is the one-dimensional nature of the laboratory combustor. Mixing of primary and secondary gases is rapid and particle ignition occurs later in a fully mixed, gas environment; particle combustion is thus not significantly affected by the gas mixing and recirculation processes in the upper regions of the reactor.

During the first 70 cm, the particles are heated by radiation. Vaporization of moisture from the coal is very rapid and devolatilization begins at about the same time for all particles and is completed very rapidly. As the raw coal devolatilizes, gas-phase products are evolved which are further reacted in bulk gas phase and the temperature rises sharply. The devolatilization process also

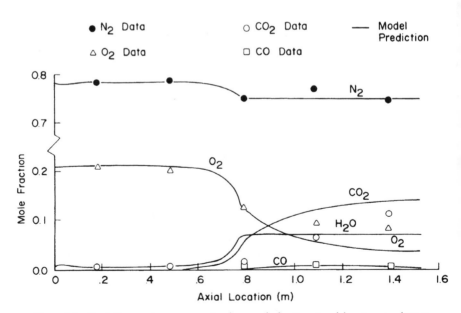

Figure 7.5. Predictions and measurements of gas mole fractions in a laboratory combustor.

forms the residual char which rises to peak quantities at the point of complete devolatilization. Heterogeneous char oxidation takes place at greatly different rates for each particle size, as illustrated in Figure 7.4. Small particles burn out quickly, while the larger particles are not burned out even at the end of the reactor.

7.4. MULTIDIMENSIONAL FLAME CODES

7.4a. Basis

Comprehensive, multidimensional modeling of turbulent combustion and gasification has been recognized as a difficult problem, due not only to the problems associated with solving the differential equation set, but also due to describing the interactions between chemical reactions and turbulence. Such a model is introduced briefly here. It is referred to as PCGC-2 (Pulverized Coal Gasification or Combustion-2 Dimensional). This description introduces the complexities of comprehensive modeling and establishes a basis for much of the material that follows in subsequent chapters. PCGC-2 is an attempt to use currently available technology to combine knowledge of the turbulent fluid

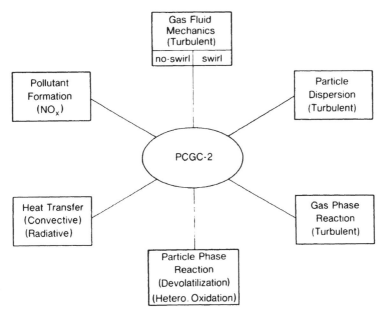

Figure 7.6. Submodels incorporated into generalized pulverized-coal gasification and combustion code (PCGC-2).

mechanics with a reasonable approach to the reaction processes. PCGC-2 is applicable to gas or pulverized, coal-fired combustion and to entrained flow gasification in axisymmetric cylindrical coordinates. Variations in the properties are considered only in the axial and radial directions. Symmetry is assumed in the angular direction. A schematic of a typical reactor modeled was shown in Figure 5.6. This particular cylindrical reactor is coaxial, with coal entering the reactor in the central (primary) stream and the majority of the oxidizer entering in the outer (secondary) stream or annulus. The pulverized-coal particles range in size from about 1 μm to over 100 μm. The model predicts the mean-gas-field properties for axisymmetric, steady-state, turbulent diffusion flames (i.e., local velocity, temperature, density, species composition, etc.). Local particle properties are also computed, such as coal burnout, particle velocity, and temperature, and coal component composition. Figure 7.6 illustrates the various submodels incorporated into PCGC-2.

7.4b.   General Description of Subroutines

Gas Phase.   The gas phase is assumed to be a turbulent, reacting, continuum field that can be described locally by general conservation equations. The flow is assumed to be time steady. Gas properties (i.e., density, temperature, species composition) are assumed to fluctuate randomly according to an assumed shape of a probability density function (PDF), characteristic of the turbulence. Turbulence is modeled by Favre-averaging of the gradient diffusion processes and with the two-equation $k - \varepsilon$ model for closure (see Chapter 10). The effect of particles on the gas-phase turbulence is modeled with an empirical correlation. Gas-phase reactions are assumed to be limited by mixing rates and not by kinetic limitations. Gaseous properties are calculated assuming local instantaneous equilibrium.

Radiation.   The pulverized-coal flame radiation field is a multicomponent, nonuniform, emitting, absorbing, scattering gas–particle system. The coal particles cause anisotropic and multiple scattering. The flame may be surrounded by nonuniform, emitting, reflecting, absorbing surfaces. An Eulerian framework is used to model the radiation, which facilitates incorporation of radiation properties into gas-phase equations and also specifies a radiation field for the Lagrangian particle calculations (see Chapter 14).

Particle Behavior.   Pulverized-coal flames of interest have a void fraction of near unity and the individual particles are very dispersed. This results in few particle–particle interactions, and hence, the particle phase has not been considered a continuum like the gas phase. Different particles at the same location may exhibit completely different properties due to the different particle paths, giving rise to a particle history effect. A Lagrangian treatment of the particles is utilized, representing the particles as a series of trajectories. Particle properties are obtained along these trajectories (see Chapter 12).

Particle Reaction. The detailed effects of particles on the gaseous flow field in turbulent combustion and gasification are unknown. The coal reaction rates are assumed to be slow compared to the turbulence time scale. This allows the particle properties to be calculated from the mean gas properties instead of the fluctuating gas properties. Without this assumption, the particle reaction rates would be extremely difficult to calculate. Particle reaction processes are considered in a manner much like that of 1-DICOG, described above.

PCGC-2 assumes that the off-gas from the coal is of constant species composition. Particles are defined to consist of coal, char, ash, and moisture. Ash is taken to be inert. Any volatile mineral matter is considered as part of the volatile matter of coal. Moisture is currently given to the gas phase through drying before the coal enters the reactor. Coal reaction rates are characterized by parallel reaction rates with fixed activation energies. Char particle swelling is accounted for empirically. The interior of the particle is assumed to have the same temperature as that of the particle surface.

Devolatilization occurs more rapidly than the char combustion reactions. The efflux of off-gases from the coal reactions can affect diffusion to the particle for the heterogeneous char combustion reactions, heat conduction from gas to particle, and momentum exchange between gas and particle; these effects are accounted by corrections from stagnant film theory. The successful prediction of particle reactions is dependent upon the available coal reaction rate data.

7.4c. Code Development

Formulation, numerical solution, and evaluation of PCGC-2 has been completed over the past several years. Development has reached the point where the code is being applied to specific problems. Appropriate use of this comprehensive code and others like it require a thorough understanding of the basic code foundations. Chapters 10–15 present technical information used in treating turbulent, particle-laden diffusion flames.

7.5. OTHER PULVERIZED-COAL CODES

Attempts to calculate the detailed performance of turbulent, pulverized-coal reaction processes have only been undertaken during the last decade. Prior to 1970, the best computations available were based on overall global calculations (Field *et al.*, 1967). The details of the mixing processes were not quantitative.

TABLE 7.2. Review of Models for Pulverized-Coal Combustors/Gasifiers in Past 10 Years[a,b]

Author(s)	Year	Description	Major results
Gibson and Morgan	1970	Two-dimensional, axisymmetric, elliptic, stream function–vorticity formulation; local gas equilibrium; particle diameter change with time based on measurement.	With simplistic radiation and coal combustion model, the predicted wall heat flux, temperature, and coal burnout agrees reasonably well with measured values.
Richter and Quack	1974	Two-dimensional, axisymmetric, elliptic, stream function–vorticity formulation, two-equation turbulence model; four-flux radiation.	Technique similar to Gibson and Morgan (see above) with improved radiation turbulence submodels. Heterogeneous reaction rates fitted to experimental furnace data.
Mehta	1976	Phenomenological model combination of PFR and PSR calculations; gas-phase chemical equilibrium; overall char gasification with Arrhenius-type rates.	Describes the outlet conditions for a three-stage entrained flow gasifier.
Lewellen et al.	1977	Phenomenological model; swirling flow reactor divided into four regions of interest; dominant physical processes identified for each region.	Predicts effects of four performance parameters on certain design variables.
Sprouse	1977	One-dimensional hydrogasification; freestream equilibrium chemistry; one-step devolatilization and hydrogasification model.	Extensive study of the boundary layer around the particles; compared well with outlet measurements for coal gasifier tests.
Finson et al.	1978	Steady, one-dimensional model, with gaseous kinetics; kinetics from data for coal pyrolysis; heterogeneous oxidation; considers char structure and reactivity.	Developed code; predictions compared with laboratory gasification measurements of axial temperature and gas composition.
Ubhayakar, Stickler, and Gannon	1977	One-dimensional, steady-state, two-step devolatilization; combined diffusion and surface reaction rates; no radiation; equilibrium gas-phase reactions.	Applied to gasification combustion and hydropyrolysis with good agreement.
Blake et al.	1977 and 1979	Three-dimensional, transient, mixed finite-element, finite-difference scheme with separate Eulerian particle equations; two-equation gaseous turbulence; separate turbulent kinetic energy for particles.	Under development; illustrative predictions shown for gasification.

	Year		
Barnhart et al.	1979	Combination of PFR and PSR elements; gas phase in equilibrium with two-equation kinetics for pyrolysis, char oxidation, and CO oxidation.	Developed code; predicts carbon burnout, product gas composition, and temperature; reported fair to good comparison between experiment and theory for cyclone gasifier.
Smith and Smoot	1979 and 1980	One-dimensional, steady-state, two-step devolatilization; combined diffusion and surface reaction rates; one-dimensional zonal radiation; equilibrium gas-phase reactions.	Good agreement with one-dimensional combustor data; less agreement with gasifier measurements.
Smith et al.	1981	Two-dimensional, axisymmetric, with recirculation, $k-\varepsilon$ turbulence model, Lagrangian particles, two-step devolatilization; diffusion and kinetic heterogeneous rates, four-flux radiation with scattering.	Gas-phase components reported and compared with data; coal code recently completed; example computations.
Wen and Chuang	1978	Single-step pyrolysis; global rate for gaseous CO combustion or gaseous equilibrium; data based rate expressions for CO_2, O_2, and H_2O reaction with char.	Predictions for Texaco entrained gasifier compared with limited pilot plant temperature and composition data.
Lockwood et al.	1980	Two-dimensional, axisymmetric, with gas recirculation, $k-\varepsilon$ turbulence model; single-step pyrolysis, mixing-controlled volatiles reaction; four-flux radiation with scattering; Lagrangian particle motion.	Computations for pulverized-coal furnace with concentric burner.
Chan et al.	1980	Two-dimensional, time-dependent, axisymmetric code, applied to high-pressure gasification, lagrangian-particle motion, global gas kinetics, and devolatilization and char oxidation in coal reaction model; first-order (algebraic) turbulence closure.	Treats multistage process; no comparisons with data; potential for three dimensions.

aData from Smoot (1980).
bRecent work by Oberjohn et al. (1982) was reported after preparation of this table on development of a two-dimensional model for pulverized coal combustion (see Table 7.2).

However, the growth in capacity and speed of digital computers and the recent awareness of the limits of energy resources have made comprehensive calculation techniques both possible and of current interest.

Modeling of turbulent reaction processes is still in a state of development. Recent reviews of this subject for gaseous systems are given by Edelman and Harsha (1978), Lilley (1979), and Williams and Libby (1980). Pulverized-coal conversion models developed since 1970 are reviewed in Table 7.2. Three codes use the perfectly stirred reactor approach in combination with plug flow; five

TABLE 7.3. Submodel Summary for Multidimensional Pulverized-Coal Reactor Models[a]

Author(s)	Date	Dimension	Turbulence	Gaseous combustion	Devolatilization
Gibson and Morgan	1970	2-D steady state with swirl	Algebraic equation for μ_{eff}	Burns instantaneously when fuel and oxidizer mix	No devolatilization; volatiles admitted separately in gaseous form
Richter and Quack	1974	2-D steady state	Two-equation $k - \varepsilon$ model for μ_{eff}	Instantaneous reaction of $CO + \frac{1}{2}O_2 \rightarrow CO_2$	Considered only low-volatile coals
Blake et al.	1977, 1979	3-D transient	Two-equation $k - \varepsilon$ model for gas, plus a k equation for particles	Global kinetics with partial equilibrium	Single reaction with distribution of activation energies
Smith et al.	1981	2-D steady state	Two-equation $k - \varepsilon$ model for μ_{eff}	Equilibrium with probability density function to account for "unmixedness"	Multiple-step devolatilization model
Lockwood et al.	1980	2-D steady state	Two-equation $k - \varepsilon$ model for μ_{eff}	Reaction rate eddy time scale	Constant release rate
Chan et al.	1980	2-D transient	Algebraic expression for μ_{eff}	Global kinetics, seven stable species	Single activated step, products specified
Oberjohn et al.	1982	2-D steady	$k - \varepsilon$ model	Various options	s_0 computed rate; single product

[a] Data from Smoot (1980).

codes are one-dimensional codes (including that described above); and six codes are multidimensional codes. Eight of the models were applied specifically to entrained gasification.

Table 7.3 reviews multidimensional models, including PCGC-2, with brief statements on some of the components included. Gibson and Morgan (1970) used techniques developed by Gosman *et al.* (1969) for steady, turbulent, gaseous flow, extended to simulate pulverized-coal combustion. The volatiles were assumed to be emitted separately in gaseous form. The carbon particles were

TABLE 7.3. (*Cont.*)

Heterogeneous reaction	Radiation	Particle dispersion	Numerical method
Mass fraction of char solved by separate transport equation for five size classes; diameter change with time based on measurement	Two-flux model (radial flux only)	Particles treated like a gas	Stream function–vorticity formulation
Arrhenius-type reaction rate with constants obtained by fitting model to data	Four-flux model	Particles treated like a gas	Stream function–vorticity formulation
Reaction with steam, CO_2, O_2, and H_2 based on Arrhenius or Langmuir expression and laboratory data	Not clarified	Separate Eulerian-particle equations with different classifications	Primitive variable, Eulerian for both phases; mixed finite element-finite difference
Diffusion rate and kinetic rate included	Four-flux model including scatter due to particles	Lagrangian particle equations with turbulent diffusion velocity	Primitive variable, Eulerian gas-phase technique/Lagrangian particle technique
Diffusion and kinetic rates included	Grey; flux method with scattering	Lagrangian particle equations with turbulent diffusion	Primitive variable, Eulerian gas-phase technique/Lagrangian particle technique
With steam, CO_2, and O_2; considers intrinsic rates	High-pressure system; diffusion approximation	Lagrangian particles, no effects of gas turbulence	Primitive variable, finite-difference, Eulerian gas-phase technique/Lagrangian particle technique
Diffusion and kinetic rates included	Flux-discrete ordnance method	By mean drag only	Eulerian gas-phase Lagrangian particle phase

represented by discrete sizes and the rate of coal reaction was based on measurements of Field (1969, 1970). Richter and Quack (1974) used a similar approach for low-volatile coals. The PCGC-2 model for two-dimensional, steady-state coal combustors (Smith and Smoot, 1980; Smith *et al.*, 1981), includes effects of gas turbulence on gaseous combustion and particle motion. Lockwood *et al.* (1980) have also recently published a steady, two-dimensional method for coal flames with computations for a pulverized-coal furnace.

Three-dimensional transient modeling of turbulent gaseous combustion has been treated by Patankar (1975) and by Gosman *et al.* (1980). A three-dimensional, transient coal gasification model by Blake *et al.* (1977, 1979) is based on a mixed finite-element/finite-difference scheme. The finite-element character of the model permits the numerical representation of a variety of geometries. An Eulerian approach for the particles requires a model for the evaluation of the particle distributions in the turbulent reacting field which change continuously for each particle property (i.e., size and temperature). The turbulence model is based on the $k - \varepsilon$ model and includes a third equation for the turbulent kinetic energy of the particulate phase. The model has been used to predict the gas and particle dynamics in a three-dimensional system. Also, a transient three-dimensional coal gasification model was developed by Chan *et al.* (1980). Work on three-dimensional codes has not progressed to the point where utility or validity have been established.

7.6. FIXED-BED COMBUSTION MODELS

7.6a. Background

It has been common to refer to moving-bed systems such as the Lurgi gasifier, illustrated in Figure 7.7, as packed beds or fixed beds. In reality, the coal in these beds and in many stokers is moving slowly. However, this kind of coal reaction process will often be referred to herein as a fixed-bed process.

In this type of process, coal is fed slowly to the top of the bed (or at one side onto a moving grate). Oxygen and other oxidants (e.g., CO_2 and H_2O) and inertants (e.g., N_2 in air) are fed into the bottom or up through the bed in this countercurrent process. Residence time of the gases is only seconds, while for the solids, it can be minutes or hours. The bed is often divided into idealized zones, as illustrated in Figure 7.8, which is a schematic of the Morgantown Energy Research Center (MERC, but now Technology Center, METC) test gasifier. The exiting hot gases dry, heat, and pyrolyze the coal. These gases then heterogeneously react with the coal char in the lower part of the bed. In direct combustion, the process is dominated by oxygen–char reactions. However, in gasification, the oxygen quickly disappears in the lowest combustion region,

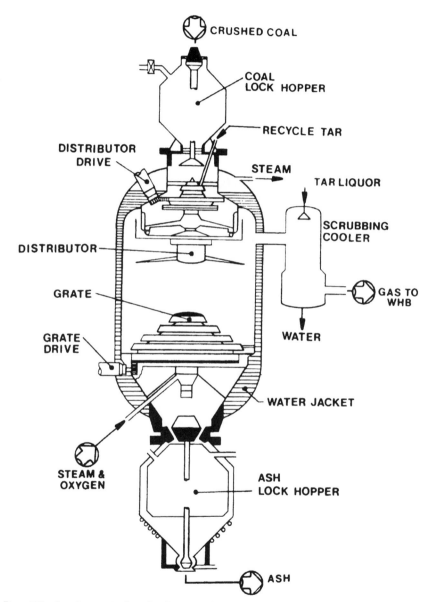

Figure 7.7. Lurgi pressurized gasifier. (Figure used with permission from Amundson and Arri, 1978.)

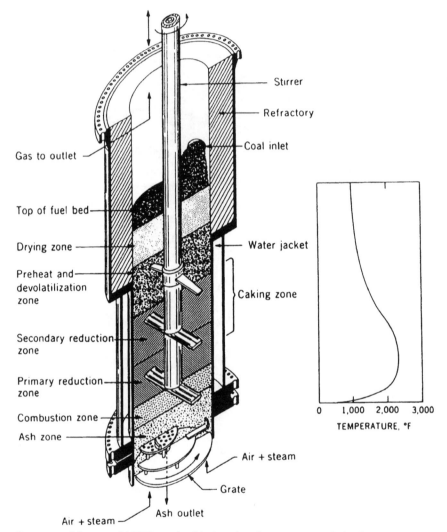

Figure 7.8 Schematic of MERC text fixed-bed gasifier, showing various idealized process sections. (Figure used with permission from DeSai and Wen, 1978, under support by the U.S. Dept. of Energy, Morgantown Energy Technology Center, Morgantown, WV.)

giving rise to slower char–CO_2 or char–H_2O reactions in the gasification section. Residual ash is removed from the bottom. Operation at very high temperatures will cause this ash to melt and be removed as slag.

The level of past research and development of these fixed or slowly moving bed combustion models is somewhat limited. No recent reviews of fixed-bed coal combustion models were identified. Further, no recent models specifically identified for direct combustion systems, such as coal stokers, were located. The majority of the recent work in this area has emphasized coal gasification in fixed-bed gasifiers. This type of gasifier is used commercially in various parts of the world. Performance characteristics of commercial [Lurgi (Figure 7.7) and Wellman–Galusha] and pilot gasifiers are shown in Table 7.4. Operation at elevated pressure in this type of process is common.

TABLE 7.4. Performance Characteristics of Moving-Bed Gasifiers[a]

Gasifier	Lurgi		Wellman–Galusha		GEGAS-D	MERC
Diameter (ft)	12	10 (OD)	10		3	3.5
Height (ft)	N/A	N/A	9		N/A	6.5
Gasifying capacity (lb/hr ft^2)		300–400	7.5–99		200	100–200
Pressure [psi (gauge)]	300–450	350	Atmospheric		200–300	15–225
Type of coal		Pittsburgh Bituminous	Anthracite		Pittsburgh	Arkwright
Coal composition						
C	57.12	74.46	N/A	73.66	67.42	75.92
H	3.93	5.1		2.25	4.98	5.70
O	8.27	4.36		6.93	7.39	4.92
N	0.83	1.43		0.66	1.35	1.38
S	4.45	2.30		0.56	3.82	2.71
Ash	13.3	7.50		9.55	15.02	8.25
Moisture	12.1	4.75		6.4	2.53	1.12
Air/Coal (lb/lb)		—	3.4	—	2.63	2.5–3.67
Oxygen/coal (lb/lb)	—	0.61	—	0.77	—	—
Steam (lb/lb)		3.10	0.4	1.4	0.45	0.5–0.74
Gas composition (mol%)						
CO	14.54	17.9	26.0	40.0	23.8	16.0–23.0
CO$_2$	16.22	30.8	4.5	16.5	6.7	7.0–12.0
N$_2$	42.85	2.4	32.0	1.5	49.2	48.0–55.0
H$_2$	22.36	38.8	13.0	41.0	17.0	13.0–17.0
CH$_4$	3.88	8.4	3.0	0.9	3.2	2.0–3.5
C$_2$H$_6$	—	0.7	—	—	—	0.3
C$_2$H$_4$	—	0.3	—	—	—	—
H$_2$S	—	0.7	0.5	—	—	0.3–0.6
O$_2$	—	—	—	—	—	0.1
Heating value of gas (BTU/SCF)	158	283.0		270	160	100.0–180.0
Steam decomposition (%)		21.1	N/A	N/A	54	30–70%
Thermal efficiency	N/A	77.0			74	65–75%
Heat loss (% of heating value of coal)						10–17%

[a]Data from DeSai and Wen (1978) and Wen *et al.* (1982).

TABLE 7.5. Summary of Recent Fixed-Bed Combustor Models

Authors	Year	Emphasis	Scope	Application	Comparison with data	Computations
Winslow	1977	Gasification packed bed, countercurrent	One dimensional, transient	In situ gasification	Peak temperature, exit gas composition, flame speed	Examples for in situ beds; subbituminous coal, 1200 K peak temperature
DeSai and Wen	1978	Gasification moving bed, stirred, countercurrent	One dimensional, steady	MERC gasifier air-blown	For three different coals; gas composition, percent conversion, temperature; results inconclusive	For several MERC gasifier cases
Yoon et al.	1978	Gasification moving bed, countercurrent, high pressure, slagging	One dimensional, steady and transient with boundary layer	Lurgi, GEGAS, slagging	Exit gas comparisons, peak temperature for pilot slagging gasifier	Effects of heat loss, ash layer oxygen rate, transient effects
Amundson and Arri	1978	Gasification countercurrent, char reaction model	One-dimensional, steady	Lurgi type	Limited comparison of steam/O_2 ratio on H_2/CO ratio	Parametric studies for several operational variables
Barriga and Essenhigh	1979	Combustion and gasification; fixed bed	Steady state	Char gasification, small-scale gasifier	15 × 40 cm laboratory gasifier with O_2, CO_2, CO temperature profiles along length; coke fuel	Sensitivity analyses on heterogeneous reaction rate, and char area with adjusted rate parameters
Wen et al.	1982	Gasification moving bed, countercurrent	One dimensional, steady with user's manual	General high-pressure gasifier (dry)	Exit temperature, carbon conversion, gas composition, heating value for four coals	Several design maps for bituminous coal

Several fixed-bed coal combustion models are summarized in Table 7.5. All six are one-dimensional models, while two of the six consider transient effects. All six emphasized gasification processes, but all of the codes presumably apply also to direct combustion. All are for slow, vertically moving packed beds with countercurrent gas flow. A brief review of the nature and state of development of these one-dimensional fixed-bed models follows.

7.6b. One-Dimensional Fixed-Bed Models

Common simplifications in modeling of fixed beds include the following:

1. Both the gas and particles are generally assumed to be in plug flow. This renders the problem one dimensional and elminates complex fluid mechanics considerations.
2. The time required to devolatilize the coal is most often neglected, although the products of devolatilization are considered.
3. Gas-phase reactions are usually assumed to be completed, to the extent that reactants are available (e.g., $CO + O_2$) or in equilibrium at the local temperature $(CO + H_2O = CO_2 + H_2)$.
4. Turbulence effects and their impact on reactions with slowly moving gases in the packed bed are neglected.

Modeling of fixed beds is, however, complicated by the possibility of layers accumulating around the large, slowly moving particles; by internal diffusion and reaction processes; by gaseous flow in porous media of changing particle size and shape; by changing oxidizers (O_2, then CO_2, H_2O), particle sizes, and temperature levels; and by changes in the controlling processes for char consumption.

The general concept used in modeling of fixed-bed gasifiers and combustors is illustrated schematically in Figure 7.9. Fixed-bed models ordinarily include:

1. Gaseous flow in porous media, for pressure drop and gas velocity.
2. Rapid drying, heating, and devolatilization of the coal to produce tars and gases.
3. Heterogeneous oxidation of the char with various oxidizers (O_2, CO_2, H_2O), and with consideration of internal and external diffusion and surface reaction, and of ash layering.
4. Mass balances for various species (CO, CO_2, H_2O, O_2, N_2, etc.), and some specification of the gas-phase reaction steps, such as equilibrium or completeness.
5. Heat balances for gases and particles including wall heat losses, conduction in porous beds, convection, and radiation.
6. Pollutant formation, particularly for oxides of nitrogen.
7. Melting of ash and slagging effects in high-temperature gasifiers.

TABLE 7.6. Summary of Code Components for Fixed-Bed Combustion Codes

Authors (Year)	Key assumptions	Basic equations	Gas flow	Gas reactions	Devolatilization	Surface reaction
Winslow (1976)	One dimensional, transient, no radiation, no tar condensation	PDE mass and energy for gas and solids	Darcy law—porous beds	Seven species and tar, water-gas equilibrium, rapid H_2/CO, CH_4 consumption	Single step, activated	Char with O_2, H_2O, CO_2, H_2; diffusion and reaction terms; first-order activated surface reaction, not coal specific
DeSai and Wen (1978)	One dimensional, steady, equilibrium gas-solid temperature, H_2O–C and CO_2–C internal reaction, finite heat losses, no radiation, neglect stirring	ODE heat and mass balances, gas species, and six solid elements	Plug flow of gas and solid: five zones	Water-gas rate with catalyst effects, 12 gas species	Fast; extent vs. final temperature; approximate products	Char with O_2, H_2O, CO_2, H_2O; coal-dependent rate constants, with kinetic and diffusion effects
Yoon et al. (1978, 1979)	Equilibrium gas- and solid-phase temperature, no heat loss to boundary layer, small radiative effect, subbituminous and bituminous coals	Mass and energy balances	Plug flow of gas and solid with adiabatic core and cooler boundary layer	Equilibrium water-gas reaction, CO and H_2 unreactive in gas phase	Instantaneous heating, proximate instantaneous volatiles release with estimate of products	Char reaction with CO_2, H_2O, H_2, O_2; internal and external diffusion effects with or without ash layer

Reference					
Amundson and Arri (1978)	Radiation included, char feed, dry bottom, equal gas and particle temperature	ODE and algebraic mass balances for C, O_2, H_2O, CH_4 with heat balance	Plug flow of gases and solids	Water–gas equilibrium with completion of other reactants with O_2	Neglected	Complicated O_2–C diffusion-controlled shrinking sphere char reaction with ash layer; $CO_2 + H_2O$ reaction inside char Carbon with O_2, CO_2
Barriga and Essenhigh (1979)	Different T_p, T_s, axial radiation, and conduction	Bed P from permeability, second order energy balances on gas and solids	Plug flow of gases and solids	$CO + O_2$ reaction instantaneous	Neglected	Carbon with O_2, CO_2
Wen et al. (1982)	Kinetic parameters for various coal types, constant pressure, radiative and convective heat transfer	Mass balances on seven gas species, particle heat balance	Plug flow of gases and solids	Catalyzed water gas shift plus H_2, CO oxidation, complete H_2 reaction; CO/CO_2 ratio temperature dependent	Extent of devolatilization, cracking carbon disposition—fast devolatilization rate	CO_2, H_2O, O_2, H_2 with kinetic and diffusion terms; internal CO_2, H_2O, H_2 reaction; shrinking core O_2 reaction with ash layer

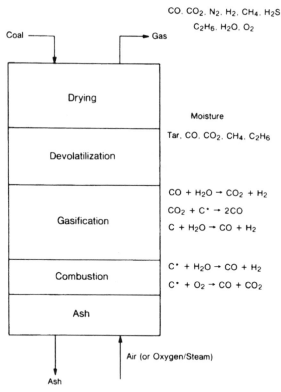

Figure 7.9. Chemical reactions occurring in a fixed-bed gasifier. (Figure used with permission from DeSai and Wen, 1978, under support by U.S. Dept. of Energy, Morgantown Energy Technology Center, Morgantown, WV.)

Table 7.6 summarizes the approaches used in recent fixed-bed gasification models. The general approach is common to all, with only secondary differences in treatment. Little attention has been given to pollutant formation or to slagging. Recent work by Wen *et al.* (1982) includes a user's manual for the fixed-bed model.

7.6c.   Fixed-Bed Models: Evaluation and Application

Solutions for the fixed-bed model are relatively straightforward for one-dimensional systems, and most investigators have performed sensitivity analyses or parametric computations, as outlined in Table 7.5. Also, most investigators have compared predicted results with observed exit properties from fixed-bed gasifiers (also see Table 7.5). Exit comparisons for three investigators are shown in Table 7.7. Most of the comparisons are from Wen and co-workers (1978, 1982), with data from the MERC fixed-bed gasifier. More recent comparisons

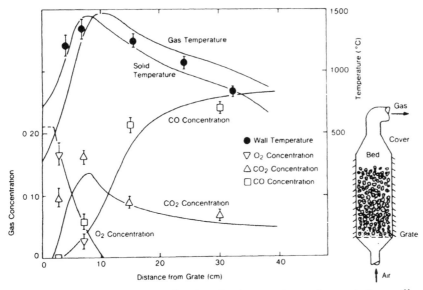

Figure 7.10. Comparison of measured and predicted temperature and concentration profiles throughout a fixed-bed gasifier. (Figure used with permission from Barriga and Essenhigh, 1979.)

(columns 7–10 of Table 7.7) with the revised model are quite good for carbon conversion, gas composition, and exit gas temperature. Since these values are interrelated, variations in one value (e.g., burnout) will certainly indicate variations in the other values as illustrated by earlier model predictions of Wen and co-workers and as shown in Table 7.7, columns 1–6. While recent results are positive, only partial confidence is gained from comparisons of exit values.

Comparisons of predictions for the fixed-bed model with detailed profile data are hampered by lack of available data. Only the measurements of Eagen *et al.* (1977) were identified for this purpose. Barriga and Essenhigh (1979) compared these profile data with model predictions as illustrated in Figure 7.10. Comparisons are encouraging but have been based on some adjustment in parameters to achieve improved agreement.

Fixed-bed codes have been applied to practical fixed-bed gasifier systems, such as *in situ* gasification (Winslow, 1977), Lurgi gasifiers (Yoon *et al.*, 1978; Amundson and Atti (1978); and Essenhigh, 1977), and several general situations with wide variation of operating parameters (see Table 7.5).

7.7. FLUIDIZED-BED MODELS

7.7a. Background

The third class of coal combustion models considered is for fluidized-bed combustion (FBC) (see Figure 5.5). Development of these models has been

TABLE 7.7. Comparison of Fixed-Bed Reactor Observations with Model Predictions

Operating conditions		Gasifier										
Bed type		MERC 1218		MERC 1097		MERC 925		MERC 1240		MERC 1371		MERC 1402
Coal feed (lb/hr)												
Coal type		Arkwright bituminous		Arkwright bituminous		Arkwright bituminous		Arkwright bituminous		UP. Freeport bituminous		West Kentucky bituminous
Pressure [psi (gauge)]		90		76		23		141		126		139
Steam (lb/hr)		576		501		342		693		537		691
Air (lb/hr)		3826		4116		2887		3756		4229		4001
Predicted quantities		Pred.	Obs.	Pred.	Obs.	Pred.	Obs.	Pred.	Obs.	Pred.	Obs.	Pred.	Obs.
Model		Desai and Wen (1978)		Desai and Wen (1978)		Desai and Wen (1978)		Desai and Wen (1978)		Desai and Wen (1978)		Desai and Wen (1978)
Dry product gas (lb mol/hr.)		200.5	200	207	205	147	148	200	206	213	218	215	201
Gas composition (mol %)												
CO		14.8	16.2	16.9	20.3	17.5	20.7	13.1	21.3	17.7	21.8	14.6	18.6
CO_2		12.7	12.2	11.4	7.7	10.6	9.2	13.9	8.9	10.8	7.0	12.6	9.2
N_2		52.2	52.2	54.5	55.2	53.7	53.4	51.4	49.4	54.4	53.1	50.9	54.3
H_2		15.5	15.7	13.3	12.9	13.6	13.1	16.8	16.9	11.4	13.9	15.7	14.3
Gas temperature (°F)		980	1253	1110	1062	1106	948	987	1065	1210	1050	1110	839
Percent carbon conversion		86.4	87.2	97.6	95.5	84.4	92.3	83.6	90.7	97.5	91.3	95.5	84.0

	MERC 1218 Arkwright bituminous		MERC 8058 Pittsburgh No. 8 bituminous		MERC 13,270 Illinois No. 6 bituminous		MERC 13,580 Montana Rosebud subbituminous		LURGI Illinois bituminous	
Operating conditions	Pred.	Obs.	Pred.	Obs.	Pred.	Obs.	Pred.	Obs.	Pred.	Obs.
Bed type										
Coal feed (lb/hr)										
Coal type										
Pressure [psi (gauge)]	90		350		335		344		368	
Steam (lb/hr)	576		25,145		33,331		13,673		$1094/ft^2$	
Air (lb/hr)	3826		5005 (O_2)		6396 (O_2)		3336 (O_2)		1140 $(O_2)/ft^2$	
Predicted quantities										
Model	Wen et al. (1982)		Wen et al. (1982)		Wen et al. (1982)		Wen et al. (1982)		Yoon et al. (1982)	
Dry product gas (lb mol/hr)	200	200	796	771	1074	1061	772	768		
Gas composition (mol %)										
CO	16.9	16.2	16.0	16.8	18.0	17.2	19.0	20.2	13.8	17.3
CO_2	10.2	12.2	31.0	31.3	30.0	31.1	28.0	28.7	32.9	31.2
N_2	52.7	52.5	1.7	1.6	1.7	1.2	1.3	1.3	—	—
H_2	16.1	15.7	40.2	39.2	38.3	38.9	36.2	36.4	42.4	39.1
Gas temperature (°F)	1169	1253	1208	1196	1119	1122	889	617		
Percent carbon conversion	87.0	87.2	98.3	98.4	98.5	99	95.5	98.7[b]		

[a] With boundary layer and tar cracking.
[b] With liquid product.

seriously pursued for the past 10–15 years, but the foundations were established by earlier research in fluidized beds. All of the fluidized-bed models reviewed can be conveniently divided into two classifications:

1. One-dimensional models
2. Two-dimensional models

Sarofim and Beér (1978) and, more recently and more extensively, Olofsson (1980) provide reviews of one-dimensional FBC model developments that have occurred principally over the past decade. No general reviews of two-dimensional FBC models were identified. Work on two-dimensional models is more recent and less complete. A brief discussion of these two model types is presented, followed by some predictions and comparisons with measurements. Historically, most of the one-dimensional models have emphasized direct combustion, while the focus of two-dimensional models has been FB gasification.

7.7b. One-Dimensional FBC Models

Foundations. Figure 5.5 showed a schematic diagram of a typical fluidized-bed coal combustor. Most of the modeling work for this type of combustor has been conducted in the past decade. This combustor is distinguished by low operating temperatures (~ 1100 K), high excess-air levels ($\sim 30\%$), intermediate particle sizes (1–3 mm), long residence times (several minutes), and vigorous particulate motion that dominates heat transfer and reaction processes. Comprehensive modeling of this process is greatly complicated by the very complex

TABLE 7.8. Summary of Selected One-Dimensional FBC models

Reference	Year	Comments
Avedesian and Davidson	1973	Early FBC model for spherical carbon-batch system
Gibbs	1975	Extended model to coal and continuous feed; diffusion-controlled char reaction
Horio *et al.*	1977	Treated sulfur capture and NO formation and added kinetic term to char consumption, polydispersed sizes, and heat-transfer tubes
Gordon *et al.*	1978	Treated gas-phase reaction in bubbles, cooling tubes, carbon fuel
Saxena *et al.*	1978b	Three-phase bed with bubbles, wake, and dispersion; ash restriction to char burnout
Wells and Krishnan	1979	Considered particle elutriation and attrition, CO_2–C gasification, volatiles release and combustion for different coals, and heat-transfer surfaces in freeboard region (user's manual)
Sarofim and Beér	1978	Considered NO_x formation, CO formation, and reaction
Lewis *et al.*	1982	Added transient and high-pressure treatment (user's manual)

fluid motion and interaction of the gas and particles. Therefore, most of the modeling reported to date for FBCs has been one dimensional, and has been based heavily on correlations of laboratory data for various controlling aspects of the process.

At least eight different one-dimensional FBC models were documented between 1973 and 1982, as summarized in Table 7.8. A continuing, consistent approach has been used in the progressive development of one-dimensional FBC models, with new components being added by various investigators. The complex motion of particles has been neglected, while gas behavior has been divided into two or three parts. One-dimensional FBC models now include the following aspects:

1. Gas fluid dynamics
2. Particle fluid dynamics
3. Coal devolatilization
4. Char oxidation
5. Volatiles combustion
6. Particle attrition (size reduction)
7. Particle elutriation (loss from bed)
8. Elevated pressure
9. Gasification reactions
10. Polydispersed particle sizes
11. Vertical and horizontal bed tubes
12. Convective and radiative heat transfer
13. Formation of oxides of nitrogen
14. Sorption of sulfur by acceptor particles (e.g., limestone or dolamite)
15. Transient systems

One-dimensional FBC models with these aspects have developed significantly over the past decade to the point of application.

Premises and Components.   Development of one-dimensional FBC models has proceeded with certain premises and assumptions common to most efforts by several independent investigators. These common premises and assumptions are noted in Table 7.9. Most of the treatments have been for direct atmospheric combustion. However, potential for application to FB gasification exists.

Simplifying assumptions have been made to avoid treatment of complex fluid motion. The predictions rely on correlations of laboratory data for model elements as suggested by Table 7.9, together with mass and heat balances.

Two of the most completely developed and documented one-dimensional FBC models are those recently published by Wells and Krishnan (1980) and Lewis and Tung (1982). Both provide user's manuals. Figure 7.11 illustrates a general schematic of the Lewis–Tung code, while Figure 7.12 illustrates the fluidized-bed and associated code flow diagram for the Wells–Krishnan code. Required components for a comprehensive one-dimensional FBC code are:

1. Fluid dynamics
2. Bed and freeboard heat transfer
3. Gas and coal combustion
4. Particle dynamics
5. Sulfur retention
6. NO_x emissions

TABLE 7.9. Premises and Assumptions Common to Most One-dimensional Fluidized-Bed Combustor Models

	Premise or assumption	Comments or justification
1.	One dimensional (algebraic and first-order differential)	Properties (e.g., O_2 concentration) do not vary with radial position in the bed. At most, they vary with axial position or are uniform.
2.	Two distinct phases: bubbles and emulsion of air and coal	Based on theory of Davidson and Harrison (1963). Excess air above minimum for fluidization is in bubble phase.
3.	Bed particles phase, well-mixed	Small particles are in such vigorous motion that the temperature, particle size distribution, and composition of particles is uniform throughout the bed.
4.	Bubble phase in one-dimensional plug flow	Bubbles grow by coalescence and expansion and by interaction with limited heat-transfer surfaces. Bubble sizes from Cranfield and Gelhart (1974).
5.	Particles are reduced in size in the bed (attrition) and are lost to the freeboard region (elutriation)	Particle attrition and elutriation treated with Highly and Merrick (1974) correlation.
6.	Bed temperature taken to be uniform throughout	Vigorous particle motion.
7.	Coal reaction treated in two parts: devolatilization and heterogeneous oxidation	Devolatilization taken to be fast or at char rate in most treatments. Recent treatments consider both rates for selected coals (e.g., Wells and Krishnan, 1980).
8.	Char oxidation is by coupled diffusion and surface reaction	Both have been shown to be important (Sarofin and Beér, 1978).
9.	Gas reaction rates such as $CO + \frac{1}{2}O_2$ modeled to account for monoxide emissions	Early treatments neglected gaseous reaction. Subsequent treatments (Wells and Krishnan, 1980) included it.
10.	Based on particle number balance, heat balances, and species mass balances for the bed	From Kunni and Levenspiel (1968), who also provide minimum fluidization velocities.

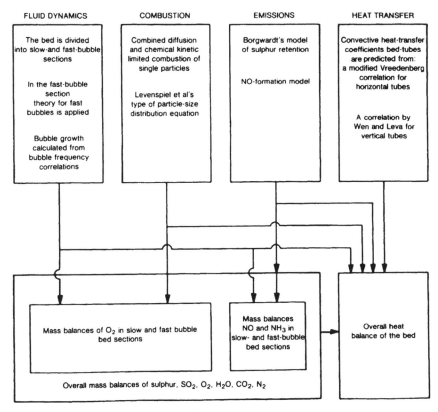

Figure 7.11. The principal parts of Lewis and Tung FBC code. (Figure used with permission from Olofsson, 1980.)

These models are largely algebraic, with ordinary differential equations to integrate along the length of the fluidized bed. Foundations for both models are very similar (see Table 7.9). Both codes can be applied to high pressure and to transient flow.

Evaluation and Application.   Comparisons of one-dimensional model predictions with laboratory and pilot scale FBC exit data are shown in Table 7.10 for carbon conversion efficiency for several fixed-bed combustors. However, given the state of development of the one-dimensional FBC codes, the extent of evaluation by direct comparison with data seems limited indeed. This is undoubtedly due, in large part, to a lack of adequate data for comparison with predictions. Submodels have often been evaluated quite extensively relative to evaluation of the comprehensive one-dimensional codes. Others have performed parametric

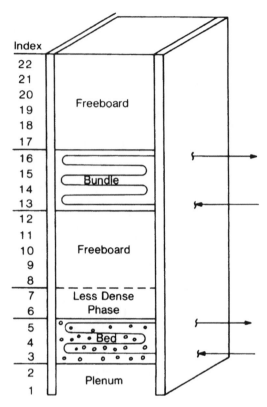

Figure 7.12. Schematic showing breakdown of AFBC (atmospheric fluidized bed combustion) cell into input indexes for ORNL (Oak Ridge National Laboratory) AFBC one-dimensional code. (Figure used with permission from Wells and Krishnan, 1980.)

analyses of these codes (e.g., Lewis and Tung, 1982), and models have also been applied to practical FBC systems (e.g., Meyer *et al.*, 1980; Wells and Krishnan, 1980).

These one-dimensional codes are somewhat restrictive for generalized application. Since model formulation relies significantly on empirical correlations, often for limited regions of operation, extrapolation of the code predictions to a variety of coals, pressures, scales, etc., may not give consistently reliable results.

7.7c. Two-Dimensional FBC Models

History. Development of two-dimensional FBC models was not initiated until the mid-1970s, with the principal focus on fluidized-bed gasification. Parallel developments with somewhat different focus have been conducted by at least

TABLE 7.10. Comparison of Experimental and Predicted Combustion Efficiencies for Atmospheric Pressure Fluidized-Bed Combustors[a]

Source of experimental data	d_p (mm)	Excess air (%)	u (m/s)	T (K)	$\eta(\%)$ (experimental)	$\eta(\%)$ (calculated)
Sheffield University 0.092 m² bed	0.6	10	0.9	993	87.5	85.4
	0.6	10	0.9	1073	91.2	93.0
National Coal Board—(NCB) 0.092 m² bed	0.22	2.5	0.9	1122	91.8	83.5
	0.15	4.2	0.9	1122	93.0	91.7
NCB–CRE 0.018 m² bed	0.502	12.1	0.9	1072	96.3	95.0
	0.502	9.8	0.9	1072	95.4	94.7
Pope, Evans, and Robbins 0.079 m² bed	0.65	7	3.21	1150	85.0	85.6
	1.65	30	2.75	1053	89.3	83.3

[a]Data from Sarofim and Beér (1978).

three groups of investigators (Schneyer *et al.*, 1981; Chan *et al.*, 1982; Gidaspow *et al.*, 1983a, b). Work has reached the stage where user's manuals are available and where independent groups are evaluating the behavior and characteristics of these codes. No reviews of available two-dimensional codes have been reported. The information that follows gives a brief summary and comparison of two of these two-dimensional codes and their relationship to one-dimensional FBC codes.

Premises, Formulation, and Components. Characteristics of these two codes are compared in Table 7.11. The codes are very similar in approach and structure, and the key components are also similar:

1. Gas fluid mechanics
2. Particle fluid mechanics
3. Gas reactions
4. Particle reactions
5. Heat transfer
6. Agglomeration

In the two-dimensional codes, the transient, partial differential equations of conservation (mass, energy, momentum) serve as the starting point. Gases are treated in the fixed, Eulerian coordinate system while the particles are tracked through the gaseous flow in the Lagrangian framework. The formulation and nature of this type of code is very different from the one-dimensional FBC codes reviewed above. The two-dimensional codes are more general, providing direct computation of bubble formation and growth, two-dimensional property profiles through the bed, effects of swirling flows, transient bed effects at various pressure, bed expansion, etc. The two-dimensional codes are more predictive and rely less on laboratory data correlations than the simpler one-dimensional codes. However, the experimental information required by the two-dimensional codes, such as solid-phase pressure, turbulent viscosity, particle collision and adherence, while more basic, is not well established. Thus, while the predictions are more fundamental, the expected accuracy is presently less certain.

The two codes reviewed in Table 7.11 (CHEMFLUB, FLAG) are similar. Both are based on the mixed Eulerian–Lagrangian framework, are transient, and are developed for FB gasification. Treatments of particle and gas reactions are similar. Both consider aspects of agglomeration, while neither has yet considered internal heat-transfer surfaces or pollutant formation. However, in CHEMFLUB, gas and particle temperatures are taken to be equal, and a small discrete bubble model is included, contrary to FLAG.

Gidaspow and associates (1983a, b) have also recently published work on modeling and fluidization measurements in two-dimensional beds. Modeling focus was on solids circulation near the inlet of a FB gasifier. Code predictions

TABLE 7.11. Comparison of Two-Dimensional FBC Codes

Item	CHEMFLUB	FLAG
Reference	Chen *et al.* (1981) (4 vols.)	Chan *et al.* (1982) (2 vols.)
Dimensions	Two dimensional (planar and axisymmetic), transient	Two dimensional (axisymmetric), transient
User's manual	Yes, limited (Vol. 3)	Yes, comprehensive (Vol. 2)
Numerical scheme	Implicit finite difference	Semi-implicit, finite difference
Systems	Combustion, gasification, high pressure	Combustion, gasification, high pressure
Particle fluid mechanics	Lagrangian; particle gas drag; particle–particle interactions	Lagrangian; effects of gas turbulence on particle drag; Monte Carlo particle collisions model
Selected assumptions	Equal gas and particle temperature; no turbulence-gas reaction interaction; no gas turbulence particle drag interaction; reaction rates not coal specific	No turbulence-gas reaction interaction
Comparison with data	Low-pressure bubble volume velocity; exit gas composition, jet penetration, and gas velocity	None noted
State of development	On-going work at METC	Formulated; illustrative solutions
Applications	Synthane, IGT UGAS, and Westinghouse gasifiers	None noted
Gas fluid mechanics	Eulerian conservation equations; turbulent viscosity	Eulerian conservation equations; turbulent viscosity and swirling flow
Heat transfer	Equal gas and particle temperature; convective and radiative; no in-bed cooling surfaces	Radiation by diffusion approximation; different particle and gas temperature
Pollutants	No considered	Not considered
Bubbles	McGill (modified) discrete bubble model in regions of small bubbles	No submodel consideration
Gas reactions	Water–gas shift (equilibrium)	Instantaneous (complete or equilibrium with O_2, CO, CO_2, H_2, H_2O, and N_2)
Coal reactions	Instantaneous devolatilization; based on char reaction with O_2, CO_2, H_2O, and H_2; shrinking core with internal and external diffusion effects and surface reaction	Instantaneous devolatilization; char reaction with O_2, CO_2, and H_2O; first-order rates, not coal specific; shrinking core with external diffusion and surface reaction
Particle dynamics	Fast sintering model based on Siegall's observations; no attrition	Agglomeration through particle collision (Monte Carlo collisions with binding forces); no attrition

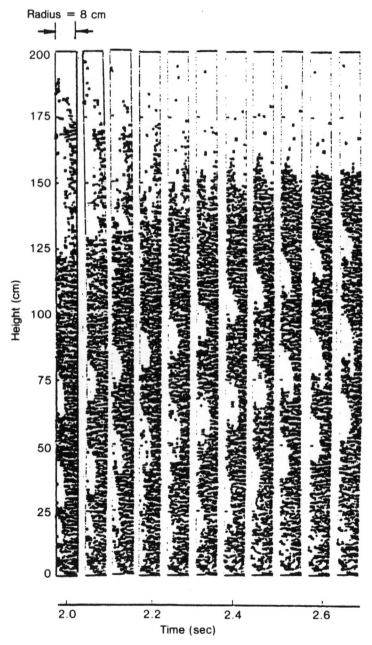

Figure 7.13. Time sequence of bubble evolution during steam–oxygen gasification. (Figure used with permission from Chen *et al.*, Vol. 1, 1981.)

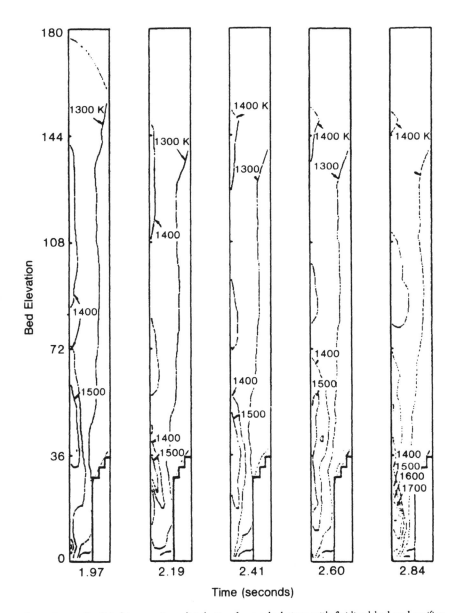

Figure 7.14. Predicted temperature distribution from calculations with fluidized-bed coal gasifier model in hot, high-pressure Westinghouse reactor from two-dimensional fixed-bed combustor code CHEMFLUB. (Figure used with permission from Chan *et al.*, Vol. 11, 1980.)

were compared with measured bed porosity data with some success. However, the code was not developed for reactive systems.

State of Development and Application. Based on the reports reviewed, CHEMFLUB is developed to the greater extent. The general nature of a two-dimensional FBC prediction (CHEMFLUB) is illustrated in Figure 7.13 for steam–oxygen gasification. The figure illustrates a time sequence of bubble evolution over a half-second period above a single perforation in the distributor plate. Transient effects of bubble behavior are illustrated as well as bubble growth, radial variations, and freeboard behavior.

Predictions of the CHEMFLUB code have been compared to FB measurements of low-pressure bubble volume and velocity, exit gas composition, gas velocity, and jet penetration depths (Chen, 1981). CHEMFLUB has also been applied to selected FB gasification systems, including Synthane, UGAS, and Westinghouse processes. Figure 7.14 illustrates computed temperature profiles for the high-pressure Westinghouse gasifier. Results show radial and axial variations in temperature that include changes of 300 K or more in lower entry regions.

7.8. SOME OBSERVATIONS ON MODELING

Mathematical modeling of complex conversion coal systems, including the direct combustion and/or gasification of coal and coal–water mixtures, requires inclusion of many physical and chemical processes which are not fully understood. It is principally because of this lack of information with respect to the fundamental processes that it is mandatory to link the mathematical modeling strategy with associated, experimental data.

Mathematical modeling at all levels can provide useful information. Table 7.12 shows a comparison of the levels of treatment of three different coal combustion or gasification models. Each of these models has different potential uses. The details of the turbulent mixing process are not needed for all applications. One argument for lower levels of sophistication is obvious by examining Table 7.12. Computational time increases exponentially with the increased level of sophistication. For many applications, the additional sophistication is not warranted.

For some applications, for example, burner design, local details are required. Multidimensional computer models can provide insight into the controlling processes. The use of these models requires not only sophisticated computer equipment, but also qualified, experienced users. Currently, these models cannot be used as a "black box" calculation of coal reaction chambers. Simplified input and color graphic output display improves the usability of these computer

TABLE 7.12. Levels of Treatment of Coal Combustion Models

Levels	Features	Uses	Typical computation time (Digital Corp. VAX Computer)
Zero dimensional Global No space variation Algebraic	Simple formulation Ready solutions Negligible computer costs No fluid mechanics	Process features Process variables	3 s
One dimensional Unidirectional Ordinary differential equations	Simple fluid mechanics Available solutions Low computer costs Residence time	Reaction length measurement interpretation	15 min
Two dimensional, three dimensional Partial differential equations Steady or transient Point variation	Local properties Complex fluid mechanics Turbulence Difficult numerical solutions High computer costs	Scaleup Detailed structure Design Optimization	5 hr

programs, but to interpret the output requires some knowledge of the assumptions made with regard to the physical and chemical processes, as well as an understanding of the numerical algorithms involved. In many applications for coal combustion and gasification, the transient startup of the reactor is also of some interest. However, transient computer programs require more computational time.

7.9. ILLUSTRATIVE PROBLEMS

1. Starting with the macroscopic equations of change for gas–particle mixtures reported by Smoot and Pratt (1979, p. 40ff) and given the volume element shown in Figure 7.2, derive the differential equations for the plug-flow model given by Eqns. 7.3 and 7.5.

2. For a bituminous coal dust/air flame (50 μm, 8% ash, 0% moisture) with the stoichiometric ratio of 1.2 (i.e., 20% excess air), estimate the maximum error that might be expected if the thermal conductivity (k_g) and diffusivity of oxygen (D_{om}) are based on air only, rather than the entire gaseous composition. (Consider the region of complete burnout where temperature is near maximum and the concentration of gases from the coal is a maximum.)

3. For the same coal dust/air flame in Problem 2, estimate the magnitude of the error that is introduced by neglecting gaseous radiation compared to particulate radiation. (Consider again complete combustion where temperature is near maximum and only ash particles remain.)

4. During the devolatilization process, the evolving gases from the coal particles act to reduce the mass-transfer rate of oxidizer from the bulk gas to the particle surface. For the coal dust/air mixture of Problem 2, estimate the magnitude of the reduction that takes place in the mass-transfer coefficient assuming that devolatilization takes place uniformly at a temperature of 1273 K (1000°C).

5. With the one-dimensional code (1-DICOG) and for the coal dust of Problem 2, compute the extent of burnout of the coal dust particles as a function of particle size:

    (a)   assuming particles are 50 μm
    (b)   assuming particles are 25 μm
    (c)   assuming particles are distributed as shown in Table 2.7 (utility boiler)(approximate with five discrete sizes with a mass mean of 50 μm).

Dimensions of the reactor and flow conditions are given in Figure 7.3 while coal composition (0% moisture) was given in Table 2.5. The reactants are initially well mixed and the wall temperature is uniform at 1773 K (1500°C). How important are particle size and size distribution?

6. For the result in Problem 5a, how long does it require (reactor length and ms) to complete devolatilization? How much of the coal (wt %) is devolatilized? How long does it take to consume the char? Is the char reaction controlled by oxidization diffusion or surface oxidation?

7. From Problem 5, what level of oxides of nitrogen (NO_x) is predicted at the combustor exit? Does it vary with particle size? Compare the result with the measured value of about 1100 ppm for a premixed coal dust/air flame. What can you conclude from this analysis about NO_x formation processes?

8. For a lignite coal of Table 2.5, what is the impact of firing the raw coal with 25% moisture as compared to that with 0% moisture (predried)? Assume the wall temperature is held constant at 1473 K (1200°C) in both cases. Reactor dimensions and primary inlet conditions are those of Problem 5, while the secondary velocity is adjusted to account for changing air quantities. Use 20% excess air and a single particle size of 50 μm. How does the extent of coal burnout compare with that for the bituminous coal (Problem 5)?

9. For an entrained gasifier with bituminous coal of Problem 5a, using 1-DICOG, resolve the following questions:

(a)   What stoichiometric mixture of coal/steam and oxygen is recommended for the gasifier at atmospheric pressure?
(b)   What is the cold gas heating value if fired with oxygen/steam as compared with steam and air?
(c)   How long does the reactor need to be at atmospheric pressure to consume the coal?
(d)   How would reactor size compare if the gasifier were operated at 10 atms? The steam is input in the secondary velocity at 600 K, while the oxygen at 300 K carries in the coal dust.

10.   A bituminous coal dust with 20% excess air is burned in a laboratory combustor. The coal contains 8% ash. Dimensions of the reactor and composition of the coal are those of Problem 5. The wall temperature is uniform and maintained at 1200 K. The coal particle size distribution is as follows:

Diameter (μm)	Mass fraction(%)
20	35.0
48	33.9
80	31.1

Other input conditions are as follows:

	Primary	Secondary
Total mass flow (g/s)	5.6	36.1
Pressure (atm)	0.925	0.925
Temperature (K)	356	591
Coal flow (g/s)	3.682	0

Compute the effect of moisture level on the combustion characteristics of the coal. In the first example, consider 0% moisture, while in the second example, consider 30% of moisture, based on the coal.

EVALUATION OF
COMPREHENSIVE MODELS

8.1. BACKGROUND AND SCOPE

In order to be used with confidence for quantitative purposes, such as combustor design, comprehensive models must be rigorously evaluated. It is generally the case that model subcomponents have been evaluated through comparison with measurements. This often takes the form of curve fitting of experimental data to provide model coefficients. Figure 3.8 and 3.10 illustrated comparisons of correlative methods and data for devolatilization. Figures 4.5 and 4.6 illustrated similar comparisons for char reaction rates. While such comparisons are vital in establishing the reliability of model components, they are insufficient in assessing the accuracy of the comprehensive codes that combine several model elements.

In this chapter methods for evaluation of comprehensive models are identified. Then a set of useful data are selected and referenced for specific evaluation of pulverized-coal models. Comparisons of predictions for a pulverized-coal model with several of these data sets are shown in Chapters 10–15.

8.2. EVALUATION METHODS

Various methods can be used to provide increased confidence in the value of solutions obtained from comprehensive codes. Four methods are discussed briefly here. Rigorous evaluation of these codes should employ all of these methods. The cost of development is sufficiently high, the reliability sufficiently uncertain, and the potential usefulness sufficiently great, that rigorous evaluation warrants major emphasis. The four methods treated are (1) numerical evaluation,

211

(2) sensitivity analysis, (3) trend analysis, and (4) comparison with comprehensive data.

8.2a. Numerical Evaluation

The uniqueness and accuracy of the numerical solutions, which often use finite-difference methods, cannot be generally evaluated analytically. However, it is useful to compare numerical solutions of limiting cases (i.e., cold flow with idealized geometry) where analytical solutions exist. Chapyak *et al.* (1982). illustrated this approach in evaluation of the SIMMER-2 code for entrained-flow gasification and combustion. Computations were performed and compared with an analytical solution of a free turbulent jet. This model, which is based on a simple algebraic representation of the turbulent viscosity, yields an analytic solution for the self-similar, far-field jet regime. These calculations furnish a direct numerical and analytic result, providing an unambiguous test of some of the numerical methods employed in the SIMMER code.

Figure 8.1 shows such a comparison for the centerline downstream velocity of a circular jet as a function of downstream distance. The analytic solution, which is valid only in the far field where the influence of the finite jet opening is negligible, is in good agreement with the calculation. Figure 8.2 shows the radial downstream velocity profile at a downstream distance of 50 jet radii. Again, excellent agreement between the analytic result and the calculation is evident. A finely zoned mesh was employed in these comparisons (i.e., 30 axial zones of variable size), with the smallest width at one-third jet radii.

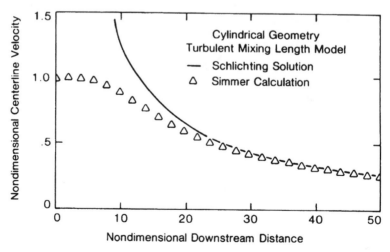

Figure 8.1. Centerline velocity *vs.* distance for a cylindrical jet. (Figure used with permission from Chapyak *et al.*, 1982.)

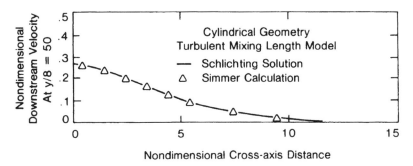

Figure 8.2. Downstream velocity profile for a cylindrical jet (in nozzle radii). (Figure used with permission from Chapyak *et al.*, 1982.)

Further numerical confidence is established by repeated computations with continual decreases in sizes of the numerical grid. The grid size must be sufficiently small such that numerical results are independent of this size. Smoot *et al.* (1984) show computations for several grid sizes using the PCGC-2 code for pulverized-coal systems. These computations were performed for a nonreactive gaseous flow and are shown in Figure 8.3. The figure shows centerline mixture

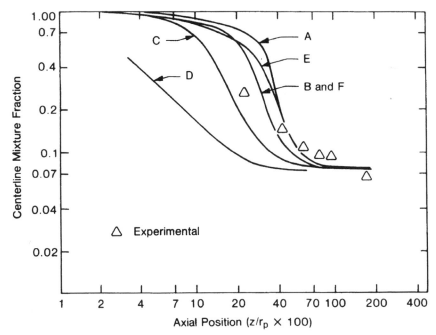

Figure 8.3. Effect of numerical grid size on predicted centerline mixture fraction decay. $A = 31 \times 31$, $B = 25 \times 25$, $C = 20 \times 20$, $D = 15 \times 15$, $E = 20 \times 31$, $F = 20 \times 25$.

fraction *vs.* axial position and experimental data are also shown. Even the finest grid of 31 axial divisions and 31 radial divisions may not be quite adequate in these computations. Although computer time increases greatly with the increased grid-size resolution, the finer mesh cells must be used to alleviate numerical uncertainties.

The lower-order differencing schemes used in finite-difference codes cause numerical error which often is of the same order of magnitude as the diffusion terms or greater. The importance of this numerical diffusion process has been

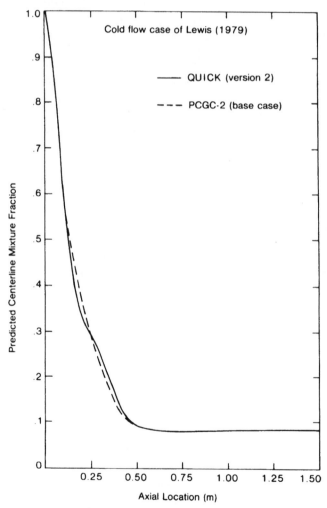

Figure 8.4. Comparison of Leonard (1979) differencing scheme with lower-order scheme.

studied by looking at higher-order differencing techniques. For example, Fletcher (1983) studied several higher-order differencing techniques and implemented one suggested by Leonard (1979) (i.e., QUICK) into PCGC-2. PCGC-2 was recoded to incorporate this higher-order differencing technique (Smoot *et al.*, 1983). Several cases were studied and comparisons were made between predictions from the higher-order and the lower-order techniques. One such comparison is shown in Figure 8.4. Although Fletcher (1983) concluded that the higher-order technique had improved accuracy, problems with numerical stability were encountered. Comparisons with the lower-order differencing scheme showed that numerical diffusion was insignificant in reactors of this geometry as long as the grid size was resolved to a fine enough accuracy to eliminate grid-size errors. Although the higher-order differencing scheme could be used with coarser cell spacing than the lower-order scheme, the increased problems with numerical stability voided any advantage.

8.2b. Sensitivity Analysis

Evaluation of the sensitivity of code predictions to various model parameters is also an important requirement in model analysis. Physical input parameters required for the particular PCGC-2 pulverized-coal model (Smith *et al.*, 1980) are summarized in Table 8.1. The list does not include fundamental physical properties such as c_p, μ, k, D_{im}, or reference enthalpy and entropy, which are used in the model. Also, required empirical mass-transfer and heat-transfer coefficients are not shown. Input parameters (i.e., turbulence intensity) add to this consideration. The parameters of Table 8.1 are the coefficients and constants for the heuristic description of various physical and chemical processes. They are most often obtained by comparison of submodel predictions with laboratory measurements. Few are established with certainty. It is very important to determine which of the parameters have significant impact on predicted results, over the range of parameter uncertainty.

Hill (1983) performed a series of computations for formation of nitrogen oxide in pulverized-coal/air systems using PCGC-2. He considered the reaction rate parameters in three chemical reactions influencing nitrogen oxide concentration. The first was the homogeneous gas-phase reaction of HCN (formed from fuel nitrogen) and oxygen. The second was the homogeneous gas-phase reaction of HCN and NO to destroy NO and form N_2. And the third was the heterogeneous reaction of NO and char to consume NO. Table 8.2 summarizes the results of decreasing each of the rate constant pre-exponential factors and each activation energy in turn by $\frac{1}{2}$.

Results show that the decrease in the activation energy of reaction 3 has the most dramatic impact on predicted nitrogen oxide level for these particular practical test conditions. This observation suggests that the homogeneous

TABLE 8.1. Summary of Key Physical Parameters Required in a Typical Multidimensional Pulverized-Coal Model[a]

Parameter[b]	Typical value	Description	Discussion (source, reference)
Turbulence model			
C_μ	0.09	Eddy viscosity coefficient	Obtained from consideration of limiting cases and by code comparison with velocity and turbulence measurements in nonreacting gaseous systems. S_t^t is the turbulent Schmidt–Prandtl number, k is the turbulent kinetic energy, and ϵ is the rate of dissipation of k.
C_1	1.44	Constant in k and ϵ transport equations	
C_2	1.92	Constant in k and ϵ transport equations	
σ_k	0.9	S_t^t for k	
σ_ϵ	1.22	S_t^t for ϵ	
Combustion Model			
C_{g1}	2.8	Constants in transport	Values generally obtained by analogy to turbulence parameters above and by code comparison with measurements in gaseous diffusion flames. σ_h and E arise from boundary conditions; f is the mixture fraction, and g is the variance in f.
C_{g2}	2.0	Eqn. for g	
σ_f	0.9	$S_t^t - f$	
σ_g	0.9	S_t^t – variance in f	
σ_h	0.9	S_t^t – enthalpy	
E	8.8	Law of wall constant	
Particle dispersion			
σ_p	0.7	Particle turbulent Schmidt number	Estimated from limited particle dispersion data; highly uncertain and may vary with particle size.

Coal model			
α_1	0.3	Stoichiometric coefficient	Subscripts 1 and 2 refer to parallel devolatilization reactions. α, E, B parameters are obtained by model comparison with coal-weight-loss data during devolatilization, and may vary with coal type.
α_2	1.0	Stoichiometric coefficient	
E_1	25	Activation energy (kcal/mole)	
E_2	40	Activation energy (kcal/mole)	
B_1	$2(10^5)$	Arrhenius coefficient (s^{-1})	
B_2	$1.3(10^7)$	Arrhenius coefficient (s^{-1})	
A	20.4	Arrhenius coefficient (g/cm^2 s atm)	A, E parameters are for surface reaction of char with oxidant. They will vary strongly with coal type and history. Parameters shown are for anthracite-oxygen from Smith (1980). Coal surface properties obtained from particle data (Smoot and Pratt, 1979).
E	19	Activation energy (kcal/mole)	
n	1	Reaction order	
γ	0.1	Coal swelling coefficient	
ξ_p	>1.0	Particle surface area per sphere area	
Φ_1	2	Moles C consumed per mole oxidizer	
Radiation			
Q_{ap}	0.85	Particle absorption efficiency	Radiative properties of gas, particles, and wall are summarized in Smoot and Pratt (1979). n varies with particle type (ash, soot, char). p is the pressure, and l is the optical length.
Q_{sr}	1.4	Particle scattering efficiency	
ϵ_w	1	Wall emissivity	
n	$1.93-1.021i$	Refractive index of particles	
ϵ_x	$f(CO_2, H_2O, p, l)$	Gas emissivity	

[a] Table from Smoot (1980).
[b] Parameters are dimensionless unless otherwise indicated.

TABLE 8.2. Effect of Chemical Reaction Rate Parameter Values on NO Formation

Variable	Coefficient value (A or E)	Peak NO value (ppm)	Integrated outlet NO value (ppm)	Change in NO values (%)
Reaction 1: Pre-exponential term A_1	1.2×10^{11}	601	226	0–10
	0.8×10^{11}	553	214	
Reaction 2: Pre-exponential term A_2	3.6×10^{11}	538	210	5–15
	2.4×10^{11}	639	236	
Reaction 3: Pre-exponential term A_3	4.9×10^{-4}	581	209	0–10
	3.3×10^{-4}	583	221	
Reaction 1: Activation energy E_1	3.4×10^8	132	81	20–40
	2.2×10^8	742	321	
Reaction 2: Activation energy E_2	3.0×10^8	1522	578	20–150
	2.0×10^8	136	78	
Reaction 3: Activation energy E_3	1.8×10^8	584	286	0–30
	1.2×10^8	564	78	

Reaction 1: $HCN + O_2 \rightarrow NO + \cdots$; $\quad w_1 = \rho A_1 X_{HCN} X_{O_2}^b \exp(-E_1/RT)$

Reaction 2: $HCN + NO \rightarrow N_2 + \cdots$; $\quad w_2 = \rho A_2 X_{HCN} X_{NO}^b \exp(-E_2/RT)$

Reaction 3: $NO + Char \rightarrow CO + \cdots$; $\quad w_3 = \alpha_p n_p A_3 \exp(-E_3/RT) A_E p_{NO} \gamma_2$

reaction of HCN and NO is particularly important in controlling nitrogen oxide reduction and that parameters for this reaction should be known with particular accuracy.

Statistical methods (Boni and Penner, 1977; Cukier *et al.*, 1978) have been identified for efficiently conducting sensitivity analysis of model parameters. Use of such methods is illustrated by the work of O'Brien and Pierce (1981), who used a nonlinear sensitivity analysis of multiparameter model systems to analyze the impact of the six parameters in the two-step devolatilization model on output from the one-dimensional code (1-DICOG).

From this brief summary, the nature of these parameters will not be entirely clear until the reader has completed subsequent chapters. However, the extensive reliance of these codes on correlation of basic laboratory measurements is dramatically illustrated. A review of reported computations for multi-dimensional codes summarized in Table 7.3 provided very little information on the sensitivity of computed results to variation in parameter values. Future work must include computations of parameter sensitivity if the models are to have any general acceptibility.

8.2c. Trend Analysis

A third useful method for evaluating comprehensive models is through comparison of model predictions with data where a test parameter has been systematically varied. For example, the effects of stoichiometric ratio variation on percentage of coal consumed or the effects of swirl number on gas-phase mixture fraction might be examined. The capabilities of the predictive method to correctly predict the observed trends provides a useful evaluation of the method.

Hill (1983) has examined the effects of stoichiometric ratio and swirl number of the secondary air flow (in a pulverized-coal/air flame) on nitrogen oxide concentration. Comparisons were made for two coals and results are illustrated in Figures 15.5–15.7 of Chapter 15. In particular, comparisons are shown for bituminous and subbituminous coals and both the magnitude of nitrogen oxide concentration and its variation with swirl number are quite well predicted. Such trend comparisons add confidence in use of a model.

8.2d. Comparison with Comprehensive Data

Coal models must be rigorously evaluated by comparison with detailed test data. While "stack" data (e.g., carbon conversion at the combustor exist) provide some idea of model validity, comparisons with detailed profile data from inside a combustor provide a more definitive basis. These comparisons of detailed local data provide information on the local structure of the flame. Mixture fractions, gas composition, temperature, particle composition, and velocity are among the properties commonly compared. This type of code evaluation is the most exacting since comparisons of several properties are commonly made throughout the flame. Illustrations of such comparisons are shown in Figures 13.8–13.10 of Chapter 13. In these figures, CO_2, H_2O, and O_2 concentration values are compared for a subbituminous coal dust/air flame.

Table 7.2 summarized the limited comparisons of theory and experiment for pulverized-coal codes reported by eight investigators. The data comparison with one-dimensional code computations generally showed reasonable agreement. However, these limited comparison were largely for combustors with a significant one-dimensional nature.

Two of the earlier multidimensional codes (Gibson and Morgan, 1970; Richter and Quack, 1974) provided comparisons with radial and axial measurements from the International Flame Research Foundation (IFRF) furnace for cement kiln and boiler-type operation. Given the limitations in the physical submodels and the effects of buoyancy on measurements (three-dimensional effects), comparisons were promising. Most of the more recent multidimensional codes (Table 7.3) have not been developed to the point where detailed comparisons with measurement have been made. The information that follows in this chapter summarizes some of the detailed test data available for comparison with predictive codes for pulverized-coal systems. Comparisons of code predictions for PCGC-2 with several sets of these data follow in subsequent chapters.

8.3. DETAILED LOCAL DATA FOR MODEL EVALUATION

8.3a. Data Requirements and Categories

A collection of detailed data are available from the published literature for use in code evaluation. Of particular importance in data selection are the following:

1. Detailed profiles of several properties, in tabular form.
2. Well-characterized initial boundary conditions.
3. Geometries corresponding to model formulation.
4. Well-specified test conditions and dimensions, with values in the range of interest.
5. Evidence of data accuracy and reliability such as reproduced tests, material balances, and error analysis.
6. Measured fluctuating turbulence properties in addition to time-mean properties.

Based on these requirements, Smoot *et al.* (1983) identified several useful data cases for pulverized-coal combustion evaluation and documented results in a comprehensive report. Data were selected in each of five categories, which varied from more simple flows to more complex flames:

1. Nonreacting, gaseous flow
2. Nonreacting, particle-laden flow
3. Gaseous flames
4. Pulverized-coal combustion
5. Entrained coal gasification.

All of the cases identified were for turbulent flows and all of the flames were diffusion-type flames. Data in various categories provide for a hierarchy of comparisons. Use of gaseous, nonreactive flow provides for evaluation of the turbulent mechanics model, in the absence of particulates or chemical reactions. Subsequent comparisons with particle-laden flows, reacting flows, and a combination of the two add various complexities in sequence. For simpler flows, data on fluctuating turbulence properties are available, while for particle-laden, reactive flows, the data are generally limited to mean properties.

8.3b. Data Cases Considered and Selected

Based on the above criteria, a total of 196 different sets of data were identified and evaluated from 41 separate investigations at 19 laboratories (Smoot *et al.*, 1983). From this data base, 34 data sets were selected for inclusion in the data book. At least four or five data sets were sought in each category where a sufficient data base was available.

Fourteen of the cases were from the BYU (Brigham Young University) Combustion Laboratory, where much of the applicable particle-laden, nonreactive profile data and all of the entrained coal gasification profile data have been obtained. An additional seven cases were selected from the results of the International Flame Research Foundation, where furnace work on gaseous and pulverized-coal combustion has been conducted for many years. In the remaining 13 data sets, data from six different laboratories were included.

In combustion systems, use of swirling flow is common for achieving stable combustion. Of the 34 data sets documented, 18 are in the absence of swirling flow, while the balance are with swirling flow. Both types of flow are represented in all of the categories except gasification, where no detailed profile data have been reported in the presence of swirling flow.

Table 8.3 provides a summary of the 33 data sets that have been documented. The table shows the chemical system, key test facility dimensions, mass flow rates, and swirl number. Measured properties and key variables are also summarized. All of these data are shown in detail (Smoot et al., 1983) in the referenced data book, including input quantities and measured properties. Several of the cases are compared with predictions for the pulverized-coal model in Chapters 10–15.

8.4. DATA NEEDS

While the detailed profile data reviewed above represent a useful source of information for code evaluation, several additional needs are apparent. Little data for fluctuating properties are available for particle-laden flows and flames. Key initial properties, such as turbulence intensity, have most often not been provided. Much of the data is obtained from intrusive probes that may interfere with the flow and combustion processes. More reliable data for more complex, highly loaded, particle-laden flows and flames with optical methods are required.

8.5. ILLUSTRATIVE PROBLEMS

1.   Based on statistical methods, such as those by Boni and Penner, (1977), Cukier et al. (1978), O'Brien and Pierce (1981), and the PCGC-2 model parameters of Table 8.1, design a computation program for performing a sensitivity analysis of various model parameters. Some judgment could be exercised regarding the most important parameters in order to reduce the number of computations required.

2.   In Chapters 10–15, several comparisons of predictions from PCGC-2 (i.e., generalized model for pulverized-coal combustion) with measurements are

TABLE 8.3. Data Sets

Investigator/ Reference	Chemical system	Diameters (m)		Chamber	Chamber length (m)
		Primary	Secondary	Chamber	
A.1 *Nonreacting Gaseous Flow without Swirl*					
Owen (1975) (ARO, Inc.)	Air/air	0.0635	0.0889	0.127	1.219
Takagi *et al.* (1981) (Osaka Univ.)	H_2/air	0.0049	1.04	1.04	2.5
Webb (1982) (BYU)	Air/air	0.025	0.130	0.206	0.914
Jones (1983) (BYU)	Air/air	0.025	0.0762	0.206	0.914
A.2 *Nonreacting Gaseous Flow with Swirl*					
Samuelsen *et al* (1982) (UCI)	CO_2/air	0.001	0.057	0.08	0.05
Webb (1982) (BYU)	Air/air	0.025	0.130	0.206	0.914
Gouldin *et al.* (1983) (Cornell Univ.)	Air/air	0.0495	0.102	0.105	0.610
Gouldin *et al.* (1983) (Cornell Univ.)	Air/air	0.0495	0.102	0.106	0.610
Jones (1983) (BYU)	Air/air	0.025	0.0762	0.206	0.914
B.1 *Nonreacting Particle-Laden Flow without Swirl*					
Leavitt (1980) (BYU)	Coal/air (43 μm)	0.0255	0.127	0.206	0.926
Sharp (1981) (BYU)	Si beads/air (46 μm)	0.0255	0.127	0.343	0.926
Modarress *et al* (1982) (Spectron Development Lab)	Si beads/air (200 μm)	0.02	0.6	0.6	2.0
Modarress *et al.* (1982)	Si beads/air	0.02	0.6	0.6	2.0
B.2 *Nonreacting Particle-Laden Flow with Swirl*					
Leavitt (1980) (BYU)	Coal/air	0.0254	0.127	0.206	0.91
Leavitt (1980) (BYU)	Coal/air	0.0254	0.127	0.206	0.91
Leavitt (1980) (BYU)	Coal/air	0.0254	0.127	0.206	0.91

Selected for Code Evaluation

Mass flow rates (kg/s)			Stoic ratio	Swirl number	Test variables	Measured properties
Primary	Secondary	Solid				
0.0077	0.0891	—	—	—	—	u, u', v, v'
0.000274	0.0508	—	—	—	Primary jet velocity	v, u', v' (gas)
0.016	0.51	—	—	—	—	u, u'
0.0067	0.0042	—	—	—	—	u, u'
0.00121	0.0936	—	—	0.3 (0.8)	—	u, u', w, w'
0.016	0.51	—	—	0.49	—	u, u'
0.0718	0.164	—	—	$Si = 0.50$ $So = 0.56$	co-swirl/counterswirl	u, w, u', w'
0.0718	0.164	—	—	$Si = 0.50$ $So = -0.56$	co-swirl/counterswirl	u, w, u', w'
0.0067	0.0042	—	—	(3.0)	—	u, u'
0.0218	0.534	0.0357	—	—	Chamber geometry	$(Ar), f, u$
0.0223	0.520	0.0333	—	—	Particle size Chamber diameter Secondary injection angle	$(Ar), f, u$
0.005	0.034	0.0032	—	—	Particle size Solid/gas ratio	$u(p), u, u', v'$ u', v'
0.005	0.034	0.0032	—	—	Solid/gas ratio	$u(p), u, u', v'$ u', v'
0.0182	0.520	0.0014	—	(0.4)	Swirl number Chamber geometry	$(Ar), f, u$
0.0182	0.520	0.0014	—	(0.4)	Swirl number Chamber geometry	$(Ar), f, u$
0.0182	0.520	0.0014	—	(0.9)	Swirl number Chamber geometry	$(Ar), f, u$

TABLE 8.3 (cont.)

Investigator/ Reference	Chemical system	Diameters (m)			Chamber length (m)
		Primary	Secondary	Chamber	
C.1 Gaseous Combustion without Swirl					
Michelfelder *et al.* (1974) (IFRF)	C_3H_8/air	0.0189	0.176	1.00	6.25
Michelfelder *et al.* (1974) (IFRF)	CH_4/air	0.0326	0.176	1.00	6.25
Takagi *et al.* (1981) (Osaka Univ.)	H_2/air	0.0049	1.04	1.04	2.5
Hassan *et al.* (1983) (Imperial College)	CH_4/air	0.056	104.0	0.600	3.0
C.2 Gaseous Combustion with Swirl					
Michelfelder *et al.* (1974) (IFRF)	CH_3/air	0.0326	0.176	1.00	6.25
Samuelson *et al.* (1982) (UCI)	Propane/ air	0.001	0.057	0.08	0.50
Gouldin *et al.* (1983) (Cornell Univ.)	CH_4/air	0.0495	0.102	0.105	0.610
D.1 Coal Combustion without Swirl					
Beer (1964) (IFRF)	Anthracite/ air	0.054	0.27	1.5	9.1
Hein (1970) (IFRF)	Anthracite/air	0.0577	0.0889	1.0	6.25
Thurgood (1979) (BYU)	Bituminous/ air	0.016	0.054	0.203	1.52
Michel *et al.* (1980) (IFRF)	Bituminous/ air	0.0703	0.130	0.950	6.25
Rees (1980) (BYU)	Bituminous/ air	0.016	0.054	0.203	1.52
D.2 Coal Combustion with Swirl					
Hein *et al.* (1970) (IFRF)	Anthracite/ air	0.0577	0.0899	1.0	6.25

Mass flow rates (kg/s)			Stoic ratio	Swirl number	Test variables	Measured properties
Primary	Secondary	Solid				
0.0636	0.832	—	1.15	—	—	(gas), T u, v, w
0.0778	0.868	—	1.15	—	Fuel/oxide ratio, Chamber geometry	(gas), T v, u, w
0.000274	0.0580	—	0.96	—	Secondary jet velocity	u, u', v'
0.00256	0.051	—	0.77	—	Fuel/oxide ratio	T, (gas)
0.0783	0.870	—	1.15	0.5	Chamber geometry Fuel/oxide ratio Swirl number	(gas), T u, v, w
0.00122	0.0936	—	4.93	0.3 (0.8)	Fuel/oxide ratio	u, u' w, w'
0.0696	0.164	—	1.19	$Si = 0.50$ $So = -0.56$	co-swirl/ counterswirl	u, w, u', w'
0.040	0.34	0.029	1.15	—	Particle size Chamber geometry	(gas), u, Char analysis
0.0542	0.465	0.0414	1.15	—	Chamber geometry Stoichiometric ratio	(gas), u, v, w (solid), T Char analysis
0.0056	0.036	0.0038	1.15	—	Particle size Secondary jet velocity and injection angle	(gas) Char analysis
0.0708	0.573	0.0589	1.12	—	Chamber geometry Stoichiometric ratio	(gas), u, v, w (solid), T, NO_x Char analysis
0.0056	0.020	0.0038	0.70	—	Stoichiometric ratio	(gas), NO_x Char analysis
0.0542	0.465	0.0414	1.15	1.0	Chamber geometry Stoichiometric ratio, swirl number	(gas), u, v, w (solid), T Char analysis

TABLE 8.3 *(cont.)*

| Investigator/ Reference | Chemical system | Diameters (m) | | | Chamber length (m) |
		Primary	Secondary	Chamber	
Hein *et al.* (1970) (IFRF)	Anthracite/ air	0.0577	0.0899	1.0	6.25
Harding (1980) (BYU)	Bituminous/ air	0.016	0.0842	0.203	1.52
Asay (1982) (BYU)	Subbituminous/ air	0.022	0.09842	0.203	1.52
E. Coal Gasification Soelberg (1983) (BYU)	Bituminous/ air	0.0131	0.0287	0.200	1.19

shown. Comparisons are shown for most of the detailed data categories from gaseous, cold flow to pulverized-coal combustion. Analyze these comparisons and make some quantitative assessment of the reliability of this particular code in making predictions for these various flows and flames.

3. Through a review of system properties and test conditions in Table 8.3, identify areas where data are limited or inadequate, as to measured properties, range of test variables, etc.

Mass flow rates (kg/s)			Stoic ratio	Swirl number	Test variables	Measured properties
Primary	Secondary	Solid				
0.0542	0.465	0.0414	1.15	1.4	Chamber geometry Stoichiometric ratio, swirl number	(gas), u, v, w (solid), T Char analysis
0.0061	0.03	0.0038	1.00	1.4	Swirl number	(gas), NO, Char analysis
0.0061	0.019	0.0038	1.06	3.0	—	(gas), NO, Char analysis
0.0073	0.0018	0.0066	0.39	0.0	—	(gas), NO, SO, Char analysis

... *to the acknowledgement of the mystery of God,*
... *In whom are hid all the treasures of wisdom and*
knowledge.

Colossians 2:2–3

BASIC MODELING EQUATIONS

9.1. INTRODUCTION

Equations describing the conservation of mass, momentum, and energy of a flowing gas–particle suspension or gas–droplet suspension can be found in various books, papers, and reports (Bird *et al.*, 1960; Trusdell and Toupin, 1960; Soo, 1967; Drew, 1971; Smoot and Pratt, 1979). The presentation of these equations in the various sources are often very different among authors. Some equations have included too many assumptions to be of practical use for predicting combustion or gasification of coal and coal slurries. Others are in a form which are too complex to be of utility for practical systems. Although the equations are presented in many sources, very few sources present analytic or numerical solutions for the two-phase flow equations. The fundamental equations are very complex and formulating a solution requires many numerical approximations, large-scale computers, and a significant dedication of time on the part of the user. This book introduces calculational procedures for combustion and gasification of coal and coal slurries which include sufficient detail to describe the important features of the reacting flow field, but which have incorporated adequate engineering approximations to render the solution algorithm tractable. In this chapter we have focused upon the basic conservation equations for the multiphase field.

We first discuss the assumptions inherent in writing the conservation equations for each of the different phases. We will then proceed with a very brief description of a typical approach for deriving any of the conservation equations. These derivations are usually based on the Reynolds transport theorem, which will be discussed briefly. We will not derive the equations themselves but will present the equations with some discussion of the approach used to derive each of them. We will focus on the importance of the particulate and droplet phases in the conservation equations. Different techniques and methods for incorporating the coupling among phases will be discussed. This chapter is not meant to be a definitive derivation of the conservation equations but only a brief discussion of these equations with references to sources where they are derived. These basic equations will be the basis for much of the discussion which follows in subsequent chapters.

Throughout this and subsequent chapters we will refer to systems that are composed of particles or droplets entrained in a gas phase. The treatment is intended for focusing on particles and droplets that are small (i.e., less than 300 μm) and are typically lightly loaded (i.e., void fraction >0.99). In some locations and some reactions the systems will be more densely loaded. For example, near the nozzle in coal–water mixtures, the flow field will contain high densities of coal–water slurry and very low void fractions. In some high-pressure, entrained-flow gasifiers, particle loading can be increased significantly above typical combustion applications. These special cases will be treated in the text with separate discussions.

9.1a. The Continuum Hypothesis

In gas-phase fluid mechanics, without the presence of the particle or droplet phases, the formulation of the governing equations of motion are nearly always derived by assuming a continuum field. In treating a fluid as a continuum, we postulate that variables such as velocity, pressure, density, etc., are continuous point functions. In fact, this is not true, since all fluids are composed of individual molecules. The molecules of a gas are separated by vast regions with linear dimensions much larger than those of the molecules themselves. The mass of the material is concentrated in the nuclei of atoms composing a molecule and is very far from being smeared uniformly over the volume occupied by the gas. Other properties of the fluid, such as the composition or the velocity, are likewise a highly nonuniform distribution when the fluid is viewed on such a small scale as to reveal the individual molecules. In most applications of gas-phase fluid mechanics, we are concerned not with the molecular level but with the more macroscopic scale, which is large compared with the distance between the molecules. Thus, it is not often that the molecular structure of a fluid need be taken into account explicitly. The basis for most conservation equations, including those discussed in this chapter, is the assumption that the macroscopic behavior of the fluid is the same as if it were perfectly continuous in structure. Physical properties such as the mass and momentum associated with the fluid and contained within a given small volume are regarded as being spread uniformly over that volume instead of being concentrated in small fractions of it.

In the continuum hypothesis, as far as the equations of change are concerned, a point may be defined as being large when compared with the microscopic molecular scale, but small when compared with the macroscopic spatial distributions of the various fluid properties. This hypothesis is most easily observed by considering Figure 9.1. When very small volume elements are considered, the density of matter enclosed within the control volume will change dramatically depending on the number of molecules enclosed within the volume. As this control volume is expanded in size until the number of molecules enclosed within the volume may be considered essentially constant, a "point" density is observed. As the volume is continually expanded, macroscopic variations in the fluid density change the observed global density within the control volume.

The continuum hypothesis for the gas phase generally holds true except for rarified flows, such as the motion of gases in the upper atmosphere or the flow of gases through small pores such as those encountered in small coal particles. All of the gas-phase equations discussed in this book are based upon a continuum.

The most significant feature of the particulate or droplet phases with respect to their transport equations is that they do not constitute a continuum. Thus, strictly speaking, one cannot define particle or droplet density at a point in the flow. This is particularly true for dispersed flow systems of interest to this book. Consider recirculation in a particle-laden reactor as shown in Figure 9.2. Select an arbitrary control volume, small compared to the macroscopic scale and large

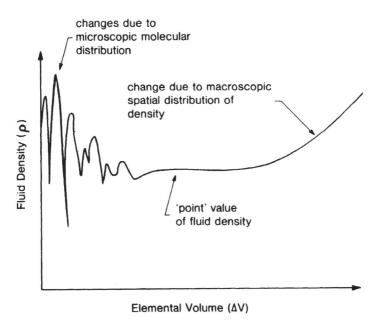

Figure 9.1.   Effect of size of volume element on fluid density.

compared to the microscopic molecular level. The gas molecules within this control volume undergo many collisions due to their random kinetic motion. When the control volume is appropriately chosen for the continuum hypothesis to be valid, we may say that the gases all have one "point" density, velocity, and temperature.

The same is not true for the particle phase. There are relatively few particles within the system when compared to gas molecules. The particles do not interact with each other in the same way as the gas molecules do. Particle–particle interactions are relatively few. Each particle has its own unique history. Many particles within the control volume can have radically different histories and thus radically different properties such as mass, composition, velocity, and

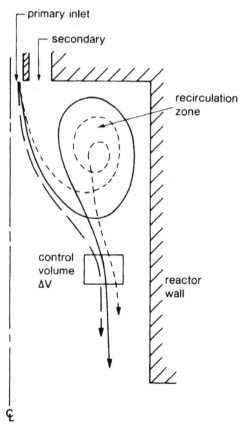

Figure 9.2. History effect of particles or droplets in dispersed flow. Particles within the control volume ΔV do not have the same properties.

temperature. In general, the particles do not communicate with each other by direct collisions as much as they do by interactions with their surrounding gas field, which in turn interacts with adjacent particles. Thus, the only meaningful definition of particle density is bulk density or mass of particles per unit volume of mixture. The control volume which is considered must contain many particles in order to define an appropriate average. Yet if Eulerian conservation equations are used to describe the system, the volume must be small enough to be considered a mathematical point. The same argument applies to the other properties of the particle including composition, velocity, and temperature.

It is clear that the particles or droplets cannot be generally considered a continuum. In many applications, each particle or droplet has its own history which cannot be neglected.

9.1b. Eulerian and Lagrangian Reference Frames

In writing the conservation equations for a system there are two fundamentally different ways of considering the problem: Lagrangian or Eulerian. From the Lagrangian point of view, the flow field is regarded from a moving reference frame associated with the fluid element itself. The motion of each fluid element behaves according to Newton's second law of motion. In contrast to following individual fluid elements, the Eulerian approach considers all fluid elements which pass a given point for all time. That is, the flow properties (density, velocity, temperature, etc.) are described at each point as a function of time. Flow-field solutions are obtained by integrating the governing equations over all points in the flow field. The reference frame is generally fixed in space and chosen to describe the boundary conditions. The Eulerian description requires the fluid properties to be defined at a point in space and thus relates all fluid elements. The Lagrangian reference frame does not assume a continuum but follows each individual particle or droplet and describes its interactions with its surroundings. The Langrangian approach is generally impractical to describe the flow of a continuum because of the large number of mass elements needed to achieve a reasonably accurate description. On the other hand, the Lagrangian approach is appealing for dispersed two-phase flows, since each particle or droplet naturally constitutes a discrete mass element.

By far the most common approach to fluid mechanics analysis involves the Eulerian description of the flow field. Because of the vast source of literature for Eulerian reference frames and continuum mechanics the approach is often applied to gas–particle flows as well. Drew (1971) has presented a very detailed mathematical derivation of the Eulerian form of the two-phase flow equations and indicates the magnitude of error to be expected by assuming a continuum. Other techniques (Smoot and Pratt, 1979) have been derived in order to combine an Eulerian gas-phase description with a Langrangian particle-phase treatment.

We recommend this latter approach for dispersed flow systems of entrained coal particles.

In this chapter, we will next discuss the Reynolds transport theorem, which relates the Lagrangian and the Eulerian frameworks in a formal mathematical sense. We will then discuss the Eulerian equations for a gas and for a two-phase system. The multiphase equations for gas–particle/gas–droplets will be presented in their Eulerian framework. In addition, we will examine the gas-phase equations for the multiphase system when the particles are treated in the Lagrangian framework. In this representation, the source terms for the particle-gas coupling will be included. We will also briefly examine the Lagrangian particle conservation equations.

9.2. REYNOLDS TRANSPORT THEOREM

Essentially all the laws of mechanics are written for "a system." A system is defined as an arbitrary quantity of mass with an affixed identity. Everything external to the system is typically referred to as the surroundings and thus the system is separated from its surroundings by its boundaries. The laws of mechanics typically take the Lagrangian reference frame. These laws then stipulate what happens when there is an interaction between the system and its surroundings. Such is the case for Newton's laws and the first and second law of thermodynamics. In order to relate these fundamental laws to a continuum for application to problems in fluid mechanics, we must relate the Lagrangian reference frame to the Eulerian form. The Reynolds transport theorem provides this relationship. It is through this theorem that we are able to convert the fundamental laws of mechanics to an Eulerian representation for application to fluids.

Figure 9.3 dramatizes the difference between the Lagrangian and Eulerian reference frame. When we are analyzing the system in Figure 9.3a, we must apply the basic laws of mechanics to the system and account for all interactions with the surrounding. For multiple systems, each individual system must be treated separately. For the two systems of Figure 9.3a, each must be followed indefinitely or until they are no longer of interest. In contrast, the hypothetical control volume of Figure 9.3b has the same two systems of Figure 9.3a passing through it at a moment in time. We are interested in observing what happens to the fixed control volume as these systems and others pass through it. The basic laws of mechanics are usually written for the Lagrangian reference frame denoted in Figure 9.3a. However, the Eulerian reference frame represented in Figure 9.3b is usually of most interest to us in fluid mechanics. The Reynolds transport theorem is the mathematical formalization for converting the funda-

(a)

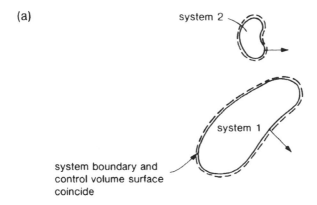

system 2

system 1

system boundary and
control volume surface
coincide

system 2

control surface

(b)

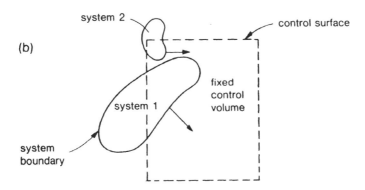

fixed
control
volume

system 1

system
boundary

Figure 9.3. Lagrangian *vs.* Eulerian reference frames. (a) Two systems passing through space with attached, moving control volumes (Lagrangian). (b) The same two systems passing through a fixed control volume (Eulerian).

mental laws from the Lagrangian reference frame of Figure 9.3a to the Eulerian frame of Figure 9.3b.

Consider the mass element which is passing through the fixed control volume shown in Figure 9.4. We defined the control volume such that at time t, the system just fills it. At time t, the control surface dA corresponds with the system boundaries. The system is moving with velocity v and has a density ρ. In Figure 9.4 such a system is shown just prior to entering the control volume at time $t - dt$. At time $t + dt$, this system will have moved out of the control volume. The Reynolds transport theorem is written for any extensive property of the fluid denoted by B (energy, momentum, etc.). Now let β be the

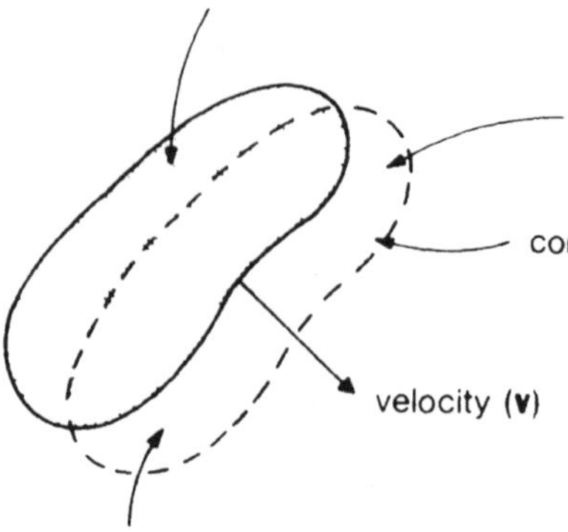

system at time t - dt (just
about to enter control volume)

control volume (dV)

control surface (dA)

velocity (**v**)

system at time t
(just occupying control volume)

Figure 9.4. Physical representation of Reynolds transport theorem—system passing through a fixed control volume and just filling it.

corresponding intensive property or extensive property (*B*) per unit mass:

$$\beta = dB/dm \qquad (9.1)$$

It is thus apparent that the total of *B* in the control volume is

$$B_{cv} = \int_{cv} \beta\rho \, dV \qquad (9.2)$$

We want to relate the Lagrangian rate of change of B_{cv} to the amount of *B* in the system which happens to coincide with the control volume at time *t*. This is the Reynolds transport theorem and may be written as

$$dB/dt = (d/dt)\int_{cv} \rho\beta \, dV + \int_{cs} \rho\beta\mathbf{v} \cdot d\mathbf{A} \qquad (9.3)$$

where ρ is the mass density, **v** is the velocity with respect to the control surface, and the differential area vector $d\mathbf{A}$ is normal outward from the control surface. The theorem simply states that the rate of change of some extensive property of a Lagrangian system (a given amount of mass) is equal to the rate of change of the property within the fixed Eulerian control volume (first term on right hand side in Eqn. (9.3)) plus the net efflux of the property across the fixed controlled surface (second term on right hand side in Eqn. (9.3)) at the very instant the Lagrangian mass element occupies the fixed Eulerian control volume.

9.3. EULERIAN EQUATIONS

9.3a. Formulation

The basic set of conservation equations encompasses conservation of mass, momentum, and energy. The fundamental laws of mechanics associated with these three conservation principles are the conservation of mass, Newton's second law of motion, and the first law of thermodynamics. Typically, these three laws are written for a Lagrangian system or mass element and are, respectively,

$$dm/dt = 0 \qquad (9.4)$$

$$\sum \mathbf{F}_{ext} = d(m\mathbf{v})/dt \qquad (9.5)$$

$$dQ/dt - dW/dt = dE/dt \qquad (9.6)$$

These Lagrangian statements of the fundamental laws of mechanics are cast into an Eulerian framework for use in fluid mechanics applications by applying the Reynolds transport theorem.

The simplest application of the Reynolds transport theorem is to the conservation of mass in order to derive the equation of continuity. We will apply Eqns. (9.3) and (9.4) to obtain the desired result. In this case the extensive property B is the mass m of the system. Thus the intensive property or extensive property per unit mass β is unity:

$$\beta = dB/dm = dm/dm = 1 \qquad (9.7)$$

Thus, for the conservation of mass, Eqn. (9.3) becomes

$$dm/dt = (d/dt) \int_{cv} \rho \, dV + \int_{cs} \rho \mathbf{v} \cdot d\mathbf{A} \qquad (9.8)$$

The left-hand side of Eqn. (9.8) refers to the Lagrangian mass element or system that has just filled the control volume at time t. Equation (9.4) indicates that this mass must be conserved and thus we substitute Eqn. (9.4) into Eqn. (9.8) to obtain the Eulerian form:

$$0 = (d/dt) \int_{cv} \rho \, dV + \int_{cs} \rho \mathbf{v} \cdot d\mathbf{A} \qquad (9.9)$$

The remainder of the derivation is simply mathematical manipulation to obtain the partial differential form of the equation of continuity desired. First, we recognize that Gauss's theorem permits us to convert the surface integral of Eqn. (9.9) to a volume integral. Gauss's theorem states:

$$\int_{cs} \psi \cdot d\mathbf{A} = \int_{cv} \nabla \cdot \psi \, dV \tag{9.10}$$

where ψ is any arbitrary vector. Thus we can rewrite Eqn. (9.9) as

$$0 = (d/dt) \int_{cv} \rho \, dV + \int_{cv} \nabla \cdot (\rho \mathbf{v}) \, dV \tag{9.11}$$

The first volume integral in Eqn. (9.11) allows the volume elements to distort with time; thus, the time derivative is applied after integration. Since we are considering a fixed, nondistorting control volume, the time derivative may be taken inside the integral to arrive at

$$0 = \int_{cv} (\partial \rho / \partial t + \nabla \cdot \rho \mathbf{v}) \, dV \tag{9.12}$$

If the integral is zero, the integrand must be equal to zero and thus we arrive at the homogeneous equation of continuity for a continuum:

$$\partial \rho / \partial t + \nabla \cdot (\rho \mathbf{v}) = 0 \tag{9.13}$$

This derivation was performed for the conservation of mass for the gas phase without the presence of the particles or droplets. The derivation was only illustrative of the technique used to derive the appropriate conservation equations for the Eulerian reference frame. The basic derivation always involves the conservation of some appropriate extensive property substituted into the Reynolds transport theorem and with substitution of the appropriate fundamental law of mechanics for the Lagrangian term.

9.3b. General Continuum Equations

When the continuum includes a reacting gas phase and reacting particles or droplets, the resulting conservation equations have many added complexities. The derivation of such equations follows steps analogous to those shown in the last section. Derivations of the governing equations for gas–particle/gas–droplet mixtures have been presented by Smoot and Pratt (1979). The final form of the continuity momentum and energy equations for the gas phase, particle phase, and mixture are summarized in Table 9.1. The presence of the

TABLE 9.1. Summary of Conservation Equations for Gas–Particle Mixtures[a]

Continuity equation

Gas phase: $\partial\rho'_g/\partial t + \nabla \cdot (\rho'_g\mathbf{v}_g) = r_p$

Particle phase: $\partial\rho'_p/\partial t + \nabla \cdot (\rho'_p\mathbf{v}_p) = -r_p$

Mixture: $(\partial/\partial t)(\rho'_p + \rho'_g) + \nabla \cdot (\rho'_g\mathbf{v}_g + \rho'_p\mathbf{v}_p) = 0$

Momentem equation

Gas phase: $(\partial/\partial t)(\rho'_g\mathbf{v}_g) + \nabla \cdot (\rho'_g\mathbf{v}_g\mathbf{v}_g)$
$= -\nabla p + \nabla \cdot [\theta\boldsymbol{\tau} + (1 + \theta)\boldsymbol{\tau}_a] - \mathbf{F}_p + \rho'_g\mathbf{g} + \mathbf{v}_p r_p$

Particle phase: $(\partial/\partial t)(\rho'_p\mathbf{v}_p) + \nabla \cdot (\rho'_p\mathbf{v}_p\mathbf{v}_p) = \mathbf{F}_p + \rho'_p\mathbf{g} - \mathbf{v}_p r_p$

Mixture: $(\partial/\partial t)(\rho'_g\mathbf{v}_g + \rho'_p\mathbf{v}_p) + \nabla \cdot (\rho'_g\mathbf{v}_g\mathbf{v}_g + \rho'_p\mathbf{v}_p\mathbf{v}_p)$
$= -\nabla_p + \nabla \cdot [\theta\boldsymbol{\tau} + (1 - \theta)\boldsymbol{\tau}_a] + (\rho'_g + \rho'_p)\mathbf{g}$

Energy equation (total)

Gas phase: $(\partial/\partial t)[\rho'_g(i_g + v_g^2/2)] + \nabla \cdot [\rho'_g\mathbf{v}_g(h_g + v_g^2/2)]$
$= -\nabla \cdot [\theta\mathbf{q} + (1 - \theta)\mathbf{q}_s] - q_{cp} + q_{r_g}$
$+ r_p(\bar{h}_s + v_p^2/2 + w'^2/2) - \nabla \cdot [(1 - \theta)p\mathbf{v}_p]$
$+ \nabla \cdot [\theta\boldsymbol{\tau} \cdot \mathbf{v}_g + (1 - \theta)\boldsymbol{\tau}_a \cdot \mathbf{v}_p] - \mathbf{v}_p \cdot \mathbf{F}_p + \rho'_g\mathbf{g} \cdot \mathbf{v}_g + \bar{p}_s s_c$

Particle phase: $(\partial/\partial t)[\rho'_p(i_p + v_p^2/2)] + \nabla \cdot [\rho'_p\mathbf{v}_p(i_p + v_p^2/2)]$
$= -r_p(\bar{h}_s + v_p^2/2 + w'^2/2) + \mathbf{v}_p \cdot \mathbf{F}_p + \rho'_p\mathbf{g} \cdot \mathbf{v}_p + q_{cp} + q_{r_p} - \bar{p}_s s_c$

Mixture: $(\partial/\partial t)[\rho'_g(i_g + v_g^2/2) + \rho'_p(i_p + v_p^2/2)] + \nabla \cdot [\rho'_g\mathbf{v}_g(h_g + v_g^2/2)\rho'_p\mathbf{v}_p(i_p + p/\rho_p + v_p^2/2)]$
$= -\nabla \cdot (\theta\mathbf{q} + (1 - \theta)\mathbf{q}_s] + \theta q_{r_g} + q_{r_p}$
$+ \nabla \cdot [\theta\boldsymbol{\tau} \cdot \mathbf{v}_g + (1 - \theta)\boldsymbol{\tau}_a \cdot \mathbf{v}_p] + \mathbf{g} \cdot (\rho'_p\mathbf{v}_p + \rho'_g\mathbf{v}_g)$

[a]Data from Smoot and Pratt (1979).

particles in the gas phase requires a new definition for gas-phase density. The volume occupied by the gas per unit volume of gas–particle/droplet mixture is identified as the void fraction ϕ. Thus, the mass of gas per unit volume of gas–particle mixture (the bulk gas density) is given by

$$\rho'_g = \phi\rho_g \qquad (9.14)$$

where ρ_g is the material density of the gas. Similarly, ρ'_p is the bulk density of the particulate or droplet phase. The particulate-phase equations shown in Table 9.1 correspond to locally uniform particle velocities. The terms accruing due to a distribution of particle velocities within the control volume for differing particle sizes can be found by referring to Smoot and Pratt (1979).

9.3c. Decoupled Gas and Solid Equations

Table 9.1 shows the conservation equations from a completely Eulerian reference frame. We have discussed the advantages of using an Eulerian reference frame for the gas phase and a Lagrangian reference frame for the particulate of droplet phases. In such a split approach, the two phases must be identified separately and equations solved separately for each phase. Since the presence of the second phase is important to the solution of each of the equations for the corresponding phases their effects must be incorporated. One convenient approach for doing so is to incorporate source terms in the Eulerian gas-phase equations to account for the presence of the particulate phase.

In this Lagrangian approach for the particle/droplet phases, the individual particle or droplet trajectories are followed through the flow field and solid properties are changed continuously for each trajectory. Interactions between the gas and particle phases are accounted for by source terms in the Eulerian gas field. These source terms are updated each time the Lagrangian trajectories are computed. This approach is based on the PSI-CELL technique developed by Crowe (1977). Particle velocities, trajectories, temperatures, and composition are obtained by integrating the equations of motion, energy, and component continuity for the particles in the gas-flow field while recording the momentum, energy, and mass of the particles on crossing cell boundaries. The cells are simply computational control volumes and the net difference in the particle properties between leaving and entering a given cell provides the particle source terms for the gas-flow equations.

The Lagrangian conservation equations for a single particle or droplet are discussed in some detail in Chapter 12 and thus will not be presented in this chapter. In this mixed Lagrangian–Eulerian approach, the Eulerian gas-phase equations are those shown in Table 9.1. The source terms mentioned match precisely with the corresponding particulate terms in the gas-phase equations of Table 9.1. To illustrate the mixed Lagrangian approach, consider the Eulerian conservation equations for the gas-phase continuum presented in Table 9.2. For simplicity, these equations have been simplified to their steady-state axisymmetric form. In the next chapter, we will discuss the necessity of performing local, time averaging to account for turbulent fluid mechanics. In Chapter 11, we will introduce the concept of Favre averaging for fluctuating density flows. Table 9.2 shows the Favre-averaged conservation equations for a typical combustion or gasification system. Immediately, we notice that there are more equations than those for simple conservation of overall mass momentum and energy. The added conservation equations are needed to include the effect of the turbulence and the chemistry in such systems. Such is the case for k, ε, f_p, g_p, η, and g_η. The meaning of these equations will be introduced in subsequent chapters. The point that is to be made here, is that the derivation of such equations follows directly the derivation for conservation of mass, momentum and energy discussed in this chapter.

Another observation regarding the equation set is the common form of the conservation equations themselves. The general ϕ equation (shown in Table

TABLE 9.2. Gas-Phase Conservation Equations for Gas-Particle/Gas-Droplet Eulerian Reference frame[a]

$$\frac{\partial}{\partial x}(\bar{\rho}\tilde{u}\phi) + \frac{1}{r}\frac{\partial}{\partial r}(r\bar{\rho}\tilde{v}\phi) - \frac{\partial}{\partial x}\left(D_\phi\frac{\partial\phi}{\partial x}\right) - \frac{1}{r}\frac{\partial}{\partial r}(rD_\phi\,\partial\phi/\partial r) = S_\phi$$

ϕ	D_ϕ	S_ϕ
$\tilde{u}$	μ_e	$\dfrac{-\partial\bar{p}}{\partial x} + \dfrac{\partial}{\partial x}\left(\mu_e\dfrac{\partial\tilde{u}}{\partial x}\right) + \dfrac{1}{r}\dfrac{\partial}{\partial r}(r\mu_e(\partial\tilde{v}/\partial x)) + S_p^x + \tilde{u}S_p^m$
$\tilde{v}$	μ_e	$\dfrac{-\partial\bar{p}}{\partial r} + \dfrac{\partial}{\partial x}\left(\mu_e\dfrac{\partial\tilde{u}}{\partial r}\right) + \dfrac{1}{r}\dfrac{\partial}{\partial r}(r\mu_e(\partial\tilde{v}/\partial r)) - \dfrac{2\mu_e\tilde{v}}{r^2} + \dfrac{\bar{\rho}\tilde{v}\tilde{w}}{r} + S_p^c + \tilde{v}S_p^m$
$\tilde{w}$	μ_e	$\dfrac{-\bar{\rho}\tilde{v}\tilde{w}}{r} - \dfrac{\tilde{w}}{r^2}\dfrac{\partial}{\partial r}(r\mu_e)$
k	μ_e/σ_k	$\Phi - \bar{\rho}\varepsilon$
ε	μ_e/σ_l	$(\varepsilon/k)(C_1\Phi - C_2\bar{\rho}\varepsilon)$
$\tilde{f}_p$	μ_t/σ_l	0
g_l	μ_e/σ_g	$C_{g1}\mu_e\left[\left(\dfrac{\partial\tilde{f}}{\partial x}\right)^2 + \left(\dfrac{\partial\tilde{f}}{\partial r}\right)^2\right] - \dfrac{C_{g2}\bar{\rho}\varepsilon g_l}{k}$
$\tilde{\eta}$	μ_e/σ_n	S_p^m
g_η	μ_e/σ_g	$C_{g1}\mu_e\left[\left(\dfrac{\partial\tilde{\eta}}{\partial x}\right)^2 + \left(\dfrac{\partial\tilde{\eta}}{\partial f}\right)^2\right] - \dfrac{C_{g2}\bar{\rho}\varepsilon g_\eta}{k}$
$\tilde{h}$	μ_e/σ_h	$q'_{rg} + \tilde{u}(\partial\bar{p}/\partial x) + \tilde{v}(\partial\bar{p}/\partial r) + S_p^h + \tilde{h}S_p^m$

where

$$\Phi = 2\mu_e\left[\left(\frac{\partial\tilde{u}}{\partial x}\right)^2 + \left(\frac{\partial\tilde{v}}{\partial r}\right)^2 + \left(\frac{\tilde{v}}{r}\right)^2\right] + \left(\frac{\partial\tilde{u}}{\partial r} + \frac{\partial\tilde{v}}{\partial x}\right)^2 + \left[r\frac{\partial}{\partial r}\left(\frac{\tilde{w}}{r}\right)\right]^2 + \left(\frac{\partial\tilde{w}}{\partial r}\right)^2$$

$$\mu_e = \mu_t + \mu_l$$

$$\mu' = C_\mu\bar{\rho}k^2/\varepsilon$$

9.2) shows that the variable ϕ is transported by convection in the axial direction as indicated by the first term and the radial direction as indicated by the second term. The last two terms on the left-hand side of the general ϕ equation show diffusion of the variable ϕ in the axial and radial directions, respectively. The right-hand side of this equation shows the source or sink terms for the general variable ϕ. Each of several conservation equation variables is shown in this table. D_ϕ shows the transport coefficient for the appropriate variable and S_ϕ shows the corresponding source or sink terms. Each of the S_p terms in the table shows sources or sinks to the corresponding property due to the presence of the particulate or droplet phase. Each of these source terms has a corresponding

Eulerian representation in Table 9.1. For example, by comparison, it is obvious that

$$S_p^u = -F_p^u \tag{9.15}$$

$$S_p^m = r_p \tag{9.16}$$

where F_p^u is the axial component of the aerodynamic force on the particles per unit volume of mixture by the gas phase. Such source terms must be provided by the Lagrangian description of the particulate or droplet phase. The source terms are discussed in detail with appropriate equations in Chapter 12.

In summary, these gas and solid equations are useful for the approach which incorporates an Eulerian description for the gas phase and a Lagrangian description for the particle phase (i.e., pulverized-coal or coal–water mixtures). The equations for the continuum are the gas-phase Eulerian conservation equations depicted in Table 9.1 with all particle-interaction terms being provided as source terms from the Lagrangian particle description. This approach makes it easier to treat the gas phase as a continuum and a particulate or droplet as a discrete mass element, and is useful for dispersed flow systems. The history effect and noncontinuum nature of the particles or droplets is explicitly incorporated. However, this method makes it much more difficult to include particle–particle interactions and thus is probably not appropriate for densely loaded particle or droplet systems.

9.4. ILLUSTRATIVE PROBLEMS

1. From the Reynolds transport theorem derive the equation of motion (or momentum equation) for an isothermal, gas-phase system. This is accomplished by application of Newton's second law of motion and results in a vector equation.

2. Compare the differences between the gas-phase equation of motion (or momentum equation) in a system with no particles derived in Problem 1 above to the gas-phase momentum equation in a system with particles. The later equation is given in Table 9.1. List the differences and discuss the physical significance of each of these differences.

3. Determine a criterion for the number density of particles above which it is possible to use the continuum hypothesis for the particulate phase without any appreciable error. Compare this criterion with the number density of molecules in air at standard temperature and pressure.

4. From the gas-phase equation of energy in Table 9.1 derive the gas-phase equation of energy in terms of enthalpy ($\tilde{h}$) shown in Table 9.2. Ignore the Favre-averaging. From your derivation, describe the physical significance of the terms S_p^h and hS_p^m.

TURBULENCE

10.1. INTRODUCTION

In the previous chapter we formulated the basic equations for describing the conservation of mass, momentum, and energy in environments typical of coal combustors and gasifiers. In this chapter we will find that due to some very practical considerations, these equations must be modified for application to turbulent combustion. We will then identify different turbulence models and briefly discuss their range of applicability. We will introduce the effects of the turbulent carrier gas on the particles or droplets and finally present a recommended approach for describing turbulence in coal combustion and gasification.

The research topic of turbulence and turbulent modeling is a highly developed field. This scientific study spans approximately 100 years and the researchers have included some of the greatest names in physics, mechanics, and engineering. Experimentalists and theoreticians alike have tackled the turbulence problem and progress in many directions has been made. However, problems still remain. A great deal of current research continues, in an attempt to understand the turbulent transport of mass momentum and energy. This chapter is designed to introduce the problems, present some of the current strategies in computational fluid dynamics, and suggest an approach for use in computing coal combustion and gasification operations with present state-of-the-art methods. As new research suggests better alternatives, those suggested in this chapter should be updated.

The flows of importance to combustors and gasifiers of pulverized coal and coal slurries have unique problems which add to the complexities of the turbulence problems. The homogeneous, confined-gas jet without any entrained particles or droplets is not completely understood. The added complexity of the second particulate or droplet phase makes it extremely difficult to use fundamental theories to describe the mean motion. A great deal of empiricism must be included in the turbulence model. In this chapter we will focus mostly on the homogeneous gas field. The introduction of the particles or droplets will be discussed briefly towards the end of this chapter and then in greater detail in Chapter 12.

Consider two streams of gases, each with different levels of momentum, mixing in a shear layer. When the external forces causing the fluid motion are very small, then the fluid is in a laminar flow. In such a state, the imposed shear stresses are absorbed through viscous forces. Thus, it is at the molecular level that the imposed shear forces are dissipated. As the velocity of the fluid increases, the imposed shear forces become larger and larger and eventually become too great for the fluid elements to absorb through molecular processes. The fluid is then forced into a different state of motion in which larger coherent regions or turbulent eddies can rotate and thus relieve the shear forces caused by the imposed velocity differences. At sufficiently high Reynolds numbers, the deceleration of the fluid always produces vorticity, and the resulting vortex interactions are apparently so sensitive to the initial conditions that the resulting flow pattern changes in time, usually in a stochastic fashion. Thus, the turbulent eddies are undergoing a chaotic sequence of events to produce sufficient interfacial shear to absorb the imposed stress.

The fluid mechanics solution of interest consists of the momentum and continuity equations for incompressible flow. It is important to realize that these equations were derived from the basic momentum principle (Newton's second law) and from conservation of mass for a continuum (see, e.g., Eqns. (9.5) and (9.13)). As such, they apply to turbulent as well as laminar flows. In other words, a complete, time-dependent solution of these equations for a set of given boundary conditions will yield an exact and complete solution for any turbulent flow. Numerically, such a solution would require extremely fine mesh spacing in order to resolve the fine structure of the fluid mechanics, since the important details of turbulence are of the order of 0.1 μm. The storage capacity required for such a computation is not available at the present stage of computer development. If such solutions did exist, they would give the instantaneous values of the velocity and pressure fluctuating in time and space.

Finding an alternate approach is the topic of turbulence modeling. No comprehensive theory exists short of solving the complete set of coupled equations of motion and energy at the scale of the smallest turbulent eddies. Nevertheless, modeling assumptions and approximations have allowed for the development of computational tools that have been more or less successful for specific restricted applications.

In selecting a suitable turbulence model, it is important to closely analyze the physical applications being considered. Such careful initial analysis will help to avoid an ill-considered selection of a turbulence model. In turbulence, the equations do not give the entire story. For example, in many situations, such as strong shear flows, a simple augmentation of the molecular viscosity by some appropriate constant or a simple algebraic expression may sufficiently describe the fluid motion. However, for recirculating streams, this assumption is completely inadequate, mainly due to its failure to account for the convective transport of energy. These criteria will be discussed when selecting the suggested turbulence model for coal combustion and gasification applications.

10.2. THE CLOSURE PROBLEM

Despite the rapid advances made in computer technology, a direct solution of the time-dependent conservation equations for turbulent flows is not currently practical. The only economically feasible way to solve practical turbulent flow problems continues to be the use of statistically averaged equations governing mean-flow quantities. This conventional technique consists of taking each of the equations of change and averaging them over a short time interval. In this way, the properties of the flow are expressed in time-mean and time-fluctuating components. The resulting equations describe the time-smoothed velocity and pressure distributions, but cross correlations involving the fluctuating velocities and pressures also appear. Expressing these correlations in terms of the mean-flow-field variables is the subject of turbulent modeling and is known as the classical closure problem of turbulent field theory.

The concept of time averaging is introduced by defining $\bar{\phi}$ as the time-averaged value of any variable ϕ. Thus

$$\bar{\phi} = (1/T)\int_0^T \phi \, dt \tag{10.1}$$

where T is large with respect to the large turbulence time scale. Then ϕ' is defined as the fluctuating component of the arbitrary variable ϕ. Thus

$$\phi = \bar{\phi} + \phi' \tag{10.2}$$

Note that $\overline{\phi'} = 0$, but $\overline{\phi'\phi'} \neq 0$.

In time averaging the transport equations of interest to combustion and gasification, all of the dependent variables are fluctuating and must be decomposed into their time-mean and fluctuating quantities. When these values are resubstituted into the equations of change, the problem is immediately obvious. The equations reduce to terms identical to the instantaneous form of the equation only in the time-mean variables, but there are a large number of extra terms involving the fluctuating components. Terms appear such as $\overline{\rho u_i' u_j'}$, $\bar{u}\,\overline{\rho' u_i'}$, $\overline{\rho' u_i' u_j'}$, etc. A constant density, constant viscosity system simplifies these problems greatly since $\rho' = 0$ and $\mu' = 0$. The only remaining terms are of the form $\overline{\rho u_i' u_j'}$. These correlations are referred to as the Reynolds stresses simply because they have the units of stress. These are the terms which are usually modeled in current turbulence models.

What about the case involving terms with fluctuating density and viscosity? The laminar viscosity is not as strongly dependent on temperature as density and will be involved with gradient fluctuations in terms like $\mu' \partial u_i'/\partial x$. Since μ' and u_i' are not likely to be strongly correlated, these terms are usually neglected. More often than not, it is also conveniently assumed that ρ' is not correlated with u_i', Thus, terms involving ρ' are neglected and the equations reduce to those for the constant density case. However, the equations are still written in terms of the time-mean density which must be appropriately computed. There is a way to circumvent this difficulty. Favre-averaging is a viable alternative to time-averaging in order to avoid neglecting the density fluctuations. This is discussed in Chapter 11.

A survey of the mean-turbulent-field closure models by Mellor and Herring (1973) gives an excellent discussion of what has taken place in mathematical models of turbulence up to 1973. Spalding (1975) gives a discussion of solved and unsolved problems in turbulence modeling. He focuses on the $k - \varepsilon$ model and enumerates its advantages and shortcomings. Indeed, over the last decade, the $k - \varepsilon$ model (Launder and Spalding, 1972) has proven to be the most widely used of all the turbulence models.

It is the deficiences in the $k - \varepsilon$ model which have encouraged the most recent developments in turbulence modeling. Rodi (1976) gives an excellent review on progress in turbulence modeling during this recent period.

Three principal directions are being taken in the latest generation of turbulence models. All of them focus on the Reynolds stresses and avoid the use of the Boussinesq hypothesis.

The first direction is a new generation of stress transport closure techniques. Launder (1979) reviews this latest direction. The second direction is an attempt to retain the numerical simplicity of the $k - \varepsilon$ model (as opposed to the complexity of the full Reynolds stress closure) by using algebraic approximations for each of the Reynolds stresses which are expressed in terms of k and ε (Launder et al., 1975; Rodi, 1976; Leslie, 1980). Finally, the third direction for recent models is a promising attempt at describing the details of the turbulence by solving for larger eddy simulations and modeling only the subgrid scale turbulence. Herring (1979) gives a short introduction and overview to subgrid modeling and Love and Leslie (1979) give more details.

In this chapter we briefly examine some of the general turbulence models. The overall philosophy is to use the simplest closure level that will give adequate agreement with the experiment for the problem of interest. The reason for emphasizing this simplicity is that most engineering flows are so complex that computer storage imposes severe constraints on the fineness of the computational mesh. In the selection of any turbulence model there are several questions which deserve careful scrutiny. Is the turbulence sufficiently developed to the point where an algebraic model for the turbulent viscosity is satisfactory? Is the turbulence sufficiently shear-layer dominated and of sufficiently high turbulent Reynolds number to permit the use of a one-equation or two-equation turbulence model? Is enough known about multi-phase turbulence to permit adequate assumptions to be made? Are there any regions in the computational domain where low Reynolds number turbulence might be a problem? Is the Boussinesq hypothesis appropriate throughout the flow field, or do bouyancy and other similar forces, which have their own characteristic length and time scales, dominate in certain regions? Is there a reasonable basis upon which the reliability of the turbulence model can be evaluated? For example, are there any simple applications where an analytical solution, numerical solution, or experimental data base are available for comparison? These kinds of questions should be addressed before final selection of any appropriate turbulence model is made.

10.3.   PRANDTL MIXING LENGTH

The Prandtl mixing length hypothesis was one of the first turbulence models ever proposed. This simple model relies on the eddy diffusivity concept which relates turbulent transport terms to the gradient of the mean flow quantities. This basic approach to the closure problem is credited to Boussinesq (1877) and is done simply by analogy to the laminar Newtonian viscosity; thus, it is assumed that

$$-\overline{u'v'} = -\nu'_g \left(\frac{\partial \bar{u}}{\partial r} + \frac{\partial \bar{v}}{\partial z} \right) \tag{10.3}$$

Or, in more general terms, the mean strain rates are

$$-\overline{u'_i u'_k} = -\nu'_g \left(\frac{\partial \bar{u}_i}{\partial x_k} + \frac{\partial \bar{u}_k}{\partial x_i} \right) - \tfrac{2}{3} (\nu'_g \operatorname{div} \mathbf{u} + k)\, \delta_{ik} \tag{10.4}$$

where ν'_g is the eddy diffusivity, which is sometimes called the eddy viscosity or effective turbulent viscosity. This is often coupled with a modified Reynold's analogy assuming that the turbulent scalar transport is closely related to the momentum transport through a turbulent Prandtl or Schmidt number.

With this Boussinesq hypothesis, the Prandtl mixing length theory calculates the distribution of the eddy diffusivity ν'_g by relating it to the local mean velocity gradient,

$$\nu_g^t = l_m^2 \left| \frac{\partial \bar{u}}{dr} + \frac{\partial \bar{v}}{\partial x} \right| \qquad (10.5)$$

where Eqns. (10.3) and (10.5) have been written for axisymmetric geometries. The closure is not yet complete, since Eqn. (10.5) still contains a single unknown parameter. This parameter, the mixing length (l_m), is typically described with the aid of empirical information.

The Prandtl mixing length theory contains many theoretical shortcomings, but despite these shortcomings has proved extremely useful for many practical flow systems. Most of the useful applications belong to the class of thin-shear-layer flows. In the process of applying this model, a vast amount of experience has been obtained in prescribing the mixing length distributions for wall boundary layers. The mixing length hypothesis has also proven to be practical for cases dominated by the local shear. In systems that are unbounded, turbulence transport considerations dominate and thus the governing characteristic length scale must be evolved by the turbulence itself.

The main shortcomings of the Prandtl mixing length theory are due to the assumed isotropy and equilibrium of the turbulence. The eddy diffusivity ν_g^t should be highly anisotropic and in general is a property which must be represented by a fourth-order tensor rather than an isotropic scalar. The Prandtl mixing length implies that the turbulence is in local equilibrium; that is, at each point in the flow field, turbulence energy is dissipated at the same rate that it is produced. On the contrary, it is well-established that both history and action-at-a-distance play essential roles in established local turbulence fields.

Attempts to account for transport and history effects of turbulence have resulted in the higher-order closure models. In general, the combustion flows of interest to this book are not dominated by wall shear forces and thus more complexity must be introduced beyond the simple Prandtl mixing length theory.

10.4. ONE–EQUATION CLOSURE MODELS

In order to relate the eddy diffusivity to the mean-flow-field properties and thus account for the transport and history effects of the turbulence, the Prandtl mixing length hypothesis must be abandoned. Principally, two approaches have been used to relate the turbulent Reynolds stresses to the mean flow field.

The first and most widely used approach employs the eddy diffusivity concept, but replaces the Prandtl mixing length theory with the Prandtl–

Kolmogorov relationship:

$$v'_g = C_\mu k^2 / \varepsilon \tag{10.6}$$

where C_μ is a constant, k is the specific turbulent kinetic energy, and ε is the rate of dissipation of k.

The second approach was suggested by Bradshaw and co-workers (1967), who did not employ the eddy diffusivity concept but converted the kinetic energy equation into a transport equation for the shear stresses $u'v'$ by assuming a direct link between $u'v'$ and k. Most of the work by Bradshaw has been intended for wall boundary layers. In these flow types, experiments have suggested that $u'v'$ is approximately equal to $0.3k$.

Both of these approaches relate the turbulence properties to the kinetic energy of the turbulent motion. This turbulent kinetic energy

$$k = \tfrac{1}{2}\overline{(u'_i)^2} \tag{10.7}$$

is a measure of the intensity of the turbulent fluctuations in the three directions. In most models the following transport equation for the turbulent kinetic energy is employed:

$$\underbrace{\frac{\partial k}{\partial t}}_{\substack{\text{rate of}\\\text{change}}} + \underbrace{\bar{u}_i \frac{\partial k}{\partial x_i}}_{\text{convection}} = \underbrace{\frac{\partial}{\partial x_i}\left(\frac{v'_g}{\sigma_k}\frac{\partial k}{\partial x_i}\right)}_{\text{diffusion}} - \underbrace{\overline{u_i u_j}\left(\frac{\partial \bar{u}_i}{\partial x_j}\right)}_{\substack{\text{production}\\(P)}} - \underbrace{\varepsilon}_{\text{dissipation}} \tag{10.8}$$

This equation can be derived in an exact form from the simple conservation principle. In this case it is found by multiplying the momentum equation for each coordinate direction by its corresponding fluctuating velocity, time averaging, and then taking one-half of the sum of the resulting equations. However, the same closure problem still exists, since this transport equation for k contains terms with fluctuating correlations (Hinze, 1967; Tenekes and Lumley, 1972). Tenekes and Lumley (1972) gave a particularly interesting discussion of the significance of each term and the order-of-magnitude estimates of their contributions. In Eqn. (10.8) the diffusion term has been assumed to be proportional to the gradient of k. The turbulence Schmidt number σ_k for the kinetic energy is simply an empirical constant on the order of unity. Bradshaw et al. (1967) do not approximate this diffusion of k by gradient model, but assume instead that the diffusion flux of k is proportional to a bulk velocity. It is obvious from Eqn. (10.8) that such a model accounts for the transport of the kinetic energy of the turbulence itself. The rate of change or accumulation of k is balanced by convective transport by the mean motion, diffusive transport by turbulent motion, production by interaction of turbulent stresses and mean velocity gradients, and destruction by the dissipation ε. In flows where buoyant forces are important, an additional term to account for this buoyancy production and destruction must be incorporated.

These one-equation closure models usually incorporate a transport equation for the turbulent kinetic energy. The dissipation rate ε must be determined from some algebraic expression. The most common expression is

$$\varepsilon = C_D(k^{3/2}/l_\varepsilon) \tag{10.9}$$

Again C_D is the semiempirical constant and the appropriate dissipation length scale l_ε must be prescribed for the flow field of interest. Prescribing l_ε is no easier than prescribing the mixing length l_m.

It is important to realize that the k equation (Eqn. (10.8)) and the associated closure assumptions are typically valid only in very high Reynolds number flows. For this reason, the solution of the equation in the viscous sublayer near walls is not even appropriate. Typically, there is little interest in solving the minute details near the walls because of the computational requirements necessary to resolve this structure. For this reason approaches have been developed for bridging over the semilaminar sublayer near walls and boundaries. These so called wall functions become an integral part of the overall turbulence models. A discussion of these wall functions is made later in this chapter.

10.5. THE $k - \varepsilon$ MODEL (TWO-EQUATION CLOSURE MODELS)

In order to predict the length scale from the mean flow properties, several two-equation turbulence closure models have been proported. Most of these models include the transport equation for the turbulent kinetic energy, and then employ a second transport equation for some form of the length scale while retaining the Prandtl–Kolmogorov relationship for the eddy diffusivity as shown in Eqn. 10.6. Of all the two-equation models presented to date, the $k - \varepsilon$ model has the widest application for confined turbulent jets. It has been tested in several reacting and nonreacting flow environments. It accounts for convective transport of energy, while retaining computational efficiency. Defects include the assumption of the Boussinesq gradient-diffusion hypothesis and the assumed isotropy of the eddy diffusivity. This method has not accurately simulated recirculation zones which occur behind blunt boundaries, and its applicability to the gas motion in a particle-laden stream is questionable. It is also somewhat sensitive to the values of the initial turbulence level. Strongly swirling flows are not accurately predicted by this technique but must include certain refinements. Yet, it continues to be the dominant technique being applied to fluid mechanics in combustion systems. For pulverized-coal systems without swirl, the $k - \varepsilon$ model seems to be the most practical choice. The impact of swirling flows on coal combustion is discussed later in this chapter.

In the two-equation $k - \varepsilon$ model, the flow field values of the turbulent kinetic energy (k) are calculated from the transport equations (10.8). In addition, another transport equation for the dissipation rate of this turbulent kinetic energy is employed. Typically, the form of the equation for ε is

$$\frac{\partial \varepsilon}{\partial t} + \bar{u}_i \frac{\partial \varepsilon}{\partial x_i} = \frac{\partial}{\partial x_i}\left(\frac{\nu_g^t}{\sigma_r}\frac{\partial \varepsilon}{\partial x_i}\right) + C_1\left(\frac{\varepsilon}{\kappa}\right)P - C_2\left(\frac{\varepsilon}{k}\right)^2 \qquad (10.10)$$

| rate of change | convection | dissipation | production | dissipation |

where P is the production term of Eqn. 10.8. This equation has many more modeling assumptions than is required for the transport of the turbulent kinetic energy. Since these modeling assumptions are heavily empirical by nature, this equation is immediately suspect when theoretical predictions fail to match experimental data. Perhaps a greater limitation of the applicability of the $k - \varepsilon$ model is the assumption of isotropy for the eddy diffusivity. Since the individual stress components may develop quite differently in the flow field, the shear and normal stresses should have different values for different directions. Thus, each eddy diffusivity should depend on the particular stress component considered.

As noted earlier, the $k - \varepsilon$ model, as formulated, is applicable only for high Reynolds number flow. Thus, special consideration must be made for both the k and the ε equations in low-level turbulence. Some turbulent combusting flows involve regions of low Reynolds number flow. Jones and Launder (1972) have proposed a version of the $k - \varepsilon$ model which includes laminarization and thus allows the model to be extended to near the wall regions and to flows where a large range of Reynolds numbers from low to high values may be important.

Although problems are prevalent, researchers over the last 10 years have applied the $k - \varepsilon$ model to many different geometries and flow regimes. Without question, the $k - \varepsilon$ model is now capable of predicting the velocity profiles in the near-jet region of several mixing chambers and combustors. Such predictions are compared with experimental measurements from two different investigators in Figures 10.1 and 10.2. Figure 10.1 shows a free-shear-layer calculation. Figure 10.2 shows comparisons in a confined jet with laser-doppler velocity measurements. This figure shows comparisons at four experimental sampling stations nearest the inlet. This is the region which is the most difficult to match. Comparisons of measurements and predictions at all downstream positions were very good and thus not shown here. Figure 10.3 shows a comparison between prediction and measurement of rms velocity, which is typically much more difficult to accurately predict. Although the rms velocity is not predicted

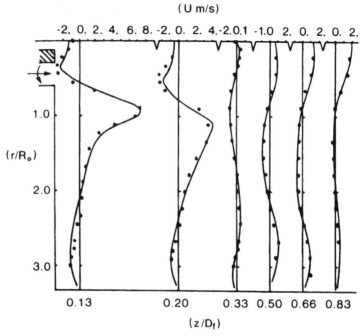

Figure 10.1.   Velocity profiles in model combustion chamber: (———) predictions and (●) laser-doppler anemometer measurements (both predictions and measurements from Hutchinson *et al.*, 1976). Conditions were for isothermal air flow with central jet Reynolds number of 5500 and 47,000 for the annular jet. The annulus included swirl with a swirl number of 0.52. (Figure used with permission from Hutchinson *et al.*, 1976.)

precisely, the resulting turbulent mixing process is predicted quite adequately. This later point will be discussed in subsequent chapters.

Strongly swirling flows are of particular interest to combustion and gasification of pulverized coal and coal slurries. Such flows are characterized by a highly nonisotropic nature, especially within the vicinity of the recirculation zone, and are therefore susceptible to the predictive inadequacies of the $k - \varepsilon$ model. Two possible approaches to the problem are: (i) incorporate a more sophisticated turbulence model like the turbulence stress models which will be discussed in the next section; or (2) adapt the $k - \varepsilon$ model to swirling flows by the addition of the appropriate source terms or modifications of the source-term coefficients in the dissipation equation. Past efforts have focused on the modification of the source-term coefficients through the Richardson number:

$$Ri = \left(\frac{k^2}{\varepsilon^2}\right)\left(\frac{w}{r^2}\right)\left(\frac{\partial rw}{\partial r}\right) \qquad (10.11)$$

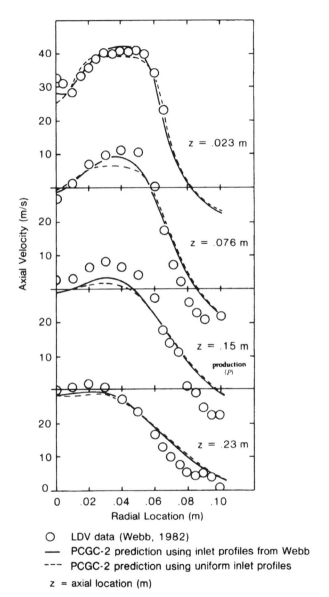

Figure 10.2. Comparison of model predictions (from Fletcher, 1983) of axial velocity with LDV data from nonreacting flow measurements (from Webb, 1982). The mixing chamber consisted of two coflowing air streams in an enlarged mixing chamber (see Chapter 8 for detailed conditions).

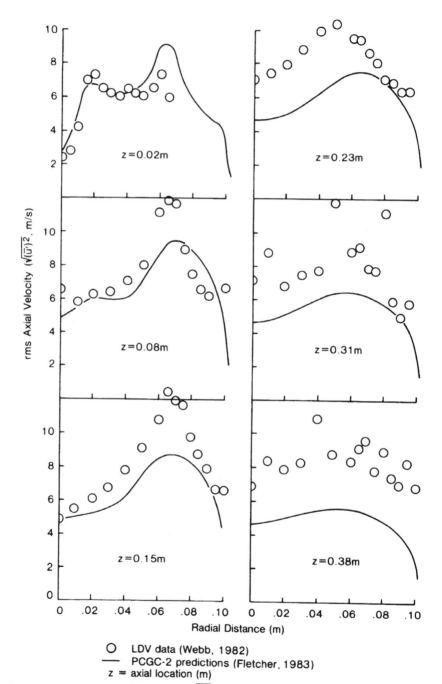

Figure 10.3. Comparisons of rms velocity $\overline{(u')^2}^{1/2}$ measurements and predictions. Conditions are those of Figure 10.2.

The Richardson number, by analogy with the streamline curvature approach, seeks to alter the local value of the dissipation rate and hence the local turbulent intensity in accordance with the degree of tangential momentum present locally in the flow field. To make the Richardson number compatible with the $k-\varepsilon$ model, one or more of the coefficients of the turbulent equations is forced to depend on the local value of Ri within the flow field. Furthermore, the denominator, which was originally expressed in terms of mean flow quantities, is replaced by a length-scale characteristic of turbulence (k/ε).

The functional dependence of the Richardson number in the turbulent transport equation is somewhat uncertain. A common form which has been utilized in the dissipation equation is

$$C_2^* = C_2(1 - C_3 Ri) \tag{10.12}$$

The corresponding steady-state, axisymmetric dissipation equation is

$$\frac{\partial(\rho u \varepsilon)}{\partial x} + \frac{1}{r}\left(\frac{\partial(r \rho v \varepsilon)}{\partial r}\right) - \frac{\partial[(\mu_t/\sigma_\varepsilon)(\partial \varepsilon/\partial x)]}{\partial x} - \frac{1}{r}\frac{\partial[(r\mu_t/\sigma_\varepsilon)(\partial \varepsilon/\partial r)]}{\partial r} = \frac{(C_i g - C_2^* \rho \varepsilon)\varepsilon}{k} \tag{10.13}$$

where

$$g = \mu_t \left\{ 2\left[\left(\frac{\partial v}{\partial r}\right)^2 + \left(\frac{\partial u}{\partial x}\right)^2 + \left(\frac{v}{r}\right)^2\right] + \left(\frac{\partial(w/r)}{\partial r}\right)^2 + \left(\frac{\partial v}{\partial x} + \frac{\partial u}{\partial r}\right)^2 + \left(\frac{\partial w}{\partial x}\right)^2 \right\} \tag{10.14}$$

Note that if the angular momentum of the mean flow increases with radius, then the Richardson number will be positive and C_2^* will decrease, causing the sink term to go to zero. Hence, ε will increase with a corresponding reduction in the turbulent kinetic energy and the turbulent viscosity. If the angular momentum decreases with radius and the derivative is negative, then ε will decrease. The constant C_2^* may be negative for high swirl, which is unrealistic.

An illustration of the predictive capability of the $k-\varepsilon$ model and the Richardson number is provided by the nonreacting experimental data taken by Vu and Gouldin (1982). The geometry consisted of a primary or inner flow passage with an inner diameter of 0.0372 m. The swirl generator upstream of the primary discharge was vaned swirler which produces a forced vortex. The swirl number was approximately 0.49 and the average axial velocity was 30.3 m/s. The secondary swirl generator was an adjustable vane swirler which produced a free or combined vortex at the discharge. The swirl number for the counterswirl case being considered was approximately −0.51 and the average axial velocity was 20.2 m/s. The diameter of the downstream dump region was the same diameter as that of the secondary duct. This case includes a combination of free *vs.* forced vortex and counterswirl.

In order to ensure that any discrepancies between the experimental data and the theoretical predictions are due to the inadequacies of the $k - \varepsilon$ model alone, computations were initiated at a distance of 0.004 m downstream of the discharge. At that point, sufficient measurements of the three velocity components, the component turbulent intensities, and the energy dissipation (qualitative only) were provided with which to formulate adequate inlet boundary conditions. Experimental data for the axial and tangential velocities for the counterswirl case, along with the theoretical predictions, are shown in Figures 10.4 and 10.5. The constant C_3 for the Richardson number function was chosen as 0.10 for the case shown.

The unmodified $k - \varepsilon$ model fails to predict the centerline recirculation zone. The theoretical predictions demonstrate that a recirculation zone is on the verge of forming much farther downstream than actually occurs. The $k - \varepsilon$ model modified by the Richardson number correction produces a recirculation

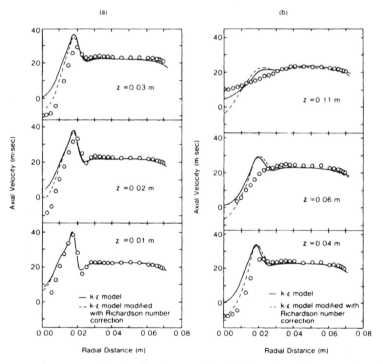

Figure 10.4. Comparison of predicted and measured axial velocity. (Data from Vu and Gouldin, 1982; prediction from Sloan, 1984). Conditions were those for a nonreacting coaxial, counterswirl jet. The swirl number of the inner jet was 0.49 and that of the annulus was −0.51.

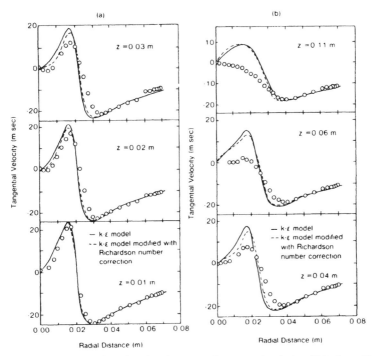

Figure 10.5. Comparison of predicted and measured tangential velocity. (Data from Vu and Gouldin, 1982; predictions from Sloan, 1984.) Conditions are those of Figure 10.4 and discussed further in text.

zone, but it also forms downstream of the experimental location. In addition, the radial profiles in the vicinity of the recirculation region do not damp out quickly enough. The discrepancies between the theoretical and experimental results have been attributed to the anisotropy of the flow near the recirculation region. In such a region, the $k - \varepsilon$ model underpredicts turbulent diffusion (i.e., radial transport of axial and angular momenta) because it associates a single length scale (assumption of isotropy) with each of the coordinate directions. In reality, the region of the recirculation zone is characterized by a large amount of turbulent kinetic energy generation due to the high local strain rates and gradients, which in turn is associated with a nonisotropic turbulence structure.

The Richardson number helps to alleviate a portion of the problem, but does not provide a complete solution to complex swirling flow fields.

10.6. TURBULENT STRESS CLOSURE MODELS

All of the models discussed so far rely on the Boussinesq hypothesis noted in Eqn. (10.3). We have already discussed the inadequacies of the Boussinesq hypothesis. A more fundamental approach to the closure problem is to eliminate the Boussinesq hypothesis and derive transport equations for $(u_i' u_j')$ directly. Such a transport equation is readily derived by taking the equation for the fluctuating velocity u_i', multiplying by u_j', adding to this the same equation with subscripts i and j reversed, and time averaging the entire result. Although 15 new unknown correlation terms arise, terms accounting for bouyancy, rotational, and other effects are introduced automatically. Such a closure scheme is often called second-order closure and a variety of such models has been proposed in the literature particularly in the last few years (Donaldson, 1969; Lumley and Khajeh-Nouri, 1974; Launder *et al.*, 1975; Gibson and Launder, 1978).

Models employing transport equations for the individual turbulent stresses and fluxes comprise a large number of differential equations whose solution is not a trivial task and is also rather costly. Hence, for practical applications, it would be desirable to use simpler models whenever possible. For this reason, so-called algebraic stress/flux models have been developed by simplifying the differential transport equations such that they reduce to algebraic expressions but retain most of their basic features. The simplest approximation is to neglect the convection and diffusion terms, but a more generally valid approximation was porposed by Rodi (1976), who assumed that the transport of $u_i' u_j'$ is proportional to the transport of k. With such approximations introduced, the transport equations yield algebraic expressions for $u_i' u_j'$ and $u_i' \phi'$ which contain the various production terms appearing in the $u_i' u_j'$ and $u_i' \phi'$ equations, respectively, and thus the gradients of the mean flow quantities. Also k and ε appear in the expressions, so the k and ε equations (10.18) and (10.10) must be added in order to complete the turbulence model. Algebraic expressions, together with the k and ε equations, form an extended $k - \varepsilon$ model.

Leslie (1980) has evaluated one of the main approximations required to develop the algebraic expressions by comparing measurements of the convection

TABLE 10.1. Comparison of Convection of Each of Four Components of the Reynolds Stresses as Measured and as Modeled in Algebraic Closure[a]

Component	Experimental term $\bar{u}(d/dx_1)(\overline{u_i' u_j'})$	Model term $(\overline{u_i' u_j'}/k)\bar{u}(dk/dx_1)$
$\overline{(u_1')^2}$	$1.96 \text{ cm}^2 \text{ s}^{-3}$	$2.09 \text{ cm}^2 \text{ s}^{-3}$
$\overline{(u_2')^2}$	0.88	0.83
$\overline{(u_3')^2}$	1.32	1.24
$-\overline{u_1' u_2'}$	0.62	0.62

[a]Data from Leslie (1980).

of each component of the Reynolds stress with that computed from the assumption in the algebraic model. Table 10.1 compares the measurements of $\bar{u}\,(d/dx_1)\,u_i'u_i'$ with the assumption used in the algebraic stress models, where this term is modeled by $(u_i'u_i'/k)\bar{u}\,(dk/dx_1)$. This term was computed from the measured data.

Algebraic stress models show promise for incorporating the effects of anisotropic turbulence with computational efficiency. They have not been tested as extensively as the two-equation $k - \epsilon$ model. The future will probably show more use of the algebraic stress models. Model evaluation of the next few years will be crucial in determining the extent of its long-range acceptance.

10.7. EFFECTS OF PARTICLE-LADEN JETS

So far in this chapter we have discussed turbulence in a homogeneous gas phase. The presence of the particulate or droplet phase will unquestionably affect the gas-phase turbulent fluid mechanics process. In addition, the homogeneous gaseous phase affects the motion of the particles or droplets themselves. This latter effect is discussed in some detail in Chapter 12. This section addresses only briefly the effect of the particulates on the homogeneous gas turbulence.

Some basic research is being done in this area. Elghobashi and Abou–Arab (1983) have incorporated the influence of the particles directly into the turbulence model and more research must be conducted in the future to understand the effects of the particle or droplet phases on the gas-phase turbulence.

Currently, fundamental models are not available for incorporating this effect on the turbulence. Melville and Bray (1979) have proposed a heuristic model for incorporating these effects. They note that the presence of particles in lightly loaded systems has only a slight effect on the gas-phase turbulence. They suggest a correlation using the ratio of the mean particle bulk density to the mean gas density as follows:

$$\nu'_{g(\text{particle})} = \nu'_{g(\text{no particle})}\left(1 + \frac{\overline{\rho_b}}{\overline{\rho_g}}\right)^{-0.5} \tag{10.15}$$

This equation shows that the amount of turbulence decreases as the bulk particle density (ρ_b) increases. This semiempirical correlation for correcting the gas-phase turbulence due to the presence of particles seems to be a reasonable approach for the present time. Unquestionably, this area needs additional work.

10.8.   BOUNDARY CONDITIONS

10.8a.   Boundary Conditions for k

To solve the transport equation for the turbulent kinetic energy k, appropriate boundary conditions must be specified for all boundaries around the computational domain. In this section, suggestions are made for each of these boundaries for k.

At inlets to the flow field, the turbulent kinetic energy of the incoming flow must be specified. This information is most preferably available from the problem specification (i.e., experimentally measured). It can be estimated from an approximate inlet turbulent intensity (I_{in})

$$I_{in} = \overline{(u_i' u_i')}_{in}^{1/2} / \bar{u}_{in} \tag{10.16}$$

Therefore

$$k_{in} = \tfrac{1}{2} I_{in}^2 \bar{u}_{in}^2 \tag{10.17}$$

At outlets from the flow field, the turbulent kinetic energy boundary condition should be some appropriate continuous outflow boundary condition. For example, one appropriate approximation is to extrapolate the upstream conditions either linearly or quadratically.

At walls to the flow domain the normal derivative of the turbulent energy is zero:

$$\partial k / \partial x_i = 0 \tag{10.18}$$

However, the turbulent production of kinetic energy (P) and the dissipation (ε) near the walls require modification. The suggested modification to P is demonstrated by remembering:

$$P = \tfrac{1}{2} P_{ii} = -\overline{u_i' u_k'} \left(\frac{\partial \bar{u}_i}{\partial x_k} \right)$$

$$= -\overline{u' u'} \left(\frac{\partial \bar{u}_i}{\partial x_i} \right) - \overline{u_i' u_k'} \left(\frac{\partial \bar{u}}{\partial x_k} \right) (1 - \delta_{ik}) \tag{10.19}$$

Near a wall this can be rewritten as

$$P = -\overline{u_i' u_i'} \left(\frac{\partial \bar{u}_i}{\partial x_i} \right) + \frac{\tau_W}{\rho_k} \left(\frac{\partial \bar{u}_i}{\partial x_k} \right) (1 - \delta_{ik}) \tag{10.20}$$

When the wall is parallel to the u_i direction, $\partial \bar{u}_k/\partial x_i$ is assumed to approach zero; for a wall prependicular to u_i, then $\partial \bar{u}_i/\partial x_k$ approaches zero.

For a specific example, near a wall in axisymmetric, polar-cylindrical coordinates,

$$P_W = -\overline{u_z' u_z'}\left(\frac{\partial \bar{u}_z}{\partial z}\right) - \overline{u_r' u_r'}\left(\frac{\partial (r\bar{u}_r)}{\partial r}\right) + \frac{\tau_W}{\rho}\left(\frac{\partial \bar{u}_z}{\partial r} + \frac{\partial \bar{u}_r}{\partial z}\right) \qquad (10.21)$$

and when the wall is parallel to the u_z direction, $\partial \bar{u}_r/\partial z = 0$; for a wall perpendicular to it, $\partial \bar{u}_z/\partial r = 0$.

The wall shear stress (τ_W) can be estimated from turbulent Couette flow near a wall. The modified log-law is used (Patankar and Spalding, 1970; Launder and Spalding, 1972):

$$u^+ = (1/\kappa) \ln (Ey^+) \qquad (10.22)$$

where

$$y^+ = y\rho(C_\mu^{1/2}k)^{1/2}/\mu \qquad (10.23)$$

and

$$u^+ = \bar{u}(\tau_W/\rho)^{-1/2} \qquad (10.24)$$

Thus, the wall shear stress can be calculated directly:

$$\tau_W = \frac{\bar{u}k(C_\mu^{1/2}k)^{1/2}\rho}{\ln[E\Delta r(C_\mu^{1/2}k)^{1/2}\rho/\mu]} \qquad (10.25)$$

This wall shear stress from turbulent Couette flow is suggested for substitution into Eqn. (10.21) for the production of turbulent energy near a wall (P_w).

Finally, for any flow symmetry boundary conditions, the derivative may be set to zero and no modification of P is needed.

10.8b. Boundary Conditions for ε

The dissipation rate of turbulent kinetic energy must also have boundary conditions specified on all computational domain boundaries.

At inlets to the flow field, the dissipation rate will not generally be known. It is suggested that ε be estimated from length-scale approximations:

$$l_m = C_\varepsilon k^{3/2}/\varepsilon \qquad (10.26)$$

The constant C_ε has been estimated by many authors and most typical values are $C_\mu^{0.75}$ and $8.3C_\mu$. Some have proposed even larger values for C_ε for swirling flows, as high as $100C_\mu$. A typical mixing length scale for fully developed inlet streams with no swirl might be

$$l_m = \tfrac{1}{4}D_{eq} \tag{10.27}$$

where D_{eq} is an equivalent pipe diamter. With swirl imparted by swirl vanes, l_m is more likely a typical distance between vanes.

Outlet conditions for ε can be obtained from a quadratic extrapolation of upstream conditions or some other continuous outflow boundary condition.

The dissipation rate at solid boundaries is harder to evaluate. In this case, it is suggested that the last node point in the flow field near a wall be set according to the following mixing length approximation:

$$l_m = \kappa \Delta y_i \tag{10.28}$$

where k is a mixing length constant and Δy, is the distance from the wall:

$$\varepsilon = C_\mu^{3/4} k^{3/2}/k\Delta y_i \tag{10.29}$$

At symmetry planes the normal derivative is set equal to zero.

10.9. ILLUSTRATIVE PROBLEMS

1. Time average the gas-phase, total continuity equation of Chapter 9 (Eqn. (9.13)). Do this first assuming constant density flow, then assume the fluctuating density case. Remember to substitute the mean and fluctuating component for all fluctuating variables (see Eqn. (10.2)). Simplify the equations by grouping all terms in the mean flow variables, dropping all terms which are zero, and group the terms with cross correlations of the fluctuating variables.

2. Repeat Problem 1 for the momentum equation for a gas–particle mixture as shown in Table 9.1.

3. Derive the exact form of the turbulent kinetic energy equation by following the procedure as discussed in the text immediately following Eqn. (10.8).

4. Discuss the physical significance of the terms in Eqn. (10.8) which are not exact but which are modeled. Include in your discussion a list of limitations which result from the approximations required in the modeling.

5. Derive the exact form of the ε equation.

6. Derive the exact form of the transport equation for $\overline{(u_i' u_j')}$ as discussed in Section 10.6.

*. . . a whirlwind came out of the north, a great
cloud, and a fire infolding itself, and brightness was
about it, . . .*

Ezekiel 1:4

CHEMISTRY AND TURBULENCE OF GASEOUS FUELS

11.1. INTRODUCTION

In the previous chapter we introduced some of the properties of turbulent flow
and briefly examined some of its characteristics. Specifically, we looked at
computational methods for predicting the mean velocity field with the aid of
turbulence closure schemes. In combusting flows the picture is greatly complicated by the interactions between the chemistry and the turbulence. The
chemical reactions occurring locally in these flow fields add many complexities
to the overall physicochemical process.

The chemical reactions take place on the molecular level. The turbulent
motion plays an important role in mixing of the two reacting species on a
microscopic scale. The local turbulence controls the time that each of the reactants
and products are associated, allowing reaction to proceed. The local instantaneous reaction process itself, often associated with local heat release or
absorption, density changes, volume changes, etc., can impact the local turbulent
fluid mechanics. Although this impact has been recognized for some time, most
applications models for turbulent combustion do not include the effect of the
chemistry on the turbulence closure scheme. The recommended $k - \varepsilon$ turbulence
model of the last chapter has been used in many turbulent combustion calculations. Although many difficulties still exist with the $k - \varepsilon$ model, as indicated
by the last chapter, its application to turbulent gaseous combustion flames
appears adequate for many situations.

In this chapter we examine the effect of the turbulent fluid mechanics on
the chemical reaction rates. Although this book is dedicated to combustion and

gasification of coal, char, and coal slurries, this chapter deals specifically with gaseous combustion. A great deal has been written about gas-phase combustion (Libby and Williams, 1981). This chapter reviews only pertinent information necessary to understanding the coal-laden flames. In coal flames, a great deal of the reaction chemistry is occurring in the gas phase after devolatilization of the coal particles. In addition, the heterogeneous reactions of the gas-phase oxidizer with char result in gaseous products that sometimes react further in the turbulent homogeneous phase. In this chapter we will discuss turbulent, gas-phase diffusion flames. The theory developed in this chapter is an important part of the overall theory for coal combustion or gasification processes developed in subsequent chapters.

11.2. MEAN REACTION RATE

Among the most important questions in current combustion research are those regarding the role of interactions between the turbulent fluid mechanics and the chemical reactions (Smoot and Hill, 1983). The apparent random fluctuations in the turbulent mixing of reactants and products dramatically influence mean chemical kinetic rates that have reaction time scales on the order of the turbulent time scale or less. Essentially all homogeneous gas combustion lies in this critical area, including the combustion of methane and higher hydrocarbon constituents of natural gas.

Since essentially all practical applications of natural gas combustion occur in highly turbulent systems, the interactions between the chemical reaction process and the turbulent fluid mechanics are crucial to the overall process. One of the largest problems in understanding these interactions lies in the characterization of the time-mean net rate of formation or destruction of the molecular species from the chemical reactions (Pratt, 1979; Bilger, 1980). Although the kinetic mechanisms are not always known, and kinetic constants difficult to identify, the major problem lies not in these areas but in obtaining the proper time-mean rate due to the presence of the turbulence.

To illustrate the impact of the turbulence on the time-mean reaction rate, consider a simple example of an elementary reaction between reacting species A and B. In this example, assume the elementary reaction to be as follows:

$$A + B \xrightarrow{k_1} C \tag{11.1}$$

For this reaction, the activation energy is E_1 and the pre-exponential rate constant is k_0. Thus the overall chemical kinetic reaction rate is

$$k_1 = k_0 \exp(-E_1/RT) \tag{11.2}$$

To calculate the mass fraction of any of the participating species in Eqn. (11.1), a conservation equation must be solved for the species of interest. This expression for laminar flows is given in Eqn. (11.3):

$$(\partial/\partial t)(\omega_i\rho) + \nabla \cdot (\omega_i\rho\mathbf{v}) = r_i \tag{11.3}$$

This equation also holds true for turbulent flows at any instantaneous point in time. Thus, to obtain the composition field for species A in our example, we must solve a conservation equation for this species,

$$(\partial/\partial t)(\omega_A\rho) + \nabla \cdot (\omega_A\rho\mathbf{v}) = r_A \tag{11.4}$$

where the reaction rate is

$$r_A = k_1\omega_A\omega_B\rho^2 \tag{11.5}$$

After substituting the Arrehenius rate expression of Eqn. (11.2) for the chemical kinetic rate constant of Eqn. (11.5) we obtain:

$$r_A = \omega_A\omega_B\rho^2 k_0 \exp(-E_1/RT) \tag{11.6}$$

Again, it is emphasized that these expressions are valid for laminar flows or for the instantaneous turbulent flows. In order to obtain the time-mean properties as discussed in the last chapter, it is necessary to decompose each of the fluctuating variables into their mean and fluctuating components. Then time-averaging of the entire expression (Eqn. (11.4)) is required. The impact of time-averaging the left-hand side of Eqn. (11.4) will be discussed later in this chapter. We focus now on simply obtaining the time-mean reaction rate on the right-hand side of Eqn. (11.4). It is emphasized that the time-mean rate expression is not simply the reaction rate calculated from the time-mean variables:

$$\bar{r}_A = \overline{\omega_A\omega_B\rho^2 k_0 \exp(-E_1/RT)}$$

$$\neq \bar{\omega}_A\bar{\omega}_B\bar{\rho}^2 k_0 \exp(-E_1/R\bar{T}) \tag{11.7}$$

Instead, after Reynolds decomposition and appropriate simplification we obtain:

$$\bar{r}_A = \bar{\rho}^2\bar{\omega}_A\bar{\omega}_B k_0 \exp\left(\frac{E_1}{R\bar{T}}\right)\left[1 + \frac{\overline{(\rho')^2}}{\bar{\rho}^2} + \frac{\overline{\omega'_A\omega'_B}}{\bar{\omega}_A\bar{\omega}_B} + \frac{2\overline{\rho'\omega'_A}}{\bar{\rho}\bar{\omega}_A} + \frac{2\overline{\rho'\omega'_B}}{\bar{\rho}\bar{\omega}_B} + \frac{E}{R\bar{T}}\left(\frac{\overline{\omega'_A T'}}{\bar{\omega}_A\bar{T}}\right)\right.$$

$$\left. + \left(\frac{E_1}{2R\bar{T}} - 1\right)\frac{\overline{(T')^2}}{\bar{T}^2} + \cdots\right] \tag{11.8}$$

The other terms, which are not shown, involve triple correlations of the fluctuating variables. This rate expression is for the simple example given by Eqn. (11.1). For any realistic combustion scheme, scores of these equations must necessarily appear. It is clear that closure of this turbulent reacting system is a formidable problem. This problem was recognized early and different alternate paths have been followed. An excellent review of the state of the art of different combustion models for turbulent diffusion flames has been given by Bilger (1976). It has been only recently that any attempt has been made to incorporate full kinetics in a turbulent environment because of this problem. In this chapter, we will focus on one of the most frequently applied techniques for turbulent gaseous diffusion flames.

The impact of these chemistry–turbulence interactions is quite different for different types of chemical reactions. To help identify the magnitude of this impact, it is convenient to identify two hypothetical time scales. First, the reaction time scale t_r is defined as a typical time for the reacting species of interest to react completely to its equilibrium value. The turbulence time scale t_t is chosen to be a typical fine-scale mixing time for scale reduction by turbulent breakup of large eddies. This time scale must be adequate for molecular interactions to take place (micromixing). The turbulence time scale then is the time required for mixing to proceed to the molecular level before reaction can occur. The reaction time scale is the time required for these reacting species, once intimately contacted, to react completely to their products. Approaches for incorporating chemical reactions in turbulent systems can be characterized by examining the relationship between these two time scales. This discussion is limited to turbulent diffusion flames. The fuel and air enter in separate streams. They must mix both macroscopically and microscopically before they can react. A majority of practical turbulent gaseous flames are turbulent diffusion flames.

11.2a.   Type A: Reaction Time Scale ≫ Turbulent Time Scale

If the reaction time scale t_r is much greater than the turbulent time scale t_t, as defined above, then the reactions are very slow compared to changes in the local turbulence. In this case, if the fluctuations of any variable (i.e., temperature) are relatively small, the chemistry is unaware of the presence of the turbulent fluid mechanics, and the effect of the fluctuations on the reaction rate can be ignored. This results in a time-mean reaction rate for our hypothetical reaction of Eqn. (11.1) of

$$\bar{r}_A = \bar{\omega}_A \bar{\omega}_B \bar{\rho}^2 k_0 \exp(-E_1/R\bar{T}). \tag{11.9}$$

As mentioned in the last section, the mean reaction rate is highly sensitive to temperature fluctuations. Thus, even if reaction rates were slow but significant temperature fluctuations existed, then the mean reaction rate would be different from that of Eqn. (11.9). However, in this Type A flame, t_l is much smaller than t_r. Thus, reactants mixing relatively quickly then proceed to react relatively slowly. Since temperature fluctuations are usually caused by a wide variety of eddies of different extents of reaction and since Type A flames preclude this occurrence, then we realize that chemical property fluctuations will be relatively small in Type A flames.

Thus, only in this very limited case is the mean reaction rate equal to the reaction rate calculated from the mean variables. Although this approximation has been used for many years, it has been shown to be valid in only a very limited number of cases. Some heterogeneous reactions are slow enough to be of this type. These heterogeneous reactions will be discussed further in Chapter 12. However, essentially all homogeneous reactions of interest to coal combustion or gas combustion are too rapid to fit this approximation. Using this approximation when reaction rates are not sufficiently slow, compared to the local turbulence, produces appreciable error (Pratt, 1979).

11.2b. Type B: Reaction Time Scale ≪ Turbulent Time Scale

The past decade has provided particularly valuable insight into the interactions between turbulence and chemistry when the reaction rates are rapid compared to the time scale for the turbulent micromixing process (Libby and Williams, 1980). In this case the chemistry occurs very rapidly once the molecules are mixed together. In this type of diffusion flame, the assumption can be made that the micromixing process is rate limiting and not the chemical kinetic process. Thus, as far as the overall flame is concerned, the chemistry is fast enough to be considered in local instantaneous equilibrium. In this type of flame, the rate-controlling process is the turbulent mixing process. Once the reactants are mixed by the turbulent fluid mechanics, the reactions proceed to equilibrium instantaneously. For these cases, the conventional conserved scalar or mixture fraction is defined to identify the degree of "mixedness" at a point. This conserved scalar is thus a measure of the local instantaneous equivalence ratio and a unique function exists between the instantaneous conserved scalar and the instantaneous chemical properties (i.e., composition, temperature, and density). The majority of the remainder of this chapter will be dedicated to a description of this Type B flame.

Briefly, the statistics of the local turbulence fluctuations ars characterized by the statistics of the conserved scalar, most easily incorporated by the probability density function (PDF) of the mixture fraction. Convolution of the chemical properties over the PDF with appropriate incorporation of the intermittency of pure fuel or air streams allows for proper accounting of the effects of the turbulence on the chemistry. The details of this approach follow. This type of flame has been well characterized over the last decade and the required assumptions work well for flames where reaction rates are fast so that equilibrium chemistry can be used. This method is sufficient even when equilibrium cannot be assumed, but the reaction rates are sufficiently fast for the composition field to be obtained from experimental data as a function of equivalence ratio and used in the computatonal procedure. This will also be discussed later in this chapter.

11.2c.   Type C: Reaction Time Scale ≃ Turbulent Time Scale

When the reaction time scale is of the same order of magnitude as the turbulent micromixing time scale, then both the chemical kinetics and the turbulent fluctuations must be integrated. It has been shown that even CH_4 combustion has some significant species that are of this type (Smith and Smoot, 1981). It is this area that has been identified in the combustion field as being among those in the most need of specific research advances (Smoot and Hill, 1983).

In 1978, Bilger reported a significant experimental observation. Over a broad region of a given nonequilibrium, laminar, hydrocarbon diffusion flame, the molecular species composition was only a function of the equivalence ratio, the conserved scalar or mixture fraction being the measure of the equivalence ratio. Fenimore and Fraenkel (1981) tested this hypothesis for thermal NO concentrations in laminar gaseous diffusion flames even though the chemistry is far from equilibrium. Their experiments confirmed this observation for thermal NO. Both authors have shown that these observations lead to the conclusion that the species reaction rate seems to be only a function of the mixture fraction in these flames. Apart from their significance for laminar flames, the observations hold great significance for the interactions between kinetically limited chemistry and the turbulent flow-field. If the experimental data base could be extended to turbulent flames with similar relationships between local instantaneous chemical composition and the local instantaneous conserved scalar (such a mixture fraction), then current statistical techniques for calculating mean composition could be readily extended to nonequilibrium combustion. This method has been applied to NO formation in coal flames even though the functional relationship between the mixture fraction and the composition remains unproven for turbulent combustors. This approach is briefly discussed in Chapter 15. Thus, the computational tools are available. The details of this approach are not discussed in this chapter.

Diagnostic equipment for reliably resolving the local instantaneous composition in practical turbulent combustors has only recently become applicable to practical gas flames (see Chapter 15). Such measurements would be beneficial in understanding chemistry and turbulence interactions independent of specific theoretical models. Well-constructed, detailed, pertinent data and the resulting analysis hold potential for significant contributions toward the incorporation of chemical kinetics in the computation of the turbulent flames.

11.3. MIXTURE FRACTION

In the gas-phase diffusion flames under consideration, the fuel and oxidizer are initially separated in different streams. Such diffusion flames were discussed in Chapter 5. Figure 11.1 shows schematic diagrams of both simple and complex

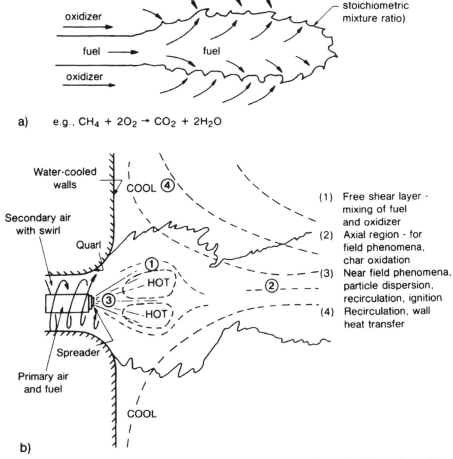

Figure 11.1. Schematic of simple and complex diffusion flames: (a) simple diffusion flame; (b) complex diffusion flame (essential features from Wendt, 1980).

diffusion flames. Whether the complexities of swirling flows are present or not, it is clear from this figure that the reactants enter the reaction chamber at the burner in separate streams, thus they exist in separate individual eddies which must be intimately contacted on a molecular level before reaction can occur. The concept of a conserved scalar or mixture fraction has already been introdcued. This variable f is defined as

$$f = \dot{m}_p/(\dot{m}_p + \dot{m}_s) = \text{mass fraction of fluid atoms} \qquad (11.10)$$
$$\text{originating in primary stream}$$

where the subscripts p and s refer to the primary and secondary streams, respectively. The primary stream is usually the fuel-containing stream; the secondary stream is the oxidizer. This is most easily understood with reference to Figure 11.1. It is clear from this definition that the mixture fraction is a measure of the local equivalence ratio. We can think of this variable as a measure of the degree of "mixedness" or "unmixedness."

In Type B flames, the rate-limiting process is the turbulent micromixing and not the chemical kinetics. It is apparent that the mixture fraction is the local measure of the extent of this micromixing process. From this variable f, any conserved scalar s with an equal diffusivity can be calculated from the local value of f:

$$s = fs_p + (1 - f)s_s \qquad (11.11)$$

where s_p is the value of the scalar in the pure primary stream and s_s is the value of that variable in the pure secondary stream.

11.4. PROBABLILITY DENSITY FUNCTION (PDF)

Although we have defined the instantaneous mixture fraction, it is recognized that in a turbulent environment this variable is fluctuating randomly about some mean value at every point in the reaction chamber. To fully describe the mixing process, more information is needed regarding the statistics of these fluctuations. The probability density function for the nondimensional mixture fraction is defined by noting:

$$P(f)df = \text{fraction of time interval } t \text{ during which} \qquad (11.12)$$
$$f(t) \text{ is in the range } (f, f + df)$$

where $P(f)$ is the probability density function of the instantaneous mixture fraction. Consideration of this definition of the PDF reveals the following two important properties:

$$\int_0^1 P(f)\,df = 1 \tag{11.13}$$

$$\bar{f} = \int_0^1 fP(f)\,df \tag{11.14}$$

The PDF is the statistical distribution of the mixture fraction at any local point in space due to the fluctuations of f in time. The first moment of the distribution $P(f)$ is simply the time-mean mixture fraction $\bar{f}$. The second moment about the mean, the variance, or the mean-square fluctuation of the mixture fraction, is defined conventionally by

$$g_t = \int_0^1 (f - \bar{f})^2 P(f)\,df \tag{11.15}$$

where it is to be remembered that f is a function of time t.

The time-mean mixture fraction $\bar{f}$ has been measured, along with its statistical PDF. These measurements have been made in cold flow as well as in gaseous combustion flames. One such example of this is shown in Figure 11.2. The conventional or Reynolds-averaged PDF is that defined by Eqn. (10.1). The Favre PDF will be discussed in Section 11.6. It can be seen from these experimentally measured distributions that fluctuations generally follow a Gaussian distribution except where intermittency is important. This intermittency is caused by eddies of unmixed gases passing through the sample volume. Thus the spikes occur in Figure 11.2 at the mixture fractions of 0 corresponding to pure secondary stream. If there was intermittency of pure primary, then spikes would occur at a mixture fraction of 1. In this context, intermittency is defined as the fraction of time the local fluid is pure primary (α_p) or pure secondary (α_s).

To further understand the implications of the mixture fraction and its associated PDF, we will consider an idealized turbulent flow field. In this flow field, a point is chosen and measurements of the local instantaneous mixture fraction are made. At this point, the mixture fraction is fluctuating about its time-mean value $\bar{f}$ in an idealized saw-tooth waveform. This is illustrated in Figure 11.3, case A. The plot on the right-hand side shows the mixture fraction as a function of time. In this idealized flow the mixture fraction spends exactly equal amounts of time at each of the mixture fractions between the maximum (f_{max}) and the minimum (f_{min}) values. The statistics of this distribution are shown on the left-hand side of case A. The probability density function is a top-hat distribution centered on $\bar{f}$ with maximum and minimum values of f_{max} and

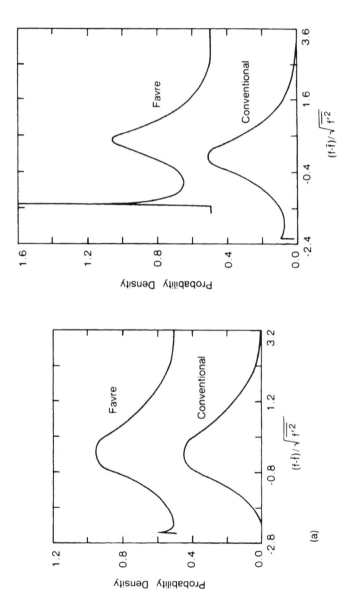

Figure 11.2. Measured PDFs in H_2–air diffusion flame. (a) conventional and Favre probability density functions of mixture fraction on flame axis at $X/D = 34$ (Favre case shifted by 0.5 on ordinate scale); (b) conventional and Favre probability density functions of mixture fraction on flame axis at $X/D = 93$ (Favre case shifted by 0.5 on ordinate scale). H_2 fuel issued horizontally from a 7.6 mm diam nozzle at 151 m/s into a coflowing air stream of 15 m/s. (Figure used with permission from Kennedy and Kent, 1980.)

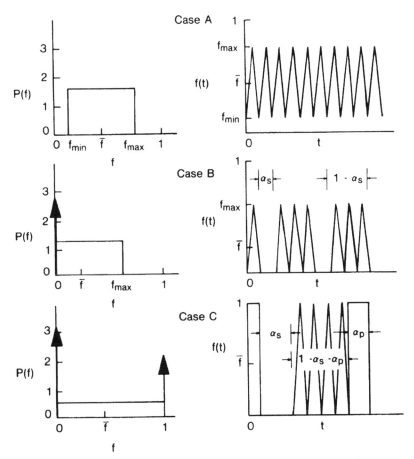

Figure 11.3. Shape of the probability density function and typical corresponding fluctuations for (A) no intermittency, (B) intermittency of secondary, and (C) intermittency of primary and secondary. (Figure used with permission from Smoot and Pratt, 1979.)

f_{min} respectively. Such would be the distribution for this idealized flow field somewhere downstream from the burner inlet such that no eddies of pure primary or secondary could reach the sample point of interest. Case B of Figure 11.3 depicts the case for this idealized flow where the sample point is located somewhere within the combustion chamber near the inlet of the secondary jet. Again, the idealized saw-tooth waveforms are depicted in the figure on the right-hand side. In this case the minimum value for the mixture fraction is at $f = 0$. A significant number of eddies reached the sample point containing pure secondary fluid atoms. Thus a fraction of time (α_s) is spent at exactly $f = 0$. The associated probability density function is shown on the left-hand side of

TABLE 11.1. PDF Parameters for Top-Hat Distribution

Case[a]	A	(a) For $\bar{f} < 0.5$ B	C
$g_{I,min}$	0	$(1/3)\bar{f}^2$	$(2/3)(\bar{f} - \bar{f}^2)$
$g_{I,max}$	$(1/3)\bar{f}^2$	$(2/3)(\bar{f} - \bar{f}^2)$	$\bar{f}(1 - \bar{f})$
f_{min}	$\bar{f} - (3g_I)^{1/2}$	0	0
f_{max}	$\bar{f} + (3g_I)^{1/2}$	$(3/2)(g_I + \bar{f}^2)/\bar{f}$	1
α_p	0	0	$2\bar{f} - 1 + \alpha_s$
α_s	0	$[g_I - (1/3)\bar{f}^2]/(g_I + \bar{f}^2)$	$3g_I - (1 - \bar{f})(3\bar{f} - 1)$
$P(f)$	$1/2(3g_I)^{1/2}$	$(2/3)(1 - \alpha_s)\bar{f}/(g_I + \bar{f}^2)$	$1 - \alpha_s - \alpha_p$

(b) For $\bar{f} > 0.5$

1. Replace $\bar{f}$ with $1 - \bar{f}$.
2. Switch subscrips p and s.
3. Switch subscripts min and max.

[a] As defined in Figure 11.3.

case B. The fraction of time spent as pure secondary stream is depicted by the δ function at $f = 0$. Case C depicts the situation at a location in the flow field where eddies of both pure primary and pure secondary can be measured. Now a fraction of the time is spent at $f = 1$ in pure primary fluid atoms. In this final case α_s is the fraction of time that the fluid elements at the measuring point are pure secondary fluid, α_p is the fraction of time that the fluid elements are measured to be pure primary fluid, and $1 - \alpha_s - \alpha_p$ is the fraction of time that the fluid elements are measured between 0 and 1.

TABLE 11.2. Parameters for Clipped Gaussian Probability Density Function

$$P(f) = (2\pi G_I)^{-1/2} \exp(-Z_I^2/2)$$

$$\alpha = (2\pi)^{-1/2} \int_L^U \exp(-Z_I^2/2)\, dZ_I$$

Intermittency	U	L	Z_I
α_p	∞	$(1 - F)/F_I^{1/2}$	$(f - F)/G_I^{1/2}$
α_s	$-F/G_I^{1/2}$	$-\infty$	$(f - F)/G_I^{1/2}$

where F and G_I come from

$$\bar{f} = \alpha_p + (2\pi G_I)^{-1/2} \int_{0+}^{1-} f \exp\left(\frac{-(f - F)^2}{2G_I}\right) df$$

$$g_I = \alpha_p - \bar{f}^2 + (2\pi G_I)^{-1/2} \int_{0+}^{1-} f^2 \exp\left(\frac{-(f - F)^2}{2G_I}\right) df$$

To describe the statistics of this mixing process mathematically, we only need the time mean mixture fraction $\bar{f}$, the variance about this mean (g_f), and the shape of the probability density function. In most applications the shape of the PDF is assumed and the other two variables $\bar{f}$ and g_f are calculated in the mathematical description. Assumed shapes of the PDF have included top hat, clipped Gaussian, β functions, multiple δ functions, as well as others. With the shape of the PDF assumed and the time mean mixture fraction $\bar{f}$ and its variance g_f known, the statistics of the mixing are completely described. The required parameters for the top-hat distribution are summarized in Table 11.1 and those for a clipped Gaussian probability density function with intermittency are described in Table 11.2. Some prediction techniques include mathematical formalism for predicting the shape of the PDF directly (O'Brien, 1980).

11.5. FAVRE AVERAGING

The need for time or Reynolds averaging has already been introduced and the problems associated with it examined. Specifically, the problem arose in variable-density flows, and terms such as $\bar{u}\overline{\rho'v'}$ and $\bar{f}\overline{\rho'v'}$ are neglected. Some measurements cited by Bilger (1976) indicate that these terms can be of the same order and sometimes greater than the momentum and mixture fraction fluxes $\overline{\rho u'v'}$ and $\overline{\rho v'f'}$. Bilger (1975) and 1976) points out how Favre averaging eliminates this problem.

In Favre averaging, quantities are weighted by the instantaneous density before averaging:

$$\tilde{\phi} \equiv \overline{\rho\phi}/\bar{\rho} \tag{11.16}$$

The tilde (˜) represents the Favre-averaged variable. This approach eliminates double correlations involving density fluctuations from the turbulent fluxes. The resulting partial differential equations are identical in form to the uniform density flow equations, except Favre-averaged variables replace the conventional Reynolds-averaged ones. The density remaining in the equations is the time-mean density. As an example, Eqn. (11.17) shows the Favre-averaged form of the species continuity equation previously given in its instantaneous form in Eqn. (11.3):

$$(\partial/\partial t)(\bar{\rho}\tilde{\omega}_i) + \nabla \cdot (\bar{\rho}\tilde{\omega}_i\tilde{\mathbf{v}}) = \bar{r}_i \tag{11.17}$$

It becomes immediately obvious that when the equations are written in conventional Reynolds average form and the fluctuating density terms are neglected, effectively, the Favre-averaged equations are being used. The same modeling terms may be used for the Favre-averaged equations as was introduced previously for the Reynolds-averaged turbulent model (Bilger, 1975; Bilger, 1976). A distorted form of the equation of state will result unless proper account is made for Favre averaging.

As already discussed throughout this chapter, the solution to all turbulent, variable-density, differential equations requires a properly averaged density. As can be seen from Eqn. (11.17), the conventional mean $\bar{\rho}$ is required; however, if a straightforward approach as outlined in Section 11.7 is followed, the Favre-averaging mixture fraction $\tilde{f}$ is predicted along with the variance $\tilde{g}_f$ for the Favre-averaged PDF. By using the local instantaneous equilibrium approximation and convoluting over the PDF, the Favre-averaged density would result $(\bar{\rho})$. The difficulties disappear when it is realized that

$$1/\bar{\rho} = (\widetilde{1/\rho}) \tag{11.18}$$

In other words, the Favre average of the inverse density equals the inverse of the Reynolds average of the density. This is easily proven by direct substitution of the inverse density into Eqn. (11.16). The calculation of $\bar{\rho}$ is discussed in the next section. The Favre probability density function of the mixture fraction f is defined as

$$\tilde{P}(f) = (1/\bar{\rho}) \int_0^\infty \rho P(\rho, f)\, d\rho \tag{11.19}$$

Bigler (1976) points out that the Favre PDF probability more closely approximates the shape of measured constant density time–averaged PDFs than does the corresponding time-mean PDF in fluctuating density flows. Figure 11.2 showed measured Favre PDFs for the mixture fraction f at two locations on an H_2–air flame axis.

To help understand the physical differences between conventional Reynolds averaging and Favre averaging, it is useful to consider what is measured by sample probes in combustion systems. If instantaneous optical measurements are being made, then either mean can be constructed. In coal flames most measurements are still made with extractive probes. In such probes Bilger (1977) shows how the Favre-averaged mole fractions will be measured if the probe is sampling at a constant velocity u, then

$$y_{i,\,\text{sampled}} = \overline{\rho u y_i}/\overline{\rho u} = \overline{\rho y_i}/\bar{\rho} \equiv \tilde{y}_i \tag{11.20}$$

On the other hand, it can also be seen from this equation, that if the probe samples at a constant mass flow rate ρu, then the Reynolds mean is measured. Anything in between would be a mixture of the two. Bilger (1977) also discusses temperature measurements with thermocouples and indicates that they probably measure the conventional mean value. Whichever mean value is measured, it should be pointed out in favor of Favre averaging that with information about the Favre-average PDF (which is only now becoming experimentally available) either the Reynolds or the Favre mean can be obtained.

11.6. MIXING

The definition and ramifications of the mixture fraction (a conserved scalar) have now been introduced. We have seen how this variable is an important measure of the degree of mixedness or unmixedness in a turbulent system. We have seen how its probability density function is a measure of the statistics of the turbulent mixing process. We have indicated that for Type B flames (see Section 11.2) it is this mixing process that is the rate-limiting step in the overall combustion process. In this section, we now examine how to calculate this progress variable, and how to use it to determine the chemical properties of the flow field.

First, the Favre-mean mixture fraction $\tilde{f}$ can be calculated throughout the combustion flow field by deriving and solving a differential transport equation for this scalar variable. Since the mixture fraction is a conserved scalar, its transport equation is simply a continuity equation with appropriate Favre averaging. Derivation of this equation is quite straightforward (see illustrative problems). The resulting mixing equation is shown here in its Favre-averaged form:

$$(\partial/\partial t)(\bar{\rho}\tilde{f}) + \nabla \cdot (\bar{\rho}\tilde{f}\tilde{\mathbf{v}} - D_{tf}\nabla\tilde{f}) = 0 \tag{11.21}$$

The first term is simply the accumulation of primary stream fluid, the second term represents the convection of the primary fluid elements by the mean velocity vector, and the last term represents the turbulent diffusive transport of the primary fluid. The turbulent diffusivity for the mixture fraction D_{tf} can be estimated from the eddy diffusivity of the fluid as discussed in the last chapter and an appropriate turbulent Schmidt number for the mixture fraction:

$$D_{tf} = \mu_r/\delta_f \tag{11.22}$$

The solution of this equation, together with the fluid mechanics model, will prescribe the mean field values for the flow and the mixing, provided the appropriate time-mean density is available. The advantage of knowing $\tilde{f}$ is that with the aid of PDF for f, the mean values may be found for any arbitrary variable β which is only a function of f. Thus, for any variable which is a unique function of the mixture fraction (i.e., depends only on the mixing) the time-mean value of this variable is simply obtained by convolution over the PDF:

$$\bar{\beta} = \int_0^1 \beta(f)P(f)\,df \tag{11.23}$$

Of course, Eqn. (11.23) includes all of the fluctuations of the mixture fraction including any intermittency of primary or secondary streams at the limits of integration. To emphasize the impact of the intermittency, and to simplify the mathematical integration of Eqn. (11.23), it is convenient to factor out the delta functions at the end points, when present:

$$\bar{\beta} = \alpha_p\beta_p + \alpha_s\beta_s + \int_{0+}^{1-} \beta(f)P(f)\,df \tag{11.24}$$

where $0+$ and $1-$ represent integration over the continuous portion of the PDF without the spike functions at the end points.

In general we would like to find Favre-averaged properties as discussed in the last section. By using the Favre probability density function, this is a straightforward process,

$$\tilde{\beta} = \tilde{\alpha}_p\beta_p + \tilde{\alpha}_s\beta_s + \int_{0+}^{1-} \beta(f)\tilde{P}(f)\,df \tag{11.25}$$

where $\tilde{\alpha}_p$, $\tilde{\alpha}_s$, and $\tilde{P}(f)$ represent the Favre intermittence and Favre PDF. With this observation we can now obtain the Favre-average density, and the time-mean density as well:

$$\frac{1}{\bar{\rho}} = \frac{\tilde{\alpha}_p}{\rho_p} + \frac{\tilde{\alpha}_s}{\rho_s} + \int_{0+}^{1-} \frac{1}{\rho(f)}\tilde{P}(f)\,df. \tag{11.26}$$

It is this Reynolds-averaged density that is required for all of the Favre-averaged equations (e.g., see Eqn. (11.21)). Of course, Eqn. (11.26) requires that the density be a unique function of the mixture fraction which is true for Type B flames as we will discuss further in the next section. With this time-mean density and the Favre statistics, it is also possible to obtain the Reynolds-averaged properties by virtue of the definition of the Favre probability density function, Eqn. (11.19):

$$\bar{\beta} = \bar{\rho}\left(\frac{\tilde{\alpha}_p\beta_p}{\rho_p} + \frac{\tilde{\alpha}_s\beta_s}{\rho_s} + \int_{0+}^{1-} \frac{\beta(f)}{\rho(f)}\tilde{P}(f)\,df\right) \tag{11.27}$$

This then completes the link between the Favre probability density function and the Reynolds PDF.

If β is a function of any number of conserved scalars [i.e., $\beta(s_1, s_2, s_3, \ldots)$], it remains only a function of the mixture fraction f. This is made possible since, as shown in Eqn. (11.11), the instantaneous value of the scalar s is only a function of the local instantaneous value of f. These conservative scalars $(s_1, s_2, s_3, \ldots)$ could be the local enthalpy, temperature, element mass fraction, etc. There are few requirements on s which must be met to make this strictly true. This approach relies on the assumption that all of the conserved scalars have equal turbulent diffusivities. In other words, the turbulent Schmidt number, Prandtl number, or exchange coefficient for local enthalpy, temperature, element mass fraction, etc. are equal. This is usually a safe assumption for turbulent flow and is commonly made. Finally, for s to qualify as a conserved scalar, it must have a transport equation of the same form as the mixture fraction, Eqn. (11.21). That is to say, there can be no source terms present, and it must have the same boundary conditions. This type of requirement is often called Crocco similarity. For example, the local enthalpy would not meet Crocco similarity if a convective or conductive wall heat flux existed (i.e., the reactor was not adiabatic). Non-adiabatic conditions will be discussed further in the next section.

Computationally, the distribution of the mixture fraction at a point is not yet completely defined. The shape of the distribution has been arbitrarily chosen and computational methods were given in Tables 11.1 and 11.2. The mean or first moment about the origin has been calculated from Eqn. (11.21), but to complete the description, the variance, second moment about the mean, or the mean-square fluctuation of the mixture fraction must be identified. The variance from the Reynolds-mean mixture fraction was given in Eqn. (11.15). Accordingly, the variance from the Favre-mean mixture fraction $(\tilde{f})$ is simply,

$$\tilde{g}_f = \frac{\overline{(\rho f - \rho \tilde{f})^2}}{\bar{\rho}}$$

$$= \frac{1}{T\bar{\rho}} \int_0^T \left(\rho f - \overline{\rho f} \right)^2 dt$$

(11.28)

where the tilde in $\tilde{g}_f$ reminds us that this is the variance from the Favre mean. Again, T is a time which is large as compared to the time scale of the local turbulence.

In order to calculate a value for $\tilde{g}_f$, an appropriate transport equation must be derived that is consistent with the closure methods of our chosen turbulence model. Launder and Spalding (1972) show how a transport equation for g_f can be derived and appropriate terms modeled in a manner analogous to, and consistent with, the other two equations in the $k - \varepsilon$ turbulence model. The resulting equation for steady-state, axisymmetric elliptic flows is

$$\frac{\partial}{\partial x}(\bar{\rho}\tilde{u}\tilde{g}_I) + \frac{1}{r}\frac{\partial}{\partial r}(r\bar{\rho}\tilde{v}\tilde{g}_I) - \frac{\partial}{\partial x}\left(\frac{D_{tg}\partial\tilde{g}_I}{\partial x}\right) - \frac{1}{r}\frac{\partial}{\partial r}\left(\frac{rD_{tg}\partial\tilde{g}_I}{\partial r}\right)$$

axial convection   radial convection   axial turbulent diffusion   radial turbulent diffusion
of I fluctuations   of I fluctuations   of I fluctuations   of I fluctuations

$$= C_{g1}\mu_e\left[\left(\frac{\partial\tilde{f}}{\partial x}\right)^2 + \left(\frac{\partial\tilde{f}}{\partial r}\right)^2\right] - \frac{C_{g2}\bar{\rho}\tilde{\epsilon}\tilde{g}_I}{k^2} \tag{11.29}$$

generation of I fluctuations   dissipation of
I fluctuations

Again the turbulent diffusivity (D_{tg}) for $\tilde{g}_I$ is calculated from the turbulent eddy diffusivity and an appropriate Schmidt number:

$$D_{tg} = \mu_e/\sigma_g \tag{11.30}$$

The two model constants appearing in Eqn. (11.29) $(C_{g1}$ and $C_{g2})$ represent additional "universal" constants. These constants with the associated appropriate turbulent Schmidt numbers for this mixing model are shown in Table 11.3.

With the $\tilde{g}_I$ equation solved simultaneously with all the others, the fluid mechanics and mixing models are complete. There still remains the issue of using the appropriate Reynolds-averaged density which will be discussed in the next section.

11.7. CHEMICAL EQUILIBRIUM

This section describes a technique for calculating the local Favre- or Reynolds,-averaged chemical properties (i.e., species mole fractions, density, temperature, etc.) for turbulent gaseous combustion of Type B flames. As has been discussed previously, for adiabatic operation of a gaseous flame, the standard enthalpy is a conserved scalar and thus with the assumption of equal diffusivities the instantaneous local enthalpy h may be calculated directly from f:

$$h = fh_p + (1-f)h_s \tag{11.31}$$

In addition, atom numbers of each element $(b_k$, defined as kilogram-atoms of element k per kilogram of mixture) are also conserved scalars. Again, with the equal turbulent diffusivity assumption, its instantaneous local value is

$$b_k = fb_{k_p} + (1-f)b_{k_s} \tag{11.32}$$

TABLE 11.3. Turbulent Combustion Model Constants for Fluctuations in f^a

Constant	Value
C_{g1}	2.8
C_{g2}	1.92
σ_f	0.9
σ_g	0.9
σ_{ht}	0.9
σ_h^b	0.8
E^b	9.793

[a] From Tamanini (1975).
[b] These constants arise from boundary conditions.

It is emphasized at this point that Eqns. (11.31) and (11.32) are not dependent on the assumption of chemical equilibrium but only on the equality of diffusivities, and for Eqn. (11.31) on the adiabatic assumption.

What if it is not realistic to assume adiabatic operation? Crocco similarity is no longer satisfied, and in this case a separate energy equation must be solved with the appropriate boundary conditions. The equation is identical to Eqn. (11.21) for the mixture fraction but with different boundary conditions. The equation is repeated as a reminder of the need to solve it for nonadiabatic operation:

$$(\partial/\partial t)(\bar{\rho}\tilde{h}) + \nabla \cdot (\bar{\rho}\tilde{h}\tilde{\mathbf{v}} - D_{th}\nabla\tilde{h}) = 0 \tag{11.33}$$

The way of obtaining the mean density, temperature, and species profiles is now prepared. If the equilibrium assumption is invoked on the grounds that micromixing is limiting rather than kinetics (as discussed in Section 11.2b), these properties are obtainable. The energy level, pressure, and the elemental composition are the only required information for a Gibbs-free-energy reduction scheme for chemical equilibrium. Since the pressure variation in most practical gaseous combustion chambers is negligible with respect to chemical reactions, the equilibrium properties are now a function of f alone:

$$T = T(b_k, h) = T[b_k(f), h(f)] = T(f) \tag{11.34}$$

$$\rho = \rho(b_k, h) = \rho[b_k(f), h(f)] = \rho(f) \tag{11.35}$$

$$y_i = y_i(b_k, h) = y_i[b_k(f), h(f)] = y_i(f) \tag{11.36}$$

In principle, the computation of the equilibrium properties can now be completed. However, calculating the complex equilibrium states and temperature at the prescribed pressure and enthalpy is not a simple task. Three schemes could be considered, namely, equating of forward and reverse rates, equilibrium constant formulation, and Gibbs function minimization. The first, dynamic equilibrium scheme encounters all the difficulties of a full kinetic scheme. Even if all the reactions could be identified, there still remains the task of finding accurate experimental rate constants, to say nothing of the complexity of the numerical scheme to solve the set of stiff differential equations. The second scheme, the equilibrium constant approach, is perhaps the most familiar, and conceptually the easiest to formulate; yet, it is known to be equivalent to the Gibbs function minimization principle for a mixture at specified energy level and pressure. This method is highly inefficient for complex chemical systems. The third scheme which determines the mixture composition corresponding to the minimum total Gibbs free energy of the reacting mixture was first described by White *et al.* (1958). This approach has been proven to be vastly superior in its computational efficiency to the other schemes. Algorithms for calculating chemical equilibrium based on this principle have been developed. Among the more well-known of these are the NASA-CEC program (Gordon and McBride, 1971) and the Edwards Air Force Base program (Selph, 1965). Both were developed for the aerospace industry.

Pratt and Wormeck (1976) have developed a code (CREK, Combustion Reaction Equilibrium and Kinetics) for either complex kinetic or equilibrium computations. This code was developed as a module for incorporation into larger fluid mechanic schemes. It was developed specifically for combustion applications and as such it has the capability of constructing its own initial estimates for practical combustion applications. The approach for equilibrium computations is the minimization of the Gibbs free energy scheme. Although CREK has the capability of computing both the kinetic and the equilibrium states, a subset of the model has been developed for computing only equilibrium compositons. Input was also restructured to receive local element atom numbers b_k as is required by the above-outlined approach. The resulting submodel was named CREE (Chemical Reaction Equilibrium for Elements). This is a useful routine for performing the computations required by the approach outlined in this chapter and is given along with user information by Smith and Smoot (1980).

The turbulent flow Favre-mean compositions are obtained by weighting with the PDF of f as in Eqn. (11.25):

$$\tilde{y}_i = \tilde{\alpha}_p y_{ip} + \tilde{\alpha}_s y_{is} + \int_{0+}^{1-} y_i(f)\tilde{P}(f)\,df \tag{11.37}$$

For the Favre-averaged transport equations the conventional time-mean density is required and can now be obtained from the equilibrium calculations in Eqn. (11.26). In this way the unmixedness is accounted for in all properties, and even with equilibrium chemistry, the time-averaged concentrations of fuel and air may both be finite at the same point in space. Indeed, experimental evidence of Hawthorne *et al.* (1949), Kent and Bilger (1971), and Lewis and Smoot (1979) indicate considerable overlap of the oxidant and fuel profiles. This is not due to sluggishness on the part of the chemical reactions but due to the turbulent fluctuations, and this unmixedness allows fuel and oxidizer to exist at a given point at different instantaneous times. The Favre values then reflect the presence of both.

This approach to find the mean properties is requisite on the enthalpy h and the element mass fractions b_k being strictly functions of the mixture fraction f. As already indicated, this is appropriate only where Crocco similarity is applicable. When heat losses are significant, the Crocco similarity must be abandoned for the transport equation in h (Eqn. (11.33)). In this situation, Eqns. (11.34)–(11.36) are no longer valid and in general the properties become functions of f and h:

$$\beta = \beta[b_k(f), h] = \beta(f, h) \tag{11.38}$$

The proper way to get the Favre-averaged properties is by convolution with a joint PDF

$$\tilde{\beta} = \text{intermittency} + \int_h \int_f \beta(f, h)\tilde{P}(f, h)\, df\, dh \tag{11.39}$$

It might be noted that this equation is valid for adiabatic operation as well, in which case the enthalpy is a function of f alone and Eqn. (11.39) reduces to Eqn. (11.25).

Now, how is $\tilde{P}(f, h)$ to be found? The problem is greatly simplified by partitioning the enthalpy into (1) the energy which is convected if there were no heat losses or gains by any other source,and (2) the residual energy due to the nonadiabatic boundary condition:

$$h = h_f + h_r \tag{11.40}$$

$$h = fh_p + (1 - f)h_s + h_r \tag{11.41}$$

In practice, the residual enthalpy can be calculated directly from the known field values of h and f, and from the inlet conditions:

$$\tilde{h}_r = \tilde{h} - \tilde{f}h_p - (1 - \tilde{f})h_s \tag{11.42}$$

Finally, it is assumed that the effects of the fluctuations of h_r are small compared with the effects of the fluctuations of h_f as implied by Eqn. (11.42):

$$h_r' = 0 \qquad (11.43)$$

Thus

$$\tilde{h}_r = h_r \qquad (11.44)$$

This means that the fluctuations are correlated by the single variable f as before, giving

$$\tilde{\beta} = \tilde{\alpha}_p \beta_p + \tilde{\alpha}_s \beta_s + \int_0^1 \beta(f, h_r) \tilde{P}(f)\, df \qquad (11.45)$$

where h_r is constant with respect to the fluctuations in f. Since h_r will sometimes be small relative to h_f, this approximation is thought to be reasonable.

This method just described for including enthalpy fluctuations is equivalent to assuming that the variations in h with respect to the mixture fraction f are parallel to the adiabatic h and passing through the location $(h = \tilde{h}, f = \tilde{f})$.

An alternative approach, used by some, is to prescribe a given heat loss from the reaction chamber of interest. With this prescribed energy deficit h_l, it is possible to perform the equilibrium calculations as before by assuming that the energy level is reduced by h_l throughout the entire computational domain. This procedure allows for heat loss, yet it maintains the simplicity of Crocco similarity for the energy equation. However, the heat loss is not calculated directly, but prescribed artificially.

11.8. NONEQUILIBRIUM CHEMISTRY FOR TYPE B FLAMES

We have noted that for flames where the chemical reaction time scale is fast compared to the turbulence time scale, the effects of the turbulence on the chemistry can be incorporated by use of the conserved scalar f through the statistics of this mixture fraction. We have discussed how the instantaneous variables (mole fractions, temperatures, etc.) can be calculated from equilibrium considerations by a Gibbs free energy minimization scheme. What if the chemistry is still rapid but does not proceed to complete equilibrium?

Most combustion does not proceed to equilibrium. Combustion of H_2 is one of the few fuels which meets the equilibrium approximation satisfactorily. Combustion of hydrocarbons includng methane (CH_4) have some species which do not generally meet the equilibrium approximation. Yet these fuels often have reaction rates that are sufficiently fast that the chemical composition is only a function of the local equivalence ratio, and thus only a function of the mixture fraction. Physically, this implies that the reactions proceed rapidly to some condition other than their equilibrium values.

Techniques for incorporating the chemical reaction rate into the turbulence calculation procedures are discussed in more detail in Chapter 15. However, one of the most practical methods is to replace the equilibrium computation for instantaneous chemical properties (i.e., mole fractions, temperature, density, etc.) with experimental data from laminar flames. As long as the properties are only functions of the mixture fraction, then the techniques outlined previously can be used to obtain the time mean (Favre or Reynolds averaged) properties. In this way the composition need not be in equilibrium.

Figures 11.4 and 11.5 show measured temperature and mole fraction data taken from a confined, laminar methane diffusion flame by Mitchell *et al.* (1980). These data show examples of the unique functional dependence of these chemical properties on the equivalence ratio. These values are generally far from the corresponding equilibrium values. Such data can be used to replace the equilibrium computation outlined earlier in this chapter. The mean values can then be obtained by convolution of these instantaneous laminar data over the

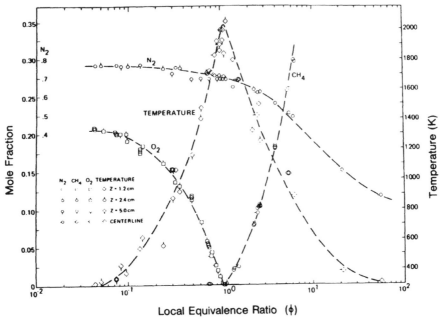

Figure 11.4. Measured temperature and reactant concentration as a function of the local equivalence ratio in a laminar methane diffusion flame. (Figure used with permission from Mitchell *et al.*, 1980.)

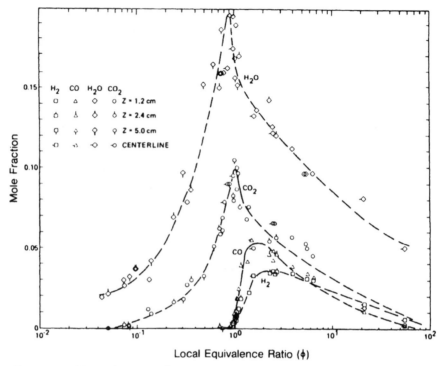

Figure 11.5. Measured major-product concentration as a function of the local-equivalence ratio in a laminar methane diffusion flame. (Figure used with permission from Mitchell *et al.*, 1980.)

appropriate PDF. This general approach is exactly that outlined in this chapter with experimental data replacing the equilibrium computation.

This method will be extended in Chapter 15. The incorporation of the experimental data has proven to be the more accurate method for hydrocarbon flames. It allows for nonequilibrium chemistry in the fast chemistry approximation . This is the recommended procedure when experimental data are available.

11.9. SAMPLE COMPUTATIONS

In order to demonstrate the capabilities of computer models based on the theoretical approach outlined in this chapter, this section presents several model predictions with associated experimental data for comparison. Gosman *et al.* (1979) and, more recently, Smith and Smoot (1980) have evaluated the capabilities of these codes for predicting gaseous mixing processes for various laboratory combustors. Figure 11.6 illustrates a comparison of theory and measurement

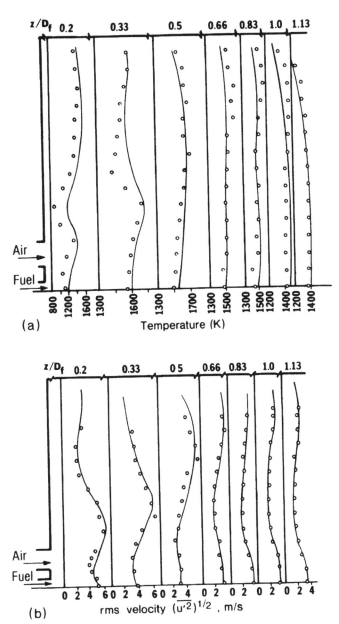

Figure 11.6. Comparison of predictions for gaseous-fired Harwell cylindrical furnace (swirl number = 0.5) with: (a) radial temperature profiles and (b) radial profile of rms velocity (points are data, lines are theory). *z* is axial distance from furnace inlet and D_f is furnace diameter. (Figure used with permission from Gosman *et al.*, 1979.)

TABLE 11.4. Parameters for Reacting, Gaseous Data Cases Used for Evaluation of Type B Flames with PCGC-2 Code

Investigator	Variables measured	Type of measurement	Inlet primary and secondary velocities (m/s)	Diameter (m)		
				Primary	Secondary	Chamber
Lewis (1981) (methane–air flame)	f, species	"Isokinetic" probe	2.3, 30	0.016	0.057	0.203
Takagi et al. (1981) (hydrogen–air flame)	$\bar{u}, \overline{u'^2}$ species, T	LDV probe, chromatograph, thermocouples	20, 5.1	0.005	0.104	0.104
Takagi et al. (1981) (hydrogen–air flame)	$\bar{u}, \overline{u'^2}$ species, T	LDV probe, chromatograph, thermocouples	70, 5.1	0.005	0.104	0.104

of temperature and rms velocity for a gaseous, reacting system. The temperature was calculated from the enthalpy where the $\tilde{h}$ equation was solved and the assumption made that the h fluctuations follow the adiabatic case. Given the complexity of this swirling, recirculating, turbulent, reacting flow, the agreement shown in Figure 11.6 is surprisingly good. Yet errors in temperature in the forward regions are particularly large. Spalding (1975) and Mellor and Herring (1973) have suggested limitations of this method. This method has not accurately simulated recirculation zones which occur behind blunt boundaries and its applicability to the gas motion in a particle-laden stream is questionable. It is also somewhat sensitive to values of the initial turbulence level.

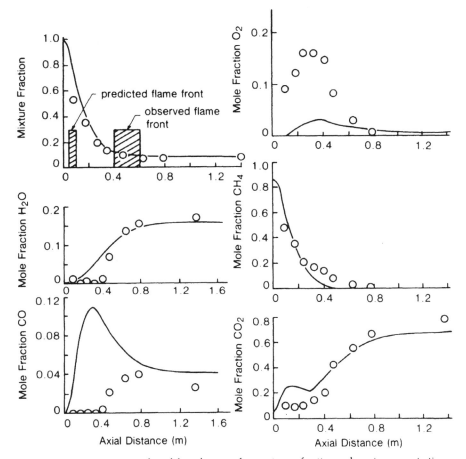

Figure 11.7. Comparison of model predictions of gas mixture fraction and species concentration with Lewis (1981) natural gas/air flame measurements. Conditions are shown in Table 11.4 (predictions from Fletcher, 1983).

The computer code PCGC-2 developed at Brigham Young University and discussed in Chapter 7 was used to predict local properties in gaseous flames for three different experimental systems shown in Table 11.4. The methane–air flame studied by Lewis (1981) has been predicted using the PCGC-2 computer program with measured inlet velocity and turbulence intensity profiles. Figure 11.7 shows centerline comparisons of mixture fraction, H_2O, CO, O_2, CH_4, and CO_2 for the Lewis case assuming local instantaneous chemical equilibrium. The agreement with the mixture fraction measurements is very good. Predictions of H_2O, CO, and O_2 do not agree as well with experimental data. The major reason for this disagreement is that the flame studied by Lewis was not stabilized at the reactor inlet but was a lifted flame. This computation is shown to emphasize a limitation of this modeling approach. The CH_4 and air mixed together in the reactor before the flame zone, so that the flame burned in a partially premixed condition. In addition, these methane flames are known to have some kinetic limitations as already discussed.

Predictions have also been made for this same case using experimental data from laminar flames for instantaneous composition vs. mixture fraction instead of equilibrium as recommended in the last section. The CO and CO_2

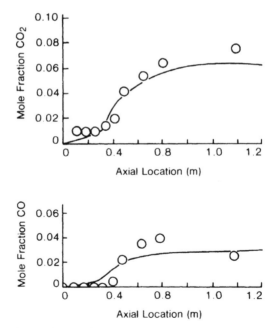

Figure 11.8. Centerline predictions of CO and CO_2 in the Lewis (1981) natural gas/air flame using nonequilibrium gas properties from Bilger (1977). Conditions are shown in Table 11.4. This figure is to be compared with Figure 11.7 (predictions from Fletcher, 1983).

concentrations are shown in Figure 11.8. This method has proven to be a very good technique for incorporating some kinetic limitations and still including the important effects of turbulence fluctuations.

Takagi *et al.* (1981) performed measurements in two different H_2–air flames. Gas chromatography was used to obtain local gas compositions, LDV (Laser Doppler Velocimetry) was used to get local mean rms velocities, and thermocouples were used to measure local gas temperatures. Comparisons of model predictions with the measurements of Takagi and co-workers are shown in Figure 11.9 for both the high- and low-velocity flames. This good agreement is to be expected, since the combustion model is best for rapid reactions that are mixing limited. The rms velocity comparisons showed that the flame is not isotropic, as assumed in the turbulence model. The lack of agreement of predictions and measurements of rms velocity data indicate the need for improvements in the model. Overall agreement with the data is considered extremely good for this H_2–air flame. H_2 combustion is one of the best examples of Type B flames.

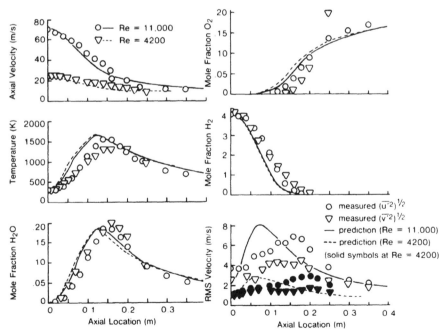

Figure 11.9. Comparison of centerline predictions of velocity, temperature, and species concentrations *vs.* Takagi *et al.* (1981) data measured in H_2–air flames ($Re = 11{,}000$ and 4200). Conditions are shown in Table 11.4. (Predictions from Fletcher, 1983.)

11.10. ILLUSTRATIVE PROBLEMS

1. Derive the steady-state, instantaneous, transport equation for the conserved scalar f (the mixture fraction), in polar coordinates in three dimensions. (*Hint:* the two most common starting points for this derivation are the Reynolds transport theorem, or the property balance around an arbitrary differential element.)

2. Starting with the instantaneous mixture fraction equation of Problem 1, derive the time-averaged equation for a turbulent reacting flow. (Do not substitute any turbulence closure approximations for the cross correlations of the fluctuating quantities.)

3. Repeat Problem 2, using Favre averaging and compare your results.

4. From the definition of a Gaussian distribution and given the physical limits on the mixture fraction f, derive the expressions given in Table 11.2 for $P(f)$, α_p, α_s, $\bar{f}$, and $\tilde{g}_f$. Discuss a numerical procedure for obtaining α_p and α_s for a given $\bar{f}$ and $\tilde{g}_f$.

5. Consider a simple turbulent diffusion flame of the type shown in Figure 11.1a, where the fuel is pure CH_4 and the oxidizer is ambient air. The inlet mass flow rates of each are such that overall stoichiometry is exactly 1. For a given location in the flame we have found that $\tilde{f} = f_{stoich}$ and $\tilde{g}_f = f_{stoich}^2$.
    a.   What is the value of f_{stoich}^2?
    b.   Derive an expression for f as a function of equivalence ratio ϕ.
    c.   Assume that the instantaneous temperature as a function of equivalence ratio is that of Figure 11.4. Assuming a top-hat Favre PDF (see Table 11.1) what is $\tilde{T}$?
    d.   Repeat (c) above for the Gaussian Favre PDF of Table 11.2.
    e.   Repeat (c) above for O_2 mole fraction.
    f.   Repeat (c) above for CH_4 mole fraction.
    g.   Explain why the mean mole fractions of O_2 and CH_4 of (e) and (f) above are both nonzero.

6. Using some chemical equilibrium program or hand calculation compute the equilibrium temperature and composition of a methane–air mixture as a function of equivalence ratio and compare to Figures 11.4 and 11.5.

7. Consider the direct combustion of natural gas and air in an axisymmetric reactor. The initial input conditions are given in Table 11.5. Use the generalized computer code PCGC–2 to predict the characteristics of this flame.

TABLE 11.5. Test Conditions of Lewis (1981)

Geometry		
Primary tube diameter	0.016 m	
Secondary duct diameter	0.057 m	
Chamber length	1.524 m	
Chamber diameter	0.203 m	
Mass flow rates		
Primary	0.0031 kg/s	
Secondary	0.0362 kg/s	
Species (mass fraction)	Primary	Secondary
Ar	0.1185	0.0128
CH_4	0.7426	0.0
C_2H_6	0.0941	0.0
CO_2	0.0071	0.0004
H_2O	0.0029	0.0002
N_2	0.0348	0.7567
O_2	0.0	0.2299
Turbulence properties		
Primary turbulence intensity	0.06	
Secondary turbulence intensity	0.06	
Physical properties		
Pressure	$8.63 \times 10^4 \text{ N/m}^2$	
	(0.85 atm)	
Primary temperature	286 K	
Secondary temperature	589 K	

Explain the general features of the predicted temperature and composition fields. From the temperature profile, identify the location of the time-mean stoichiometric mixture.

8. For the conditions of Problem 7 (Table 11.5) what is the effect of the turbulence on the chemistry and the resulting flame structure? Use PCGC-2 to predict the flame structure when the turbulence fluctuations are ignored.

9. Consider the direct combustion of natural gas and air in an axisymmetric reactor. The initial input conditions are given in Table 11.5. What differences would be expected between predictions made with a top-hat distribution for the probability density function and those made with a Gaussian distribution?

10. With the same case as Problem 1 (Table 11.5), examine the effect of the burner design by parametrically studying inlet profiles and turbulence levels as follows:

a. Use PCGC-2 to alter the shape of the inlet velocity profile and the inlet turbulence kinetic energy profile from plug flow to more realistic profiles. Velocity profiles and turbulence energy profiles have been measured for a similar geometry and are given in Figure 11.6. Use the shape of these measured profiles as model input parameters. What differences are observed? Why?

b. Use PCGC-2 to alter the turbulence level of the burner from the base case of Table 11.5. Change the inlet turbulence intensities of both the primary and secondary to 10%. What is the resulting effect on the flame structure? Why?

The glory of God is intelligence, or, in other words, light and truth.

Doctrine and Covenants 93:36

PARTICULATE AND DROPLET REACTIONS IN TURBULENT FLOWS

12.1. INTRODUCTION

In the last chapter we discussed the importance of the turbulent fluid mechanics on the homogeneous gas-phase chemistry. Different approaches for incorporating the chemistry–turbulence interactions were investigated. An approach for mixing-limited reactions with local instantaneous chemical equilibrium was examined in some detail. The importance of the homogeneous reactions in overall coal combustors and gasifiers was emphasized.

In this chapter we will stress the importance of the particulate and droplet motion and reactions in these systems. The emphasis will be on individual particles and on their heterogeneous reactions. In Chapter 13 we will examine the effect of the particulate and droplet reaction products on the overall chemistry in this turbulent environment. In both of these chapters we will discuss the physical processes occurring and present engineering procedures for calculating their performance.

There are many different levels of sophistication at which these practical reaction processes may be investigated. Some researchers have focused on the fundamental molecular level including details of the pore evolution and studies of the intrinsic kinetics involved in the heterogeneous surface reactions. Others have looked at greatly simplified assumptions including global particle surface areas correlated from experimental data and global kinetic rates for the heterogeneous reactions. Some investigators have studied the individual particle motion including the random fluctuations of individual eddies on the path of flight of each particle, while others have looked at mean properties for the entire particle cloud. It is beyond the scope of this chapter to review the entire range of particulate and droplet research. A brief review will be given of the different methods and the focus of the chapter will be on a discussion of one of the many possible approaches, but one that retains a common level of sophistication with other assumptions made in describing in this book the overall combustion and gasification of coal, char, and coal–water mixtures.

Chapters 3–5 have already discussed in some detail the physical phenomena and associated theory for coal particle ignition, devolatilization, heterogeneous char reaction, and combustion of coal slurries. The incorporation of these theories into an overall predictive model was discussed to some extent in Chapter 7. The major purpose of this chapter is to focus on techniques for calculating the particle motion in turbulent reactors and methods for incorporating the reaction scheme, previously outlined, into an overall multidimensional predictive code.

12.2. APPROACHES

In all turbulent reaction vessels involved with combustion and gasification of coal, char, and coal slurries, the location of the individual particles or droplets during the reaction time is one of the controlling factors in the overall reaction process. In these turbulent diffusion flames, large regions of both fuel-rich and oxidizer-rich environments exist. Since the source of reacting fuel originates in the particle or droplet phase, the flight history of each particle becomes important.

A distinguishing feature among current pulverized-coal models is the treatment of the dispersed particulate phase. In large-scale furnaces, large recirculation zones have a major influence on the particle motion. Gas molecules may usually be treated as a continuum field. These gases tend to equilibrate in temperature and composition over some small region of space at some point in time. This equilibration is due, in most part, to the random molecular collisions. However, in particle/droplet systems, the particles or droplets are not a continuum; particles or droplets exist locally with different properties due to their varying histories. For example, some particles of a given size class may travel through the recirculation zone one or more times, then may subsequently be found downstream in the same computational cell as particles of the same initial size class that did not recirculate. Particles or droplets in those disperse systems do not communicate with each other by random collisions but, instead, largely by interaction with the neighbouring gas field, with each particle having its own unique history. These differences between a continuum fluid and dispersed particles were discussed in more detail in Chapter 9. Procedures for handling this particle history effect are of major significance.

Effects of turbulence on particle motion are also important. Turbulence in a homogeneous mixture still evades a firm theoretical basis and relies on empiricism. The added complexity of the second phase introduces unresolved issues regarding the interaction between the two phases. A lack of understanding of the detailed processes has prevented formulation of generalized techniques for representing the governing equations.

Two basic approaches are being used in pulverized-coal codes where the particle dispersion rates are considered to be different from the gas. In one approach, the particle phase is treated analogously to the gas. The history effect is introduced by allowing for several discrete classifications of the particles. Properties such as size or temperature are established for a group of coal particles. Eulerian (fixed reference) equations are written for each classification with source and sink terms allowing for particles to change from one group to the next. A large number of added differential equations is needed to handle particle history effects in this manner.

In a second approach, particles are tracked through the flow field while the solid properties are changed continuously for each particle trajectory. The interactions between the gas and the particle phase are accounted for by source terms in the Eulerian gas field which are obtained from the particle computations. This technique was discussed in Chapter 9. The Lagrangian (moving reference) approach involves reduced computer storage but introduces problems in treating effects of gas turbulence on particle motion and in obtaining average particle properties for comparison with experimental data.

Gibson and Morgan (1970) and Richter and Quack (1974) treated coal particles as gas (i.e., always in dynamic equilibrium). Blake et al. (1979) proposed the addition of another balance equation for the transport of the turbulent kinetic energy of the particles, in addition to the gaseous turbulence properties. Lockwood et al. (1980) and Smith et al. (1981) account for turbulent gas effects on the particle Lagrangian velocity. Only limited studies of particle motion or fluid–particle turbulent interaction have been presented. Much of the work on particle dispersion has neglected the critical influence of gas turbulence. Those who have considered this issue have generally proposed extensions to the gas-phase turbulence model through empiricism. Chapter 7 reviewed these various treatments in more detail.

The approach discussed in this chapter is based on the PSI-CELL technique of Crowe et al. (1977). Care is taken to account for all modes of coupling between the gas phase and the particle phase. The method does not account for particle–particle interactions and thus would not be applicable to highly loaded particle–gas flow nor to general two-phase flow systems. It is intended to be applied to dispersed flow systems. The model is based on following the trajectories or paths of representative particles or droplets through the gas-phase field. These particles are then treated as sources of mass, momentum, and energy to the gaseous phase.

The particle velocities, trajectories, temperatures, and composition are obtained by integrating the equations of motion, energy, and component continuity for the particles in the gaseous-flow field. This is done with the Lagrangian approach, while recording the momentum, energy, and mass of the particles upon crossing cell boundaries. In this straightforward manner, the net difference in the particle properties between leaving and entering any given cell provides the particle source terms for the gas-flow equations.

We will now examine the Lagrangian equations of motion for individual particles or droplets. These equations will be the basis for extension to the mean trajectory and history of representative particles or droplets in turbulent combustors and gasifiers. This turbulent dispersion of the particles is one of the controlling mechanisms in the particle/droplet motion. Procedures for estimating this effect will be presented in some detail. Subsequently, the formulation for the particle source terms for the gas-phase equations will be discussed. Finally, we will review the particle reactions already discussed in preceding chapters as they pertain to a comprehensive model.

12.3. LAGRANGIAN PARTICLE DESCRIPTION

In order to establish the trajectories of representative coal particles or coal slurry droplets in combustors or gasifiers, it is essential to examine the formulation of Lagrangian equations for the representative particles or droplets. On following these individual trajectories, we can establish the history effect of the particles and thus examine the net source of momentum, mass, and energy between the particulate or droplet phases and the neighboring gas phase. This is accomplished by integrating the particle or droplet equation of motion and the heat- and mass-transfer equations relating to the particle or droplet temperature and size. Of course, the effect of the turbulence on these particle properties must also be included. This turbulence effect will be discussed in Section 12.4. In addition, the velocity, pressure, and temperature field of the neighboring gas is used in these calculations.

Consider the particle momentum equation for a single particle or droplet in the Lagrangian framework:

$$\alpha_p \, d(\mathbf{v}_p)/dt = \mathbf{F}_p/n_p + \alpha_p \mathbf{g} \tag{12.1}$$

This equation is simply an expression for Newton's second law of motion for an individual particle. The rate of change of momentum is equal to the sum of external forces on the particle. The first force on the right-hand side of Eqn. (12.1) is the drag force between the particle and its surrounding gas field. The second force on the right-hand side of Eqn. (12.1) is simply the gravitational force on the particle. In this equation the other terms contributing to aerodynamic forces on the particle or droplet are neglected because they are of a much smaller significance. These terms include the pressure gradient, virtual mass,

Bassett term, the Saffman lift, and Magnus forces (Smoot and Pratt, 1979). These terms are small for these applications because the gas-to-particle or gas-to-droplet density ratio is very small and the particles are not typically in a high shear region for the applications of interest.

In Eqn. (12.1), Newton's second law applied to the reacting particle has been written with the mass of the particle on the outside of the derivative on the left-hand side of Eqn. (12.1). It has been written in this fashion even though the particle is reacting and α_p is changing with time. To understand the implications of this form of the equation, it is sometimes useful to write the equation of change in its more general form:

$$d(\alpha_p \mathbf{v}_p)/dt = \mathbf{F}_p/n_p + \alpha_p \mathbf{g} + \mathbf{v}_{fp} \, d(\alpha_p)/dt \qquad (12.2)$$

This equation shows that the rate of change of momentum of the particle is equal to the sum of the external forces on the particle—the external drag force, the gravitational force, and a third force due to the mass efflux from the surface of the coal particle. Figure 12.1 depicts this process schematically. In this representation, we must realize that the system is composed of just the solid particle. This particle is changing mass as gases are evolved from its surface due to moisture evaporation, particle devolatilization, and heterogeneous char reactions. If the velocity of the fluid evolved from the surface of the particle is $\mathbf{v}_{fp}$ and the evolution rate of the gas evolved from the particle is $d\alpha_{fp}/dt$, this fluid will exert a differential impulse on the particle it is leaving:

$$d\mathbf{J}_{fp} = -\mathbf{v}_{fp} \, d\alpha_{fp} \qquad (12.3)$$

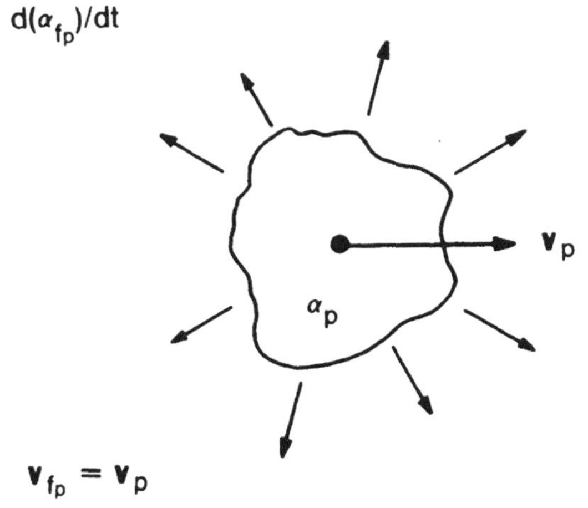

$d(\alpha_{fp})/dt$

$\mathbf{v}_p$

α_p

$\mathbf{v}_{fp} = \mathbf{v}_p$

Figure 12.1.   Effect of mass efflux on the Lagrangian equation of motion of a coal particle.

The negative sign in this equation indicates simply that the fluid is exerting the impulse on the particle. By Newton's third law, there is a reaction force of equal magnitude exerted on the fluid by the particle in the opposite direction. This efflux of mass, therefore, must exert a force on the particle:

$$\mathbf{F}_{fp}\, dt = d\mathbf{J}_{fp} = -\mathbf{v}_{fp}\, d\alpha_{fp} \tag{12.4}$$

$$\mathbf{F}_{fp} = -\mathbf{v}_{fp}\, d(\alpha_{fp})/dt \tag{12.5}$$

Since the rate of mass depletion of the particle is simply equal to the negative of the rate of the mass addition to the gas, the overall force exerted by the evolving gas on the coal particle can be written in terms of the rate of mass depletion of the coal particle:

$$d(\alpha_p)/dt = -d(\alpha_{fp})/dt \tag{12.6}$$

$$\mathbf{F}_{fp} = \mathbf{v}_{fp}\, d(\alpha_p)/dt \tag{12.7}$$

This then is the force exerted on the coal particle due to the efflux of mass from its surface as shown in Eqn. (12.2).

In reality, the velocity and rate of evolution of gases from the surface of the coal particle will be functions of the solid angle. By the use of holography, pictures have been taken of coal particles undergoing rapid devolatilization. McLean et al. (1981) and Seeker et al. (1981) have shown and presented such holographs. Figure 3.15 showed examples of these photographs. The volatiles sometimes evolve from the particle in a localized jet behavior. This jetting will obviously affect the momentum of individual coal particles. Procedures for calculating this jetting phenomena have never been developed. If, however, we are willing to assume that on the average the gases evolving from many representative particles can be considered to evolve uniformly about the particle, then the fluid velocity $\mathbf{v}_{fp}$ is represented by the particle velocity $\mathbf{v}_p$. With this assumption and by carrying out the differentiation of the left-hand side of Eqn. (12.2), it is apparent that the force exerted by the fluid on the particle does not contribute to a change in the particle velocity. Equation (12.1) is the resulting Lagrangian equation of motion.

To complete the description of the Lagrangian equation of motion for the particle, a correlation must be chosen for the drag force on the particle. Since the particles are typically small for combustion and gasification applications, the Reynolds number based on the relative velocity between the gas and the particles is typically less than 10. Empirical correlations for low Reynolds number particle or droplet flow are adequate. One such correlation is

$$\mathbf{F}_p = A_p \rho_g C_d n_p (\mathbf{v}_p - \mathbf{v}_g)\|\mathbf{v}_p - \mathbf{v}_g\|^2 \tag{12.8}$$

where

$$C_{do} = (24/Re)(1 + 0.15Re^{0.687}) \tag{12.9}$$

The mass efflux from the reacting coal particle can reduce the drag coefficient. A relationship suggested by Bailey *et al.* (1970) can be used to correct the drag coefficient without rates of high mass transfer to one with mass transfer:

$$C_d = C_{do}/(1 + B_m) \tag{12.10}$$

where B_m is the particle blowing parameter for mass transfer.

12.4. TURBULENT PARTICLE DISPERSION

One of the major factors involved in the combustion and gasification of coal and coal slurries is the turbulent dispersion of the particle phase. It has been found that the particles do not mix at the same rate as the gas phase and that this difference in turbulence mixing is of paramount importance to the coal reaction process. The environment around a given particle during heatup, devolatilization, and heterogeneous reaction controls the overall combustion process.

At the outset, it is necessary to recognize that turbulent multiphase flows present many very difficult problems. The problems are not only mathematical in nature, but even formulating the governing equations is often not possible because of a significant lack of understanding of the physical processes involved. Thus, any attempt to obtain a theoretical relationship for the mass dispersion of the solids due to the turbulent motion of the continuum requires the introduction of a significant number of assumptions in order to arrive at the resulting equation. Often the simplified form of the equation requires a high level of empiricism to obtain the correlating coefficients.

As we saw in Chapters 10 and 11, turbulence in a homogeneous gas phase has been studied extensively, and many advances have been made in recent years. It was also apparent from the discussion in those chapters that much remains to be learned about the turbulent mixing process, even in the isothermal gas phase. The complexities are compounded even more in the two-phase turbulent jet. As the mass fraction of the particulate phase increases, so too does the effect of these particles on the primary fluid flow. It is through the turbulence in this primary fluid that the particles respond to the turbulent field. It may be useful to understand the interactions of the carrier fluid with the individual particles by identifying three regimes. First, consider very small particles, so small that they are able to respond to all the turbulence irregularities.

In the limit of extremely small sizes they have no influence on the flow apart from increasing its effective density in proportion to their mass addition. These particles follow the gas closely. Indeed, it is this regime that is assumed for seed particles in laser–doppler anemometer applications. The second regime is the opposite extreme; in this case, consider the particles that are very large. Now the particles make significant contributions to the mass and energy balances of the mixture, providing additional means of storing and transferring momentum and energy, apart from those already present in the corresponding primary carrier gas. In this extreme, the particles fall according to imposed external forces but are not affected significantly by the turbulent motion of the gases surrounding them.

The third regime is an intermediate-size range in which both phases, gas and solid, interact with each other. In this regime the mean velocity and turbulent fluctuations of the gas phase are affected by the presence of the particles. In turn, the particles are carried by the gas phase and moved about by the random turbulent eddy structure of the carrier gas. It is unfortunately in the third regime where the pulverized-coal particles of interest to coal combustion and gasification are found. In this regime the mass transfer diffusion rates of the

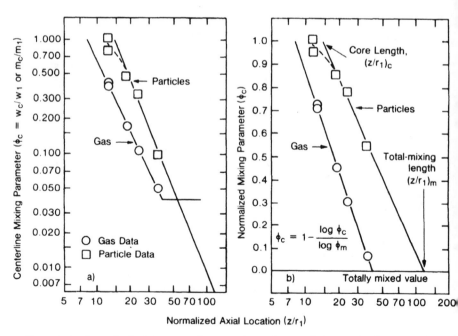

Figure 12.2. Example of mixing rates of gases and particles in turbulent, coal-laden confined jet. Primary velocity is 33 m/s, secondary velocity is 42 m/s, 61% solids loading, and 43-μm mass-mean particle diameter. (Figure used with permission from Leavitt, 1980.)

particles due to the turbulent motion of the surrounding fluid are different from the corresponding mass transfer rates of the carrier fluid itself.

It is useful to examine turbulence effects by experimental observations. Figure 12.2 compares the mixing rate of gases and particles in a particle-laden confined jet. This particular case is an example of a coal-laden jet with a typical pulverized-coal particle size distribution. The measurements were made in a nonreacting isothermal jet. The geometry of the mixing chamber is shown in Figure 12.3. The plots in Figure 12.2 show the centerline decay of the gas mixture fraction and the coal particle flux. The centerline values were normalized to the values at the primary jet exit. These values were referred to as the centerline mixture fraction or particle dispersion parameter and were indicated generally by the symbol ϕ_c. By plotting the logarithm of the centerline mixture fraction or dispersion parameter, ϕ_c, as a function of the logarithm of axial location, the data points were observed to correlate linearly over a significant portion of the jet as illustrated in the axial decay plot in Figure 12.2a. There are two axial boundary conditions for axial decay plots. The upstream value is that of the primary stream and normalizes to unity. The downstream value is the totally mixed value, ϕ_m, which can be determined from input conditions and mixing-chamber diameter. These values vary with the test conditions. The decay from the central core of pure primary stream to the fully mixed value indicates the region where mixing of the secondary fluid penetrates through to the centerline. The axial location where the regressed line and the totally mixed value intersect, has been defined as the total mixing length $(z/r_1)_m$. This value is indicative of the first axial location downstream in the mixing chamber where total mixing of the gas or particles has occurred. Figure 12.2b shows the normalized mixing parameter for both the gas-phase carrier stream and the particulate streams.

It is immediately obvious from Figure 12.2 that the primary carrier gas begins to mix on the centerline with the secondary gas much earlier than the

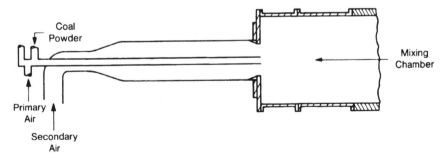

Figure 12.3. Test facility configuration coal-laden confined-jet experiments. (Figure used with permission from Leavitt, 1980.)

particles. The primary stream particle flux persists on the centerline to a normalized axial location of about 15, whereas the gases along the centerline begin dilution at a normalized axial location of about 8. The gases reach their totally mixed value at the centerline at a normalized axial location of near 40, whereas the particles disperse at a much slower rate and would not seem to be totally mixed until reaching a normalized axial location of over 100. This general trend has been observed repeatedly for particles in the size range of interest for pulverized-coal combustion and gasification applications. The particles do not mix at the same rate that the gases do. The turbulent mixing of the gas phase was much more rapid than that for the particle phase in this particular case and for all cases observed at the BYU Combustion Laboratory.

There is some experimental evidence pointing to an opposite effect, that is, the particle turbulent dispersion rate is faster than that for the gas. According to Goldschmidt *et al.* (1971), this effect increases with size of the discrete particles. Several investigators have tried to explain this effect, but without wholly satisfactory results. Obviously, this trend must reverse itself at some reasonable particle size and larger and larger particles must disperse more slowly than the associated carrier gas if the limit of very large particles (regime 2) is to be achieved.

The extensive data (Thurgood *et al.*, 1980; Memmott and Smoot, 1978; Leavitt, 1980) obtained for particle-laden isothermal jets and reacting jets with particle sizes and gas velocities in the regime of interest to pulverized-coal combustion and gasification applications, indicate the trend depicted in Figure 12.2. It appears, at least for the particle sizes and velocities of interest, that the trend is always in the direction of slow particle dispersion superimposed on rapid gas dispersion. Knowing this dispersion rate is extremely important in any calculational procedure for combustion or gasification. This dispersion is dominated almost completely by the turbulent motion of the carrier fluid and to a much lesser extent by the ballistics or mean motion of the particles themselves.

Predicting the extent of particle dispersion in a turbulent environment is extremely complex even in a nonreacting isothermal jet; but to include all of the turbulence and chemistry interactions in a reacting, particle-laden jet and to predict the overall effect from fundamental principles is currently beyond reach. Although all of the fundamental physics cannot be incorporated into an overall predictive model for such a system, engineering approximations and empirical correlations can often adequately bridge the void between theory and application. Some fundamental treatises on turbulent fluid and particle interactions are available (Soo, 1967; Hinze, 1971). We will outline an approach in this chapter which will rely heavily on empiricism and still incorporate the turbulent dispersion of the particle.

In this suggested approach for turbulent particle dispersion, the individual Lagrangian particle is recognized to have a different velocity than the local continuum carrier-gas field. The mean particle velocity is arbitrarily broken into two separate contributions—a convective velocity and a diffusive velocity:

$$\bar{\mathbf{v}}_p = \bar{\mathbf{v}}_{pc} + \bar{\mathbf{v}}_{pd} \tag{12.11}$$

The mean convective velocity $\bar{\mathbf{v}}_{pc}$ is defined as that velocity which results in the absence of turbulence, or in other words, that velocity based on the mean gas velocity. Thus, the convective velocity of the particle is as calculated from Eqn. (12.1) with mean properties substituted for each of the dependent properties. The solution to Eqn. (12.1) as calculated along a trajectory by numerical integration gives the mean convective velocity. A second integration of this velocity vector gives the position of the particle or, in other words, the particle trajectory. The mean diffusion velocity $\bar{\mathbf{v}}_{pd}$ represents the contribution to the mean particle velocity by the turbulent dispersion of the particle caused by the eddy fluctuations in the transporting fluid. This diffusion velocity is calculated by assuming turbulent diffusion proportional to the mean particle bulk density gradient:

$$\bar{\mathbf{v}}_{pd}\bar{\rho}_b = D_p^t \nabla \bar{\rho}_b \tag{12.12}$$

Equation (12.12) can be viewed as a particle turbulent diffusion equation. Since the left-hand side can be thought of as turbulent mass flux (relative to the convective velocity):

$$\bar{\mathbf{j}}_p = (\bar{\mathbf{v}}_p - \bar{\mathbf{v}}_{pc})\bar{\rho}_b = \bar{\mathbf{v}}_{pd}\bar{\rho}_b \tag{12.13}$$

Equation (12.12) defines the transport coefficient D_p^t as a turbulent mass diffusion coefficient, which can also be expressed as

$$D_p^t = \nu_p^t / \sigma_p^t \tag{12.14}$$

where D_p^t is the turbulent particle mass diffusivity, ν_p^t is the particle eddy diffusion coefficient, and σ_p^t is the turbulent particle Schmidt number, or the ratio of particle turbulent momentum transport to particle turbulent mass transport. Much more research is needed to more accurately attain the turbulent particle eddy diffusivity ν_p^t.

An expression for ν_p^t should account for the particle size, so that larger particles are not affected by the turbulence as much as the smaller particles. The ν_p^t expression should also include some term to account for the degree of turbulence. Several simplified correlations for such a term have been proposed on simple order-of-magnitude arguments (Abramovich, 1971; Owen, 1969). The recommended procedure has been suggested by Melville and Bray (1979) and relates ν'_p to ν'_g as follows:

$$\nu'_p = \nu'_g(1 + t_p/t')^{-1} \tag{12.15}$$

The particle relaxation time t_p is related to the Stokesian particle drag by

$$t_p = \alpha_p/3\pi\mu_g d_p \tag{12.16}$$

The time scale of turbulence t' is related to the local turbulence by

$$t' = \nu'_g/\overline{\mathbf{v}'_g \cdot \mathbf{v}'_g} \tag{12.17}$$

Since the $k - \varepsilon$ model is used to obtain ν'_g and $\overline{v'^2_g}$, the turbulent time scale can be expressed as

$$t' = 1.5 C_\mu k/\varepsilon \tag{12.18}$$

Particle dispersion models other than that of Melville and Bray (1979) are reported in current literature. Lockwood *et al.* (1980) use an empirical approach to get $\bar{\mathbf{v}}_{pd}$ from $\overline{u'^2}$ and $\overline{v'^2}$. Longwell and Weiss (1953) considered that the ratio of particle eddy diffusivity to gas eddy diffusivity followed a sinusoidal pattern involving t_p and t'. Lilly (1973) gives evidence that ν'_p/ν'_g decreased as t_p/t' increased, especially for small particles (6–16 μm).

In order to calculate the turbulent diffusion velocity from Eqn. (12.12), not only is the turbulent mass diffusion coefficient required but also the gradient in the bulk density of the particles. Such a bulk density gradient is easy to obtain in Eulerian particle calculation, but not so easy in a Lagrangian description. In the Lagrangian framework, this density gradient can be obtained by computing the trajectories of several representative particles and then counting trajectories in any given cell to back-calculate the cell bulk particle density. It has been our experience that such a procedure requires many trajectories to obtain smooth-gradient information. We have tried methods of calculating bulk densities in computational cells then smoothing these density profiles with a cubic spline fit, then differentiating the spline to obtain the particle gradient density. Such a procedure is quite tedious and in our experience has not provided adequate resolution for the bulk particle density gradient.

An alternate procedure is to use a mixed Lagrangian–Eulerian particle treatment. This mixed procedure uses the entire Lagrangian formulation except for the bulk density gradient term used in Eqn. (12.12). This bulk density gradient is obtained by solving an Eulerian continuity equation for the particulate phase. Such a continuity equation must be solved for each of the particle-size classifications considered in the Lagrangian treatment.

This particle dispersion model has been evaluated by comparing predictions from this technique with measurements from nonreacting, isothermal, coal-laden jet mixing experiments. The apparatus was shown in Figure 12.3. Figures 12.4 and 12.5 show comparisons of the predicted particle mass flux compared

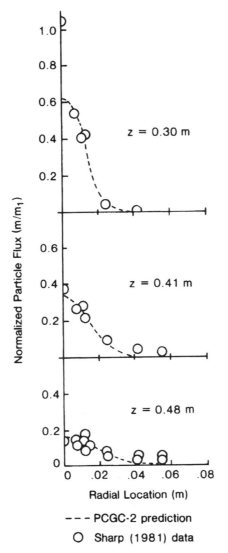

Figure 12.4. Comparisons of predictions and experimental measurements of particle mass flux for a mixing-chamber diameter of 0.21 m. Primary velocity ~33 m/s, secondary velocity ~40 m/s, 67% particle loading, and 46-μm mass-mean diameter.

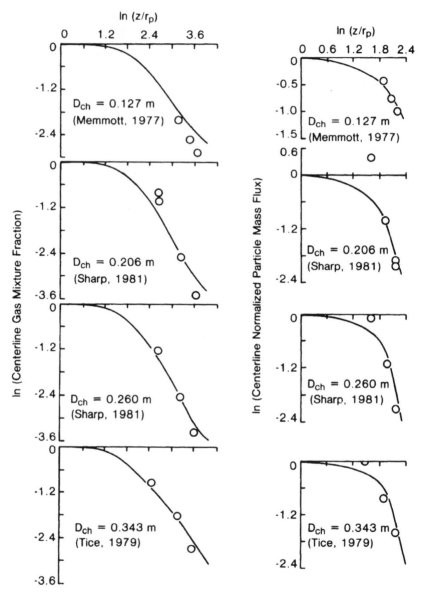

Figure 12.5. Comparison of gas mixture fraction and particle mass flux for silicon particle cases as measured (data points) and predicted by mixed Lagrangian–Eulerian particle dispersion model (predictions from Fletcher, 1983) (lines with $\sigma'_p = 0.35$). All cases were similar except for chamber diameter as noted, primary velocity ~33 m/s, secondary velocity ~40 m/s, 67% particle loading, and 46 μm mass-mean diameter.

with the measured values as obtained from extractive Pitot probes. Figure 12.4 shows radial dispersion as a function of axial location for one test condition. Figure 12.5 shows centerline mass flux for both gas and particle phase for four different conditions. Given the assumptions needed to predict this turbulent mixing phenomena, the agreement is surprisingly good. Although it is to be recognized that even though the calculated and measured results are in general agreement, it is clear that quantitative differences exist. For all of these calculations, a single value of the turbulent particle Schmidt number (defined as the ratio of the particle mass diffusivity to the particle momentum diffusivity) of 0.35 was used. Melville and Bray (1979) have suggested that this value should be somewhere around 0.7. The fact that this level of empiricism is needed to adequately describe the data indicates that much more research needs to be done in the fundamental turbulent mixing processes of particles and gases. As can be seen from the two examples given by Figures 12.4 and 12.5, the single value of 0.35 appears to be adequate to describe the turbulent dispersion in several different geometries typical of coal combustors and gasifiers.

12.5. PARTICLE REACTIONS

This section presents an overview of single coal particle model reactions that occur in pulverized-coal combustion and gasification processes. The description of coal reaction processes includes devolatilization, char oxidation, and gas–particle interchange. The resulting model describes the response of a coal particle to its thermal, chemical, and physical environment. The details of the coal reactions were discussed in Chapters 2–4. Here we interface these components with the turbulent fluid mechanics.

Development of an analytical treatment of pulverized-coal/char behavior in reacting systems is based largely on independent experimental observations and kinetic parameters deduced from these observations. Since there are still unresolved questions regarding the kinetics of coal reaction, an attempt has been made to formulate a general reaction scheme that can accommodate results of future measurements and improved kinetic parameters.

The description applies to pulverized-coal reaction processes, where particles are small ($\leq 100 \ \mu$m) and heating rates are high (10^3–$10^5 \ \mathrm{Ks^{-1}}$) for application to pulverized-coal furnaces, entrained coal gasifiers, and coal slurry combustion.

The coal particle model consists of four components: coal, char, moisture, and ash. Ash is defined as that part of the coal particle that is inert in the combustion scheme. Char is the residue left in the coal particle when the volatile products are released. The continuity equations for each coal component for the jth particle size are

$$d\alpha_{cj}/dt = r_{cj} \tag{12.19}$$

$$d\alpha_{hj}/dt = r_{hj} \tag{12.20}$$

$$d\alpha_{wj}/dt = r_{wj} \tag{12.21}$$

$$d\alpha_{aj}/dt = 0 \tag{12.22}$$

The process by which each particle reacts is schematically presented in Figure 12.6. The reacting particle is assumed to be composed of specified amounts of raw coal, char, moisture, and ash at any particular time. The raw coal, or the dry, ash-free portion of the coal, undergoes devolatilization to volatiles and char by one or more reactions (M in total number) of the form:

$$(\text{raw coal}), \xrightarrow{k_m} Y_{pm}(\text{volatiles})_{pm} + (1 - Y_{pm})(\text{char}) \tag{12.23}$$

The volatiles react further in the gas phase. The char reacts heterogeneously after diffusion of the reactant (i.e., O_2, CO_2, H_2O, and H_2) to the particle surface by one or more reactions (L in total number) of the form:

$$\phi_l(\text{char}) + (\text{oxidizer})_l \xrightarrow{k_l} (\text{gaseous products})_l \tag{12.24}$$

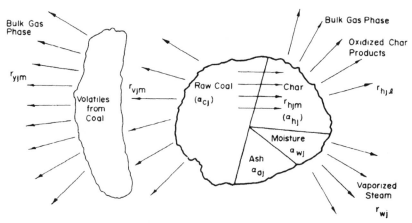

Figure 12.6. Schematic of coal particle, illustrating constituents and reaction processes. (Figure used with permission from Smoot and Pratt, 1979.)

The particle energy equation must be solved to obtain the particle temperature. The heat of reaction of the coal can be assigned to the gas phase, or partitioned between the two. We have found that gas and particle temperatures were greatly affected by the portion of the heat of reaction of coal given directly to the gas phase (Beck, 1980).

The particle is allowed to swell linearly with the extent of devolatilization. The average particle diameter increase is often on the order of 10% for highly volatile bituminous coals.

Experimental quantification of particle diameter change after devolatilization and during heterogeneous char oxidation has been discussed by Smith (1979). Two possible options regarding the diameter of the char residue have been suggested. After devolatilization, the particle could be envisioned as a porous sphere where reaction take place mainly within the pores themselves. As char combustion proceeds, the particle diameter can be taken to remain relatively constant, but with variable density until breakup or complete burnout of the fuel. This situation seems to be substantiated by visual observation of scanning electron micrographs of coal particle samples taken from combustors. Alternatively, the particle may be perceived to react mostly on the surface and thus burn out at a near-constant density, but as a shrinking particle. Methods for treating these options were outlined in Chapter 4.

Constitutive equations must be solved for heat and mass transfer to and from the particle surface. It has been found important to correct the heat- and mass-transfer relationships for high rates of mass transfer, since the devolatilization rates are particularly rapid.

The oxidation and gasification scheme suggested thus far combines the effects of film diffusion at high temperatures with chemical kinetics at low temperature. The kinetics are modeled from an nth-order global reaction rate (based on external surface area) with the oxidation mechanisms outlined in Chapter 4. The effects of this mass transfer on diffusion (Stefan flow) can be included by use of film theory, which allows a smooth transition between the rapid "blowing" characteristics of pyrolysis and the subsequent slower characteristics of heterogeneous oxidation. In this method it is possible to incorporate the option of allowing the particle to either shrink at constant density or decrease in density at constant particle size. This option, however, does not include intraparticle transport or pore structure monitoring. Little incentive has been provided for this level of modeling because experimental reaction rates and rate constants have usually been based on exterior surface area rather than on interior surface area (intrinsic kinetics). Intraparticle diffusion effects are supposedly absorbed into the global reaction expression (see Chapter 4).

Although a lack of intrinsic rate data exists in the literature and pore structure development is a little understood phenomenon, it is desirable to implement intrinsic rate expressions, because it allows the greatest potential of differentiating between the various char reactivities on the basis of fundamental kinetic parameters. (Note that while this is desirable in theory, the heterogeneous differences among various ranks of coal may render it somewhat less tractable in practice.) The derivation of such fundamental kinetics will necessitate a knowledge of the intrinsic kinetics of the char, its initial pore structure, pyrolysis temperature, heating rates, cenosphere growth, etc. Such influences seem to restrict modeling efforts to an idealized development. However, successful simulation of a generalized pore structure would greatly enhance the ability to optimize char parameters for a given reactor environment and burnoff.

Two major classifications of pore structure have evolved over the years: macroscopic and microscopic. Macroscopic modeling constitutes the majority of existing char models, including the classical unreacted shrinking core model, as discussed thus far, and the progressive conversion models. An adequate description of an effective diffusivity, which varies with burnoff and local position, appears to be the most ill-defined aspect of this model. The microscopic model attempts to describe diffusion through a single pore and then predicts the overall particle rates of an appropriate statistical description of the pore size distribution. Generally, these models may attempt to include the effects of pore coalescence, and hence must be derived from hypotheses in geometrical probability. The detailed pore description inherent in the microscopic models seems to provide a greater potential for successful predictive capability than the more simplistic macroscopic models.

Of the recent microscopic models, the three which appear to be 'most prevalent are those of Simons (1979b), Gavalas (1980), and Zygourakis et al. (1982). All three models attempt to solve either partially or completely a differential equation in the number density function or pore size distribution F:

$$\partial F/\partial t = (\partial/\partial r)(F \, \partial r/\partial t) = B - D \qquad (12.25)$$

where r refers to the pore radius and the B and D terms refer to the birth and death of pores, respectively. The model advanced by Simons (1979b) attempts to integrate an approximate form of the pore distribution function with the species continuity equation and obtain an analytical solution. A statistical derivation in conjunction with experimental data suggest an r^{-3} distribution for the pore number density. This infers that each pore that reaches the exterior surface of the char particle is the trunk of a pore tree with smaller pores as branches. Each tree trunk of radius r has an associated internal surface area within its branches on the order of r^3. The implication of this observation is that the pore tree will not parallel the behavior of a simple cylindrical pore since the trunk of the pore tree is responsible for feeding an extremely large surface area. Thus, small pore trees will be kinetically limited, while the larger trees are diffusion limited. This is the opposite of what would be concluded from the traditional Thiele approach applied to cylindrical, noninteracting pores.

Simons utilized a simplified form of a Langmuir expression to obtain analytical solutions from the species continuity equation. The analytical solutions from the kinetically limited pores, the diffusion limited pores, absorption limited pores, etc., are then patched together to encompass all possible limiting regimes. Although the model is efficient and validated for some char reactivities and appears to be relatively simple to code, it may be limited in its potential for improvement.

The random capillary model of Gavalas considers the porous structure of char as a set of straight cylindrical capillaries which intersect each other and partially overlap. The probability density function is defined as the density of intersections of the axis of any pore with a surface element and is assumed to remain constant with conversion. The exterior particle surface recedes when the pore walls merge. The species continuity equation is solved in terms of a fixed boundary problem via a semi-implicit finite-difference algorithm. The method monitors both the concentration profile and burnoff at each axial position within the pore. The computational challenge is somewhat more difficult than the more simplistic model advocated by Simons. It may also prove necessary to modify the code for gases involving a complete film diffusion limitation.

The model of Zygourakis et al., is an expansion of earlier work performed by Hashimoto and Silveston (1973a, b). The model attempts to transform Eqn. (12.25) into an equation for the moments μ_n of F where

$$\mu_n = \int_0^\infty r^n F \, dr \qquad (12.26)$$

The transformation is attractive because the surface area and porosity may be expressed as functions of the moments μ_n. The resulting differential equations, in terms of moments, may be solved directly for porous structure characteristics as a function of time. Although this approach is rather appealing, the many approximations made to render the moment equations tractable may have limited its potential drastically. Model parameters may be too simplistic to satisfy pore volume continuity equations over the entire range of burnoff.

Each of the microscopic models seems to have been derived in a detailed and complete manner. However, the disparity in the various methodologies demonstrates the lack of a common basis from which to attack the problem of heterogeneous char oxidation or gasification. Each method seems to have its advantages and limitations. It is doubtful whether any of the current microscopic models would easily lend themselves to potential improvements in range or scope as new insights become available. For that reason, it may be feasible to choose among the less complex derivations for the intermediate future (e.g., Zygourakis et al., 1982). Chapter 4 discussed microscopic models in more detail.

Perhaps a more practical limitation imposed on char modeling concerns the potentially dominating effects of other overall model operations. It is conceivable that particle interactions with fluid aerodynamics coupled with equilibrium chemistry, may dominate reaction parameters to such an extent that time-dependent char characteristics, as calculated by a refined char model, would not noticeably influence conversion.

12.6. PARTICLE SOURCE TERMS

So far we have discussed the overall gas-phase field equations in Chapters 9–11 and thus far in this chapter we have discussed different techniques for approaching the particulate and droplet phases. The recommended procedure has been a Lagrangian approach in which the droplets or particles are followed along individual representative trajectories. This section discusses the coupling between the carrier-gas field and the particulate or droplet field. This discussion is focused on an assumed Lagrangian treatment for the particulate phase. In such a procedure the Eulerian gas-phase equations are coupled to the Lagrangian particulate- or droplet-phase equations through particle source terms (S_p^m, S_p^u, S_p^v, S_p^h). The particle field is modeled as a set of discrete trajectories, where each trajectory represents a number of particles of uniform particle size and particle starting location in the primary jet. It is assumed that the particle number flow rate n_{ij} along a trajectory is constant, so that n_{ij} may be calculated from the initial mass flow rate of particles (m_{po}), the appropriate initial mass fractions representing particle size and location, and the initial particle mass α_{pio}. The particle number flow rate of the ith particle size and the jth starting location is calculated from the input conditions as follows:

$$\dot{n}_{ij} = \dot{m}_{po} X_{io} Y_{io} / \alpha_{pio} \qquad (12.27)$$

The particle mass source S_p^m to the gas phase is the change in mass of all particles that traverse the particular cell of interest. For the kth cell, this change is represented by

$$(\Delta \dot{m}_{pij})_{kcell} = \dot{n}_{ij} [(\alpha_{pij})_{out} - (\alpha_{pij})_{in}]_{kcell} \qquad (12.28)$$

The mass term is taken to be negative when the particle loses mass, in order to fit the gas-phase sign convention. The mass source term is calculated from

$$(S_p^m)_{kcell} = \left[\left(\sum_i \sum_j \Delta \dot{m}_{ij} \right) \Big/ V \right]_{kcell} \qquad (12.29)$$

The particle momentum source term S_p^u for the u component of momentum is similarly derived:

$$(S_p^u)_{kcell} = \left((1/V) \sum_i \sum_j n_{ij} [(u_{pij}\alpha_{pij})_{out} - (u_{pij}\alpha_{pij})_{in}] \right)_{kcell} \qquad (12.30)$$

The particle source term for the radial (v) component of momentum (S_p^v) can be similarly obtained.

The particle energy source term S_p^h represents the energy given to the gas phase by the particles. The heat generated from the coal reactions is given to the gas phase simply by giving the mass and enthalpy of the reacted portion of the coal to the gas phase. The heat of coal reaction does not enter into the equation for S_p^h unless this heat of reaction is partitioned between the gas and particle phases. The S_p^h equation also includes a term to account for radiation between the gas and the particles. The radiation level calculated is a function of time, and the time change across the cell must be included. The equation for S_p^h is

$$(S_p^h)_{kcell} = \left((1/V) \sum_i \sum_j n_{ij} [(h_{pij}\alpha_{pij} - tQ_{rp})_{out} - (h_{pij}\alpha_{pij} - tQ_{rp})_{in}] \right)_{kcell} \qquad (12.31)$$

where t is the elapsed time along the integrated particle trajectory. It should be noted that this equation includes any energy source or sink due to the addition or depletion of mass to the gas phase because of particle or droplet reaction.

12.7. ILLUSTRATIVE PROBLEMS

1. Examine the effect of turbulence on the particle momentum by time averaging Eqn. (12.1). Assume a constant drag coefficient and substitute the drag force of Eqn. (12.8) into Eqn. (12.1) before time averaging. Group terms together involving the mean field variables, drop the terms which are zero, and group together the cross correlations in the fluctuating variables. What is the physical significance of each term?

2. Derive Eqn. (12.18) from the definitions given in Chapter 10. List all requisite assumptions in your derivation.

3. Derive Eqn. (12.31) and explain why this equation includes addition or depletion of mass to or from the gas phase due to particle or droplet reaction.

CHEMISTRY AND TURBULENCE OF COAL REACTION PRODUCTS

13.1. INTRODUCTION

In Chapter 11 we discussed the impact of the turbulence on the gas-phase chemistry with particular application to gaseous fuels. We also examined the problems in obtaining time-mean reaction rates for chemical reactions in the homogeneous phase. We limited our discussion to turbulent diffusion flames and noted three types of turbulence–chemistry interactions. The majority of Chapter 11 focused on Type B flames when the reaction time scale was much less than the turbulent time scale and thus the turbulence mixing process controlled the chemical reactions. All of Chapter 11 focused on gaseous fuels. An appropriate mixture fraction or conserved scalar was defined; the statistics of the turbulence were characterized by an appropriate probability density function; and the mean chemical properties were obtained by convolution over this PDF of either chemical equilibrium properties or experimental data.

In the previous chapters we looked at the particulate and/or droplet phases in turbulent flows and focused on the effect of the turbulence on the motion of the particles or droplets. We also discussed the heterogeneous reactions involved with this added phase. We did not discuss the subsequent reactions of the products of the heterogeneous particulate- or droplet-phase reactions in the homogeneous phase. Throughout the book we have indicated the importance of these subsequent homogeneous reactions. The products of devolatilization, the gaseous products of heterogeneous oxidation reactions, evaporated moisture from water droplets or moist coal—all provide a potential source for further reactions in the homogeneous gas phase.

This chapter deals with the understanding and modeling of the effects of turbulence on the homogeneous reactions of the gaseous coal reaction products (i.e., off-gas) in the gas phase. This process is similar to the gas-phase reactions of gaseous fuel, but differs in that the source of fuel did not originate in the primary carrier gas but originated from the coal particles or water droplet phase. Thus, the source of fuel is continually appearing throughout the combustion or gasification process and not arriving from a simple inlet condition. The treatment discussed in this chapter will be somewhat analogous to that for gaseous fuels in Chapter 11. Limitations of the approach and alternate methods will be addressed. Finally, overall problems and long-range research needs will be identified.

13.2. COAL–GAS MIXTURE FRACTION

The concept of the conserved scalar or mixture fraction (f) as introduced in Chapter 11 was shown to be a useful progress variable to follow the progress of the mass fraction of fuel in a gaseous combuston environment throughout the turbulent flow field. This mixture fraction became a measure of the degree of mixing between the fuel and oxidizer streams. For Type B flames, the chemistry was only a function of this progress variable f. Methods were shown whereby the statistical distribution of f about its mean value could be described with the help of the turbulence model. Understanding the statistics of f proved useful in understanding the statistics and mean values of other dependent variables.

In particulate- and droplet-fired systems, the source of fuel to the gas phase originates in the particle or droplet. It would be convenient to follow the progress of some representative variable which could describe the local fuel mass fraction. Such a variable would be analogous to f in gaseous firing. This progress variable can be defined and is here called the coal–gas mixture fraction η. This variable is a measure of the local mass of gas originating from the coal particle or fuel droplet divided by the total local gas mass. That is

$$\eta = \dot{m}_c / (\dot{m}_p + \dot{m}_s + \dot{m}_c) \tag{13.1}$$

where $\dot{m}_c$ is the local mass flow rate of gas originating from the particle or droplet, and $\dot{m}_p$ and $\dot{m}_s$ are the local mass flow rates of gas originating from the primary or secondary streams, respectively. It is clear that η is simply a measure of the local mass fraction of gas originating from the particles or droplets. In most coal- or droplet-fired systems it is a measure of the local mass fraction of fuel in the gas steam. It is thus a progress variable measuring the evolution and mixing of the coal off-gas in the turbulent transporting fluid.

Since the coal off-gas is simply a component of the homogeneous continuum or transporting fluid, a transport equation of continuity can be derived for η analogous to that for f. The coal off-gas is convected and diffused through the gas flow field just as is f. The variable η is not, however, a conservative scalar like f. A source exists for the coal off-gas throughout the entire flow field. The steady-state continuity equation for η in turbulent flow field becomes

$$\nabla \cdot (\bar{\rho}\tilde{\mathbf{v}}\tilde{\eta} - D_\eta^t \nabla \tilde{\eta}) = \bar{S}_p^m \tag{13.2}$$

The convection and diffusion terms are the first and second terms, respectively, on the left-hand side of Eqn. (13.2). The source term is on the right-hand side. This source term is simply the rate of mass addition to the gas phase from the particles due to heterogeneous reactions, devolatilization, or liquid evaporation. This source term was briefly discussed in Chapter 12.

The turbulence in the transporting fluid affects η significantly. Different eddies will contain varying amounts of off-gas and thus different values for the progress variable η. Its value at any one point will fluctuate in time about some mean value. Of course, the eddies of transporting fluid also contain various amounts of coal particles or coal slurry droplets. Local temperatures and compositions will also be fluctuating due to the turbulent environment. The fluctuations undoubtedly affect the heterogeneous reactions in some nonlinear fashion. Thus, the source of coal gas coming from the droplets or particles is also fluctuating in time. It is the time-mean value of this particle mass source that is required in the right-hand side of Eqn. (13.2).

The effect of the turbulence on the mean particle mass source to the carrier gas is an important question in turbulent particle-laden or droplet-laden combustion applications. This mean source term is governed completely by the mean reaction rate of each of the individual particles or droplets. In Chapter 12 we discussed how to obtain these reaction rates for individual particles in a turbulent environment. We did not specifically address the effect of the turbulence on the reaction rate or how to obtain a time-mean reaction rate. Unlike fast, gas-phase reactions, the heterogeneous reactions, such as evaporation, devolatilization, or heterogeneous oxidation, are not mixing limited. The chemical kinetic limitations are crucial in understanding the operation of such combustion chambers. These reactions are, however, much slower than the gas-phase reactions. Indeed, most of these reactions fall within the Type A turbulent kinetics discussed in Chapter 11. In other words, many of these heterogeneous reactions may be considered slow enough to be unaffected by the turbulent fluctuations of the transporting fluid. Such reactions see only the mean properties of the mean carrier fluid. It is recommended that, initially, it be assumed that the reaction time is much longer than the turbulence time scale for heterogeneous reactions. Mean values can then be used in the kinetic expressions suggested in Chapter 12 for the heterogeneous particle reactions. The mean source term $\bar{S}_p^m$ is the resulting rate of mass addition to the carrier-gas fluid through these mean global heterogeneous reaction rates. Methods for relaxing this assumption are suggested in in Chapter 15. It is recognized from the outset that some evaporation rates and devolatilization reactions are rapid enough to cause some question concerning the validity of this assumption.

Heterogeneous reactions may sometimes be considered to be adequately slow to be considered Type A flames; however, this is not the case with the gas-phase reactions of the coal off-gas (measured by η) in the homogeneous carrier-gas phase. Once the coal off-gas leaves the vicinity of the particle surface and enters the bulk gas phase, the fluctuating fluid mechanics dramatically affects subsequent chemical reactions. When the homogeneous reactions are fast enough, the chemistry may be considered to be in local instantaneous equilibrium and mean properties can be obtained assuming mixing limited reactions as with gaseous firing. In such cases, η and f become the important progress variables and mean values of chemical properties are obtained by describing the statistics of these two variables. This fast-chemistry approximation will be described next. Later in the chapter, we will examine alternate approaches and review possible techniques for relaxing the fast-chemistry approximation.

If the only source of fuel for the combustion or gasification application of interest is from the coal particles or droplets, and if all of the remaining transporting fluid is composed of the same chemical composition with the same energy level, then η is the only required progress variable. Such is the case for practical combustors where the coal particles are transported with air in the primary stream and the secondary stream is composed of air at the same temperature as the primary stream. In such systems, the gaseous part of the primary and secondary streams have exactly the same composition and energy level. In this limiting case, the coal–gas mixture fraction η is adequate to describe the progress of the chemical field since the components of the gas can be broken into only two parts, namely, the coal off-gas and everything else.

In most applications, the primary carrier-gas fluid is different from the secondary gas in composition and/or temperature. In such cases, the gas phase must be broken into three components: (1) mass of gas originating from the primary stream, (2) mass originating from the secondary stream, and (3) the coal off-gas. In such cases, at least two progress variables are required to describe the gas field at any one point. The variable η must be used to describe the mass fraction of coal off-gas; in addition, the mixture fraction f provides the information regarding the amount of primary carrier-gas fluid. There is a significant advantage in maintaining the same definition for f as used in Chapter 11, namely,

$$f = \dot{m}_p/(\dot{m}_p + \dot{m}_s) \qquad (13.3)$$

In this case f is not the local mass fraction of gas originating from the primary stream, but instead is the mass fraction of the primary stream in the local mixture of primary plus secondary streams. This quantity f is useful for two reasons. First, η provides information regarding the partitioning of the local gas mass between gas originating from the particles and gas originating at the inlet (i.e., primary and secondary). The quantity $1 - \eta$ is the local mass fraction of gas originating at the inlet. The variable f provides information regarding the partitioning of this quantity of gas to primary or secondary sources. This partitioning is represented schematically in Figure 13.1. Second, since the mixture fraction f contains only information on the intermixing of the primary and secondary gas streams, it is somewhat insensitive to local values of η; thus, it is less dependent on the variable η than is the local mass fraction of primary fluid. This statistical independence is useful in calculating the local chemical composition. In most cases of pulverized-fuel combustion, the equivalence ratio will be independent of f, since the primary and secondary streams will have

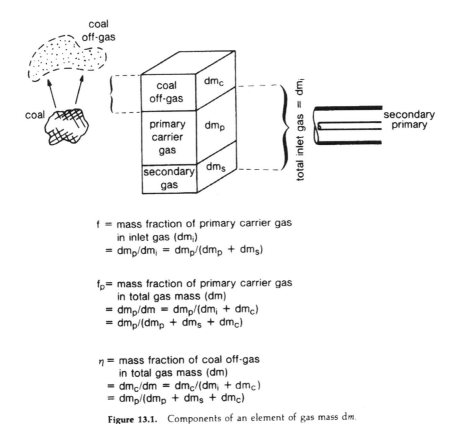

f = mass fraction of primary carrier gas
    in inlet gas (dm_i)
    = $dm_p/dm_i = dm_p/(dm_p + dm_s)$

f_p = mass fraction of primary carrier gas
    in total gas mass (dm)
    = $dm_p/dm = dm_p/(dm_i + dm_c)$
    = $dm_p/(dm_p + dm_s + dm_c)$

η = mass fraction of coal off-gas
    in total gas mass (dm)
    = $dm_c/dm = dm_c/(dm_i + dm_c)$
    = $dm_p/(dm_p + dm_s + dm_c)$

Figure 13.1. Components of an element of gas mass dm.

the same composition (i.e., air). Even in a few cases where f is required, the amount of coal off-gas at any point will be only weakly dependent on the partitioning between the primary and secondary gases; thus, to a first-order approximation, f and η can be considered statistically independent.

Although f, as defined in Eqn. (13.3), is a useful quantity for the calculational procedure, the local mass fraction of primary fluid is also an important parameter. This quantity defined as

$$f_p = \dot{m}_p/(\dot{m}_p + \dot{m}_s + \dot{m}_c) \tag{13.4}$$

is the conserved scalar for reacting particle-laden or droplet-laden systems. The variable f, as defined in Eqn. (13.3), is not a conserved scalar. The importance of f_p is obvious from this conservative principle. Its mean value can be calculated from a Favre averaged ($\tilde{f}_p$) conservation equation which includes convection and turbulent diffusion terms. There is no source term for the $\tilde{f}_p$ equation. From these definitions, it is clear that the relationship between f and f_p is

$$f_p = f(1 - \eta) \tag{13.5}$$

This equation quantifies the dependence of f_p on η.

13.3. MEAN CHEMICAL PROPERTIES

The progress variables f and η described in the last section are useful quantities for partitioning the components of the gas field. In order to obtain the mean chemical properties (i.e., temperature, density, composition, etc) of the gas field, we must understand the statistical distribution of these variables due to the presence of the turbulent fluid mechanics. In addition, each of the three components, coal off-gas, primary carrier gas, and secondary gas can react chemically to form new products. At this point, it is important to examine the properties of each of the three components, identify the statistics of each, and understand the implications of further chemical reactions among the three components.

We have discussed the properties of the primary carrier gas and the secondary gas in Chapter 11. Each has a constant chemical composition before subsequent reactions. The statistics at a given point vary about a mean value and can be described by the mixture fraction f. In particulate or droplet systems, $\tilde{f}$ can be obtained from a nonconservative (i.e., with a source term) transport equation. Alternatively, $\tilde{f}_p$ can be calculated from a conservative transport equation and $\tilde{f}$ obtained through the relationship shown in Eqn. (13.3). Of course, Eqn. (13.5) relates the instantaneous f and f_p variables. But if f_p can be considered statistically independent of η, then the relationship also holds true in the mean:

$$\tilde{f} = \tilde{f}_p / (1 - \tilde{\eta}) \tag{13.6}$$

As with the gaseous combustion case discussed in Chapter 11, the turbulent fluctuations on the mixture fraction can be obtained by solving a transport equation for the mean-square fluctuations or variance as shown in that chapter. All conserved properties of dm_i (see Figure 13.1) can be calculated from f in the same manner as shown in Chapter 11 for gaseous-fired systems. We recall, however, that the energy level is only a conserved property for a locally adiabatic system. When heat loss or other nonadiabatic conditions are significant, then the enthalpy must be obtained from an alternate approach. Considerations for this nonadiabatic operation are given later in this chapter.

The coal off-gas component dm_c of the element of gas mass dm has somewhat different characteristics from the other two components, dm_p and dm_s. While the primary and secondary gases have their own unique elemental composition, the coal off-gas is known to vary in elemental composition. This characteristic was discussed in the previous chapter. The mass evolved from the coal particles initially is enriched in hydrogen, whereas that evolved later in the particle history is enriched in carbon constituents. Indeed, each of the elements will have its own unique history for evolution from the particle or droplet. This history will be different for different coal types, and possibly for different operating conditions (i.e., type of oxidizer, heating rates, etc.). Figures 13.2–13.4 show experimental data for the fraction of three different elements released to the gas phase from the coal, for two different coal types. These data include many different samples taken from locations throughout a laboratory combustor. The analysis for composition was made on the solid-phase samples and the percent released obtained by simple difference from the original coal composition. If the composition of the coal off-gas was constant throughout its history then each of the elements released from the coal plotted against the coal burnout would lie on the 45° line in Figures 13.2–13.4. In such cases, the mass fraction of the element in the coal off-gas will be equal to the composition in the dry, ash-free coal. From these experimental data, it is apparent that carbon and nitrogen are released at about the same rate as the total coal gas. Figure 13.4 shows that the hydrogen apparently leaves the coal during the early stages of mass evolution. These three figures are representative of all the elements in coal. Of the major elements, carbon, hydrogen, nitrogen, and sulfur all seem to be released proportional to the total mass evolution except for hydrogen. These indicate that in order to completely describe the elemental composition of the coal off-gas (dm_c, see Figure 13.1) more than one progress variable is needed. Then, the coal off-gas (dm_c) depicted in Figure 13.1 could be broken

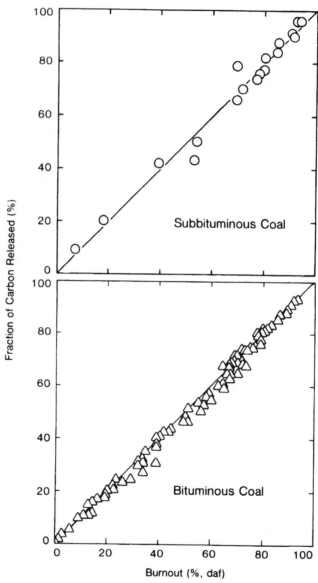

Figure 13.2. Fraction of carbon released to the gas phase from the coal particles for two different coal types (used with permission from Asay, 1982). Points are experimental data; solid line has slope of 1.0. Primary velocity ~15 m/s, primary temperature 300 K, and secondary temperature 590 K.

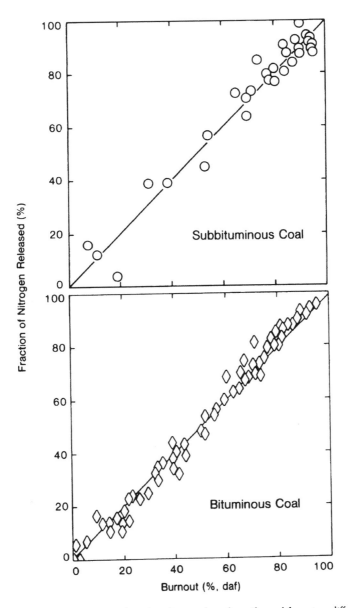

Figure 13.3.   Fraction of nitrogen released to the gas phase from the coal from two different coal types. Solid line has a slope of 1.0. (Figure used with permission from Asay, 1982.)

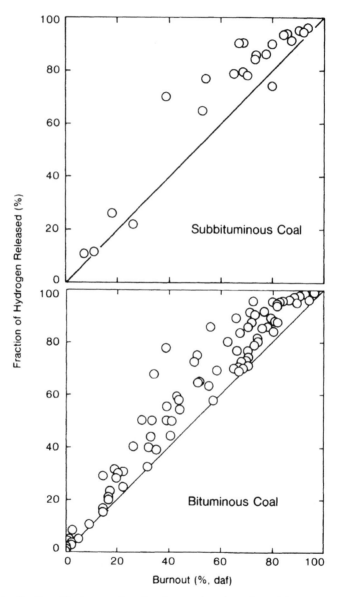

Figure 13.4. Fraction of hydrogen released to the gas phase from the coal for two different coal types. Solid line has a slope of 1.0. (Figure used with permission from Asay, 1982.)

into several subsets, each subset having a constant composition. Of course this complicates the mathematical description of the process, as we will see in section 13.4.

There are several different bases that could be chosen for subdividing the coal off-gas. Each subdivided component must have a progress variable η_m to track the evolution of that component and to describe the mass fraction of that component at any point throughout the flow field. Thus, each η_m must have its own transport equation analogous to Eqn. (13.2). The appropriate source term on the right-hand side of such an equation must be the rate of mass evolution of that component. Thus, whichever basis is chosen for η_m, an appropriate evolution rate for that component of mass must be known.

For example, one such basis might involve a separate progress variable for each heterogeneous reaction involved. There might be one progress variable η_1 for the low activation energy pyrolysis equation, another η_2 for the high activation energy pyrolysis reactions, and yet others, $\eta_3, \eta_4, \ldots$, for each of the heterogeneous oxidation reactions involved. On this basis, each reaction involved could produce products with different gaseous elemental compositions. In such a case, η_m would be the local mass fraction of gas originating from the mth reaction. The source term for the component is simply the reaction rate of each reaction.

As a second example, each η_m might represent the mass fraction of each element evolved from the coal particles or droplets. In this case there would be a different η for tracking the progress of each element (i.e., carbon, hydrogen, nitrogen, . . .) evolved from the particulate or droplet phases. Each η would require a transport equation with a mean source term which describes the mean rate of evolution of each element from the coal particle. The pyrolysis and heterogeneous reaction rates are not usually formulated on such a basis. Perhaps experimental work like that represented in Figures 13.2–13.4 might guide such a formulation.

We have so far described techniques for following the progress of the coal off-gas component of the homogeneous phase in the combustion and gasification of coal, and coal slurries. By far the simplest approach to use in combustion of multiphase flow processes would be to assume that the composition of the coal off-gas is constant. As was seen in Figures 13.2 and 13.3, this is a very good approximation for those elements evolved from the coal particles. The assumption is not as adequate for hydrogen (as seen in Figure 13.4) and is analogous to assuming that all of the data points in Figure 13.4 lie along the 45° solid line. The description which follows is for this special case when the coal off-gas has constant composition. The discussion is not encumbered with the added complexity of multiple η's. This first-order approximation may be adequate for many applications. The approach for extending this technique to many different progress variables should be apparent.

The overall objective of this section is to understand the importance of the turbulent fluctuations on the mean chemical properties in the gas phase of reacting coal, liquid fuels, and slurries. As we discussed in Chapter 11, very few gas-phase reactions are slow enough to be unaffected by the turbulent fluctuations. Many of the coal reactions are in the intermediate range where both chemical kinetics and turbulent fluctuations affect the mean reaction rates. Mathematical techniques for describing this regime of intermediate kinetics are very new and will be discussed somewhat in Chapter 15. NO_x pollutants are some of the important species which lie in this regime. In the remainder of this chapter, we will assume that the turbulent time scales are much longer than the chemical reaction times scales. It is thus assumed that the mixing process is the rate-limiting step in the chemical reaction sequence. With this assumption, local instantaneous homogeneous equilibrium can be calculated by the Gibbs free energy reduction scheme as discussed in Chapter 11. The only required information for the equilibrium computation is the local instantaneous energy level h, the pressure, and the elemental composition. For ideal, local adiabatic reactors at constant pressure, these quantities can be calculated directly from the progress variables defined in this chapter.

For example, with information known about the local instantaneous values of f and η, the corresponding elemental composition of the homogeneous phase can be calculated:

$$b_k = b_{k_c} \eta + (1 - \eta)[f b_{k_p} + (1 - f) b_{k_s}] \tag{13.7}$$

where b_{k_c}, b_{k_p}, and b_{k_s} are the kilogram-atoms of element k per kilogram of component. The components are those depicted in Figure 13.1, namely, the coal gas component c, the primary gas component p, and the secondary gas component s. Atom numbers have been used in this representation simply because of its utility in computational procedures. Mass fractions of each element would be equally conserved. The atom numbers b_k are simply an Avogadro's number of k atoms per kilogram of component.

A procedure for obtaining the mean density, mean temperature, and mean species mole fractions has now been prepared. If the equilibrium assumption is invoked on the grounds that micromixing is limiting, rather than chemical kinetics, then these properties are only functions of the two progress variables f and η. This follows from the fact that the pressure may be considered constant insofar as the equilibrium properties are concerned; thus

$$T = T(b_k, h) = T[b_k(f, \eta), h(f, \eta)] = T(f, \eta) \tag{13.8}$$

$$\rho = \rho(b_k, h) = \rho[b_k(f, \eta), h(f, \eta)] = \rho(f, \eta) \tag{13.9}$$

$$y_i = y_i(b_k, h) = y_i[b_k(f, \eta), h(f, \eta)] = y_i(f, \eta) \tag{13.10}$$

In other words for any equilibrium property β then

$$\beta = \beta(b_k, h) = \beta[b_k(f, \eta), h(f, \eta)] = \beta(f, \eta) \tag{13.11}$$

where the functional relationship of b_k or h to f and η is simply

$$b_k = b_{k_c}\eta + (1 - \eta)[fb_{k_p} + (1 - f)b_{k_s}] \tag{13.12}$$

The Favre-mean and conventional-mean gas properties (i.e., species composition, density, and temperature) can be obtained for any values of η and f by convolution over a joint probability density function:

$$\tilde{\beta} = \int_{-\infty}^{\infty} \int_{-\infty}^{\infty} \beta(\eta, f)\tilde{P}(\eta, f)\, df\, d\eta \tag{13.13}$$

Earlier, we discussed the independence of the variables f and η. The importance of this independence is now obvious. The joint probability density function $(P(\eta, f)$ is a complicated statistical function that is difficult to calculate. However, if η and f are statistically independent, then the joint probability density function can be separated:

$$\tilde{P}(\eta, f) = \tilde{P}(\eta)\tilde{P}(f) \tag{13.14}$$

In order to describe each PDF, the mean, variance, and shape must be quantified. The mean for each of these probability density functions is obtained from the Favre-mean continuity equation for η and f, respectively. The variance about that mean is obtained from the equations for g_η and g_f, which have already been discussed. A shape must be assumed for the probability density function as discussed in Chapter 11. Once the shape has been assumed, the complete mathematical description of the PDF can be formulated. This description is discussed briefly in the next section.

Sometimes the conventional, time-mean properties are desired or required, as for the density in each of the conservation equations. As in Chapter 11, this quantity can be obtained from the Favre probability density function

$$\bar{\beta} = \bar{\rho} \int_\eta \int_f [\beta(\eta, f)/\rho(\eta, f)]\tilde{P}(\eta, f)\, df\, d\eta \tag{13.15}$$

Intermittency occurs for both η and f and must be handled carefully. Physical limits on η and f range between 0 and 1 for both variables. Intermittency of pure inlet gas and pure coal gas is physically possible. In addition, the regions of pure inlet gas can be composed of eddies of mixed primary and secondary composition or could have intermittency of pure primary stream or pure secondary stream. Expansion of Eqn. (13.13) to include this intermittency results in the following expression, with intermediate steps required to obtain the final formulation shown for completeness. By substituting Eqn. (13.14) into Eqn. (13.13) and expanding:

$$\tilde{\beta} = \int_{-x}^{x} \left[\int_{-x}^{x} \tilde{P}(f)\beta(\eta, f)\,df \right] \tilde{P}(\eta)\,d\eta \tag{13.16}$$

$$\tilde{\beta} = \int_{-x}^{x} \left[\alpha_p\beta(\eta, 1) + \alpha_s\beta(\eta, 0) + \int_{0+}^{1-} \tilde{P}(f)\beta(\eta, f)\,df \right] \tilde{P}(\eta)\,d\eta \tag{13.17}$$

Finally,

$$\tilde{\beta}(\eta, f) = \alpha_i\beta_i + \alpha_i \left[\alpha_p\beta_p + \alpha_s\beta_s + \int_{0+}^{1-} \tilde{P}(f)\beta(0, f)\,df \right] + \alpha_p \int_{0+}^{1-} \tilde{P}(\eta)\beta(\eta, 1)\,d\eta$$

$$+ \alpha_s \int_{0+}^{1-} \tilde{P}(\eta)\beta(\eta, 0)\,d\eta + \int_{0+}^{1-} \int_{0+}^{1-} \tilde{P}(\eta)\tilde{P}(f)\beta(\eta, f)\,d\eta\,df \tag{13.18}$$

In this equation the intermittency of pure coal gas, pure inlet gas, pure primary stream, and pure secondary stream has been separated from the integrals. The integration over limits of $0+$ to $1-$ represents the integration over the continuous part of the probability density function, with the integral beginning at an infinitesimally small distance from the positive side of zero to the upper limit of an infinitesimally small distance from the lower side of 1. The integrals at the extremes of 0 and 1 are represented by the respective α's.

13.4. PROBABILITY DENSITY FUNCTIONS

Although, in principle, we have completely described the techniques for calculating the mean chemical properties, some mathematical representation of the probability density functions $\tilde{P}(f)$ and $\tilde{P}(\eta)$ must be defined to complete the mathematical description. A complete definition of the probability density function was given in Chapter 11. In that chapter we also discussed the physical significance of the PDF.

Table 11.1 showed the mathematical representation for a top-hat or flat-top probability density function for the mixture fraction f. As we discussed in Chapter 11, this form of the PDF is probably not the most physically realistic form. Although the turbulent fluctuations are probably not completely random, a normal or Gaussian probability distribution function can not be too far from the actual case, as long as intermittency is properly accounted for. Since the Gaussian distribution is defined from minus infinity to plus infinity, the assumption is made that all values beyond the physical limits of less than 0 or greater than 1 are representative of the intermittency at these end points. The Gaussian curve can be integrated for all values less than 0. This area can be represented by a δ function at 0 for the intermittency of fluid at 0. The same integration can be performed for the Gaussian curve for values above 1. The resulting clipped Gaussian curve can then be used to describe the continuous portion of the probability density function between 0 and 1 and the appropriate δ functions represent the intermittency of fluid at 0 and 1. It is apparent that such a clipping process changes the moments about the mean. For example, if we were examining the probability density function of the mixture fraction f, and there were intermittency of both primary and secondary streams (α_p, α_s), the clipping at 0 and 1 results in a continuous portion of the probability density function between 0 and 1 and two δ functions at each end. The continuous portion of the distribution between 0 and 1 no longer has a mean at $\bar{f}$, but a new mean which we will call F. Likewise, the original variance is no longer at g, but is at a new variance which we will call G_f. The introduction of the two new variables F and G_f requires two more mathematical expressions. These expressions are obtained by realizing that the mean value for the clipped Gaussian PDF with both δ functions, if present, must be the same as for the PDF for minus infinity to plus infinity. Likewise, both PDF's must have the same variance.

The mathematical representation of this new PDF is thus described as follows. First, the total area under the curve must be equal to 1:

$$1 = \alpha_p + \alpha_s + (2\pi G_f)^{-0.5} \int_{0+}^{1-} \exp\left[-(f-F)^2/2G\right] df \qquad (13.19)$$

The definitions of the intermittency have already been discussed:

$$\alpha_p = (2\pi G_f)^{-0.5} \int_{1}^{\infty} \exp\left[-(f-F)^2/2G\right] df \qquad (13.20)$$

$$\alpha_s = (2\pi G_f)^{-0.5} \int_{-\infty}^{0} \exp\left[-(f-F)^2/2G\right] df \qquad (13.21)$$

Finally, the new variables f and G_f are obtained implicitly from the definition of the mean and the variance:

$$\tilde{f} = \alpha_p + (2\pi G_f)^{-0.5} \int_{0+}^{1-} f \exp\left[-(f-F)^2/2G_f\right] df \qquad (13.22)$$

$$g_f = \alpha_p(1-\tilde{f})^2 + \alpha_s(-\tilde{f})^2 + (2\pi G_f)^{-0.5} \int_{0+}^{1-} (f-\tilde{f})^2 \exp\left[-(f-F)^2/2G_f\right] df$$

$$(13.23a)$$

$$g_f = \alpha_p - \tilde{f}^2 + (2\pi G_f)^{-0.5} \int_{0+}^{1-} \tilde{f}^2 \exp\left[-(f-F)^2/2G_f\right] df \qquad (13.23b)$$

With the assistance of the definitions above, this last equation can be simplified to:

$$g_f = \alpha_p - \tilde{f}^2 + (2\pi G_f)^{-0.5} \int_{0+}^{1-} \tilde{f}^2 \exp\left[-(f-F)^2/2G_f\right] df \qquad (13.24)$$

TABLE 13.1. Parameters for Gaussian Probability Density Function

$$\tilde{P}(\phi) = (2\pi G_\phi)^{-0.5} \exp\left(-Z_\phi^2/2\right)$$

$$\alpha = (2\pi)^{-0.5} \int_L^U \exp\left(-Z_\phi^2/2\right) dZ_\phi$$

Intermittency	ϕ	U	L	Z_ϕ
α_p	f	∞	$(1-F)/(G_f)^{0.5}$	$(f-F)/(G_f)^{0.5}$
α_s	f	$-F/(G_F)^{0.5}$	$-\infty$	$(f-F)/(G_f)^{0.5}$
α_c	η	∞	$(1-H)/(G_\eta)^{0.5}$	$(\eta-H)/(G_\eta)^{0.5}$
α_l	η	$-H/(G_\eta)^{0.5}$	$-\infty$	$(\eta-H)/(G_\eta)^{0.5}$

where F and G_f come from

$$\tilde{f} = \alpha_p + (2\pi G_f)^{-0.5} \int_{0+}^{1-} f \exp\left[-(f-F)^2/2G_f\right] df$$

$$g_f = \alpha_p - \tilde{f}^2 + (2\pi G_f)^{-0.5} \int_{0+}^{1-} f^2 \exp\left[-(f-F)^2/2G_f\right] df$$

and where H and g_η come from

$$\tilde{\eta} = \alpha_c + (2\pi G_\eta)^{-0.5} \int_{0+}^{1-} \eta \exp\left[-(\eta-H)^2/2G_\eta\right] d\eta$$

$$g_\eta = \alpha_c + (2\pi G_\eta)^{-0.5} \int_{0+}^{1-} \eta^2 \exp\left[-(\eta-H)^2/2G_\eta\right] d\eta$$

Again, analogous equations exist for the PDF of the coal–gas mixture fraction, η. All of these expressions are summarized for the clipped Gaussian probability density function in Table 13.1.

It is immediately obvious from Table 13.1 that the variables F, H, G_f, and G_η cannot be obtained explicitly but exist in an implicit form only. However, it is clear that for every $\tilde{f}$ and g_f there exists only one F and one G_f. This is also true for the coal–gas mixture fraction. Since the mean values of the mixture fraction and their variances are predicted from transport equations, it is most convenient if values of F, H, G_f, and G_η could be available explicitly for the calculated mixture fractions. This is most easily accomplished by tabulating values for these variables as a function of the mixture fractions and their variances. This tabulation needs to be done only once for all computations, since they are unique functions of the mixture fractions and their variances. The computation to construct the table is most easily constructed by numerical iteration.

13.5. ENERGY BALANCE CONSIDERATIONS

The overall technique discussed thus far in this chapter has dealt with strictly adiabatic, fast-chemistry reactions in the gas phase. By now it is clear that these assumptions are intimately tied to the overall approach for finding mean chemical properties in the gas phase. In Chapter 15 we will be looking at novel ways of extending these theories to include chemical kinetics. All of these approaches require that the chemical properties be functions of the mixture fractions only. This requirement is dictated by the turbulent field. Since we know very little about the history of the gas molecules in the turbulent system, the statistics of fluctuating turbulence are incorporated by predicting the statistics of the mixture fractions. All variables that have the same statistics as the mixture fractions can then be calculated.

One of the greatest limitations in this approach is the requirement that the enthalpy or energy level have the same statistics. We have seen how this assumption is valid when the local homogeneous chemistry may be considered adiabatic. In most practical combustion or gasification applications this is not the case. The purpose of pulverized-coal combustors is to obtain heat from the system. Thus the adiabatic assumption is not generally valid. The approach shown in Chapter 11 for residual enthalpies for gaseous systems can be extended to the coal-gas mixture fraction approach. Equations (11.38)–(11.45), and the adjoining discussion, indicated how the local instantaneous equilibrium properties are functions of f and h for nonadiabatic gaseous combustion. The enthalpy was then partitioned, and the fluctuations of the residual enthalpy were ignored in Eqn. (11.40)–(11.42). For reacting particles and/or droplets, the local instantaneous equilibrium properties are functions of f, η, and h. Again, the enthalpy can be partitioned:

$$\tilde{h} = \tilde{h}_{f,\eta} + \tilde{h}_r \tag{13.25}$$

$$\tilde{h} = h_c \tilde{\eta} + (1 - \tilde{\eta})[\tilde{f} h_p + (1 - \tilde{f}) h_s] + \tilde{h}_r \tag{13.26}$$

where h_c is taken as the initial particle enthalpy. Again it is assumed that the effects of the fluctuations in h_r are small compared with the effects of the fluctuations in $h_{f,\eta}$ and η. The result is that the mean properties of the gas phase are functions of f and η alone and thus can be calculated, as before, from Eqn. (13.18). The Eulerian gas-phase energy equation was given in Chapter 11, Eqn. (11.33), and the source term was shown in Eqn. (12.30).

Computationally, a table of values may be constructed for the instantaneous values of the gas-phase properties (β). The table would be a function of f, η, and h_r. In some cases the large storage requirement for the table may override the computational time required for a complete equilibrium solution at each point.

Although this technique is computationally tractable, the required assumptions are not physically meaningful in many cases. When the residual enthalpy is small, this approach is adequate, but far from exact. In the approach outlined, h_c must be taken to be constant. Selective heating or cooling of the individual particles due to radiative heat transfer obviously takes place; thus, in reality, the value of h_c will vary throughout the life history of the particles. However, it might be said in favor of this approach, that the sensible energy is not typically the dominant term. Chemical energy, which is directly accounted for in the equilibrium computation, is usually the dominant factor in computing local temperatures. It is important to realize that this technique is not adequate when significant heat loss occurs. Ignoring the fluctuations in h_r is inappropriate for reactors with significant amounts of heat transfer.

An alternate approach, which has been used by some investigators, is to approximate the amount of heat loss from the reactor and apply this fractional heat loss to every point throughout the reactor. For example, the total amount of sensible heat that could be lost from the reactor is approximated by the following equation:

$$Q_t = C_p(T_a - T_\infty) \tag{13.27}$$

where T_∞ is the ambient temperature and T_a is the adiabatic temperature. By assuming a fractional heat loss from the reactor (γ), each of the local instantaneous enthalpies are reduced by the local fractional heat loss. This fractional heat loss is assumed to be constant throughout the reactor:

$$h = h_a - Q_t = h_a - \gamma C_p (T_a - T_\infty) \qquad (13.28)$$

The enthalpy from Eqn. (13.28) is then used to calculate the local gas properties from the equilibrium computations as discussed previously. The adiabatic enthalpy h_a is that which is computed from the mixture fractions, and the adiabatic temperature T_a is that which would be calculated with no heat loss at the local values of the local instantaneous mixture fractions. Obviously, this approach is not as rigorous as solving the energy equation with appropriate boundary conditions; however, this approach does allow an approximation to the heat loss without excessive computational time and storage costs. The major effect of this heat-loss approximation is to lower gas temperatures by a certain percentage throughout the computational domain. The resulting temperature change affects the velocity field and the coal reaction rates, thereby affecting the overall flame performance.

The importance of the energy equation and appropriate heat-loss considerations varies from reactor to reactor. Figure 13.5 shows a comparison of computations made with the three different assumptions discussed in this chapter. This reactor was a pulverized-coal combustion reactor. The combustion was occurring in a cylindrical, down-fired, single-burner, laboratory reactor discussed in Section 5.4b. Figure 13.5a shows isotherms as calculated by the model with the full enthalpy equation and fluctuations ignored for the residual enthalpy (Eqns. (13.25) and (13.26)). Figure 13.5b shows isotherms with a completely adiabatic reactor assumed. Figure 13.5c shows heat-loss factor included where γ equals 0.5. The computations of Figure 13.5a used a boundary condition of a constant top wall temperature of 1000 K and a constant sidewall temperature of 700 K. In Figure 13.5a the effect of the wall on the flow-field temperatures is shown. The curvature near the walls is apparent. The profiles are more pronounced when heat loss at the wall is considered. The radial profiles appear flatter in Figure 13.5b due to the adiabatic assumption. Figure 13.5c again shows flat profiles because of the constant heat-loss factor. The large heat loss of 50% assumed for this computation shows a dramatic change in the overall flame structure. Ignition occurs late in the reactor due to heat loss earlier in the reactor. It is apparent that this reactor, even with relatively cool walls, operates near adiabatic conditions. Assuming a value for γ that is too large would result in significant error.

Other reactors may have significant heat loss and in such cases the heat-loss factor approach may be useful. It is clear that a more detailed theory for the effect of turbulent fluctuations on chemical reactions with significant heat loss could have an important impact on computational methods for combustion and gasification of coal and coal slurries.

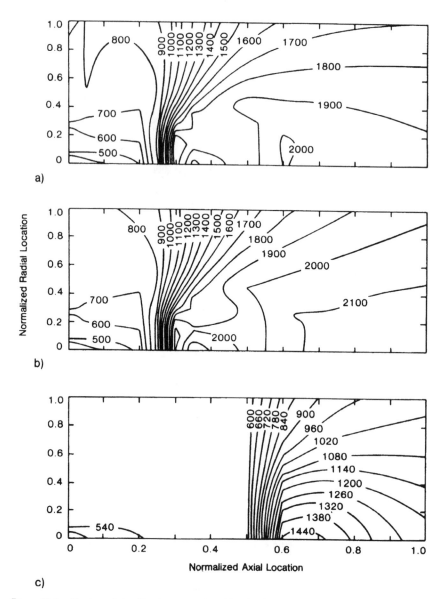

Figure 13.5. Contour plots of isotherms predicted by the three energy equation options discussed in the text: (a) full energy equation and ignoring fluctuations in h_r; (b) assuming adiabatic reactor operation; and (c) heat-loss factor γ with $\gamma = 0.5$.

13.6. PROBLEMS AND RESEARCH NEEDS

In this chapter we have discussed some techniques and options for including the turbulent fluctuations in the computations of gas-phase chemical properties. We have focused on the importance of incorporating turbulent fluctuations in the computation. We have noted that including the statistics of the turbulence on the gas-phase chemistry complicates the computational procedure significantly. The importance of these fluctuations is more intuitively obvious to combustion of gaseous fuels than it is to the combustion of coal and coal slurries. In combustion of gaseous fuels, it seems apparent that the turbulent micromixing process might be one of the rate-limiting steps. This is not so apparent with the heterogeneous fuels. In Chapter 12 and in previous chapters we saw that heterogeneous reaction, such as devolatilization, and heterogeneous oxidation were relatively slow reactions compared to fast homogeneous reactions. Since the source of all fuel in the gas phase comes from the solid particles or liquid droplets and since this source of fuel is being evolved relatively slowly from the particles or droplets into the gas phase, it might be argued that the

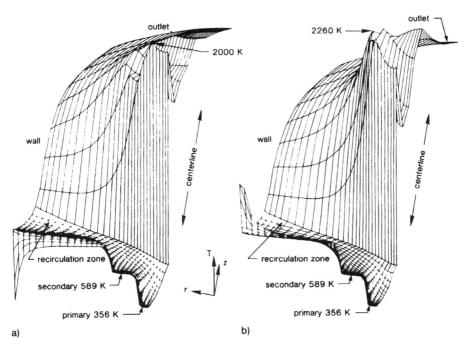

a)   b)

Figure 13.6. Three-dimensional projection of computed temperature surface for near-stoichiometric combustion of pulverized coal showing the predicted importance of turbulent fluctuations: (a) temperature surface including fluctuations; (b) temperature surface excluding fluctuations.

effect of the turbulent fluctuations on the resulting gas-phase reactions may not be significant. Is it possible to ignore turbulent fluctuations for heterogeneous fuels and still maintain the important qualitative features of such flames?

Figure 13.6 compares the mean temperature profile for two different computations testing the importance of the gas-phase turbulent fluctuations in coal combustion computations. This computation was performed for a lifted coal combustion flame in a laboratory axisymmetric, single-burner, down-fired furnace. Figure 13.6a shows a three-dimensional projection of the temperature profile throughout the combustor. The mesh superimposed upon the surface represents the computational grid. The centerline wall inlet and outlet are marked. The computation was performed using the energy equation and ignoring the fluctuations in h_r. The outer wall was kept at a temperature of 700 K and the top wall at 1000 K. These are the same conditions as for those shown in Figure 13.5. The heatup region can be seen before ignition of the coal particles. Devolatilization causes a high peak in the gas temperatures followed by a region of lower-temperature, fuel-rich, devolatilization products. It is apparent that the fluctuations in η caused a significant smoothing of temperature peaks, as

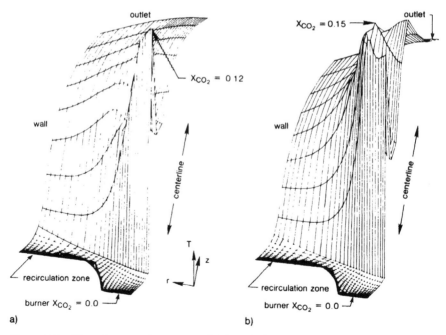

Figure 13.7. Three-dimensional projection of CO_2 mole fractions as predicted from the model discussed in the text for conditions of Figure 13.6 showing effect of fluctuation on composition: (a) CO_2 mole fraction including fluctuations; (b) CO_2 mole fraction excluding fluctuations.

would be expected. Figure 13.6b shows a high-temperature ridge near the stoichiometric point for the coal–gas mixture fraction. The temperature profiles from centerline to wall are much flatter when the turbulent fluctuations in the coal–gas mixture fractions are included. A difference of 260 K is noted in the peak temperatures. It is apparent that the turbulent fluctuations are important in predicting the structure of the flame resulting from combustion of pulverized coal. Ignoring the fluctuations causes the flame to be very localized. The turbulent fluctuations spread the reaction zone over a wider area and thus moderate high-temperature peaks significantly.

Figure 13.7 shows the effect of the fluctuations in η on one representative species mole fraction. The same conclusions can be made regarding the mole fraction of CO_2 as were made regarding the temperature. The flattening of the composition profile caused by the turbulent fluctuations is even more prevalent in the CO_2 profiles than in those for temperature. Both Figures 13.6 and 13.7 show a significant peak near the devolatilization point, even when fluctuations are included. This peak results from the rapid devolatilization reactions as discussed in Chapter 3 and seems to be even more pronounced in the predictions than in any of the experimental data. Even though the fluctuations are included in the gas-phase chemical properties, these fluctuations are ignored in the heterogeneous kinetics. The devolatilization reactions may be too rapid to be considered nonfluctuating. This seems to indicate that there may need to be more smoothing due to turbulent fluctuations than is included in the present approach.

The smoothing caused by the fluctuations has a beneficial effect on the numerics of the solution algorithm. The steep gradients caused by the predictions with no turbulent fluctuations result in numerical instabilities. The smoothing caused by the fluctuations helps relax these instabilities to make the computational algorithm much more stable, even though an added equation must be solved.

Selected coal combustion and gasification comparisons of theory with experimental data are shown in Figures 13.8–13.10. Agreement for these coal reaction predictions is not as good as for gaseous combustion but given the complexity of the physical and chemical processes, agreement is acceptable. These few figures demonstrate that coal processes can be simulated with acceptable accuracy to predict overall trends.

One of the overriding observations of modeling reacting coal systems is the many approximations and assumptions that have to be incorporated in order to describe the systems with current levels of understanding and computer capabilities. With the many assumptions required, some would propose that modeling these types of systems should not even be attempted. They argue that the level of assumptions required makes any predictions simply speculative. We have attempted to rely heavily on experimental data where inadequate

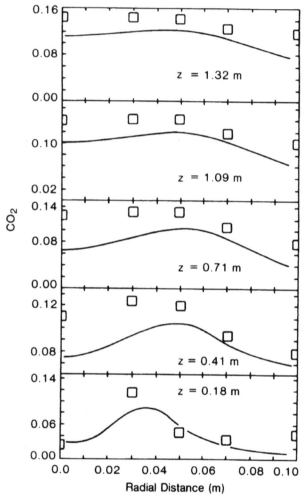

Figure 13.8. Comparison of measured and predicted radial profiles of CO_2 mole fractions at various axial locations for laboratory axisymmetric combustor with subbituminous coal. (Data from Asay, 1982; predictions from Hill, 1983.) Reactor conditions are discussed in Chapter 8.

knowledge was available about the physical and chemical processes. With this engineering approach, we have shown that many phenomena can be described with relatively simple approximations. It is true, however, that unjustified proliferation of simplifying assumptions can cause the predictions from these kinds of models to be misleading or nonuseful. The same is true of extending realistic assumptions to regions beyond the levels of validation.

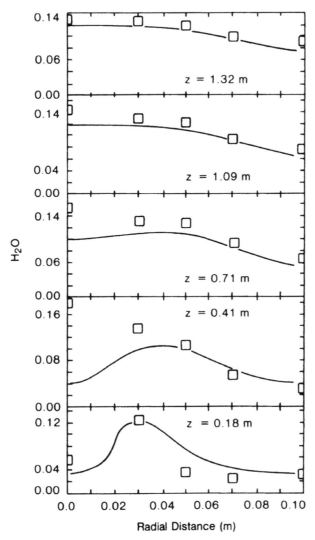

Figure 13.9. Comparison of measured and predicted radial profiles of H_2O mole fractions at various axial locations for a laboratory axisymmetric combustor with subbituminous coal (data from Asay, 1982; predictions from Hill, 1983). Reactor conditions are discussed in Chapter 8.

A great deal of effort remains to be accomplished in evaluating these theories and approximations in simple burner configurations. Applying these techniques to multiburner, transient, large-scale furnaces is beyond the scope of current validated code capabilities. Essentially, no validation studies have been performed with any transient coal codes. Reaching a steady state with

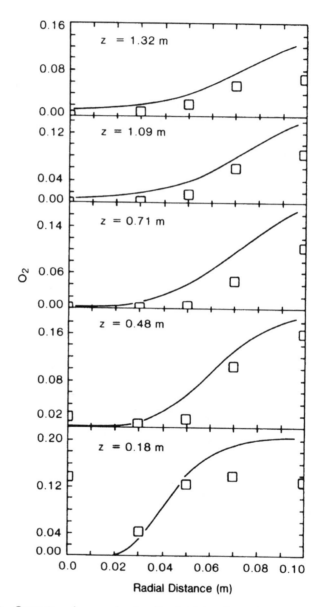

Figure 13.10. Comparison of measured and predicted radial profiles of O_2 mole fractions at various axial locations for a laboratory axisymmetric combustor with subbituminous coal (data from Asay, 1982; predictions from Hill, 1983). Reactor conditions are discussed in Chapter 8.

such codes currently seems to be impractical. Resolving the computational problems caused by scale in large, multiple-burner boilers also appears to be beyond current computer technology. In addition, there are many remaining issues for simple burner configuration which require more research for adequate resolution.

A promising application of any multidimensional code to larger-scale systems is in individual burner design. In the near-burner region, the turbulent mixing process is of paramount importance. Theories such as those discussed in this chapter could be invaluable in evaluating burner options in the flame of these burners, and to some extent, in evaluating the pollutant output from such burners (see Chapter 15).

The approaches outlined in this chapter are useful at the current state of the art, and are also a foundation for adding new submodels as further insight is obtained in the fundamental processes involved in coal combustion and gasification.

Mathematical modeling of pulverized-coal systems, including the direct combustion or gasification of coal and coal slurries, requires inclusion of many physical and chemical processes which are not fully understood. It is principally because of this lack of information with respect to the fundamental processes that it is mandatory to link the mathematical modeling strategy with an associated, experimental program. It is the underlying philosophy of this book that through this joint effort greater insight will be obtained to the basic combustion processes. Current predictive techniques can provide a quantitative description of local jet behavior in the near-burner field, providing a valuable interpretative tool for experimental data.

Mathematical modeling at all levels can provide useful information. Table 7.12 showed a comparison of the levels of treatment of three different coal combustion or gasification models. Each of these models has different potential use. The details of the turbulent mixing process are not needed for all applications. One argument for lower levels of sophistication is obvious by examining Table 7.12. Computational time increases exponentially with the different levels of sophistication. For many applications, the additional sophistication is not warranted.

For some applications (e.g., burner design) local details are required. Multi-dimensional computer models can provide insight into the controlling processes involved. The use of these models requires not only sophisticated computer equipment, but also qualified, experienced users. Currently, these models cannot be used as a "black box" calculation of coal reaction chambers. Simplified input and color-graphic output display improves the usability of these computer programs; but to interpret the output requires some detailed knowledge of the assumptions made with regard to the physical and chemical processes, as well as an understanding of the numerical algorithms involved. In many applications for coal combustion and gasification, the transient startup of the reactor is also of some interest. However, transient computer programs require more computational time. It is questionable whether transient coal combustion codes can reach a steady state in realistic computational run times.

The discussion of each of the submodels involved in the coal combustion and gasification processes indicates that much more research needs to be done with each component involved in the overall theory. Two aspects of the reaction sequence are of particlar concern: (1) coal particle turbulent dispersion and (2) interactions between turbulence and chemistry. New *in situ* laser measurement techniques and the availability of larger and faster computers, together hold great potential for significant advances in these areas in the near future. Individual laboratories must focus on both the modeling and experiments to provide new insights to these processes. With the combined effort in experimental technology and mathematical modeling efforts, greater insight to the complex phenomena occurring in combustion and gasification processes can be obtained. This knowledge undoubtedly will contribute to better design and operation of coal combustors and gasifiers in the near future.

13.7. ILLUSTRATIVE PROBLEMS

1. Calculate the value of η for complete combustion of the coal shown in Table 15.4 with air at an overall stoichiometric ratio of 1.2 (i.e., 20% excess air).

2. Calculate the elemental composition of the gas phase for coal gasification of the coal shown in Table 15.4, when $f = 0.5$ and $\eta = 0.1$. The primary carrier is O_2 at a flow rate such that 20% of the coal can react with the available O_2. The secondary stream is all steam such that after O_2 combustion, enough H_2O is available to react exactly with the remaining carbon in the fuel.

But the path of the just is as the shining light, that
shineth more and more unto the perfect day.
Proverbs 4:18

14

RADIATIVE HEAT TRANSFER

14.1. INTRODUCTION

We have focused on the turbulent fluid mechanics, the chemical reactions, and
the interactions between these two aspects of combustion and gasification. In
this chapter we change emphasis and briefly examine the dominant heat-transfer
mechanism in coal-fired combustors and gasifiers; namely, radiative heat transfer. Radiation is electromagnetic energy in transport. A photon of energy
interacts with an individual molecule to excite it to increased energy states. As
the molecule relaxes to lower-energy levels, different photons are emitted.
Radiation can be transmitted at wavelengths anywhere from 1 nm to 1 km
corresponding to x-ray radiation and radiowave radiation, respectively. Of
interest to heat-transfer calculations is thermal radiation within the range of
0.1–100 μm in wavelength. In combustion and gasification processes, the most
significant contributions to thermoenergy arrive from radiation having
wavelengths between 0.5 and 10 μm.

When dealing with the fluid mechanics and the chemistry, we have studied
the importance of including the turbulence and its effect on various properties.
It was essential to examine the dominant time scales for the different rate
processes and then determine which physical processes control the overall
combustion and gasification processes. Radiative heat transfer is a rate process.
The photons are traveling at the speed of light which is a rate much faster
than any other process occurring in the combustion and gasification environment.
However, the rate at which various surfaces react to the photon flux is a more
important consideration. Typically, it is found that irradiated surfaces experience
no difficulty in maintaining equilibrium among the energy states associated with
the absorption process at any given wavelength. Most of the theories for
radiative heat transfer within enclosures are based on the assumption that, with

respect to the radiative heat transfer, the transfer of energy is a pseudoequilibrium process. Thus, although surfaces and volumes may be changing in temperature with time, the radiative heat transfer is assumed to respond to these changes as if they were in equilibrium.

The details of equilibrium thermal radiation in a multiphase reaction chamber is a complex physical process that involves a massive compilation of knowledge. This one chapter is intended only to survey the issues involved and outline different approaches which have been taken to develop a computational procedure for incorporating radiative heat transfer into a comprehensive predictive coal combustion and gasification model. The restriction that the radiative-heat-transfer equations must be incorporated into a model which includes the turbulent fluid mechanics, chemistry, and other physical and chemical processes forces added restrictions on the radiative-heat-transfer model used. The calculational model for the radiation component must be computationally efficient to permit its inclusion in the comprehensive model. As with other submodels discussed in this book, it seems apparent that there is a significant disadvantage in formulating a treatment which is so mathematically elaborate as to obscure the physical nature of the problem. Finally, the numerical procedure used for solving the radiative-heat-transfer equations must combine well with the transport equations for the other processes. This last requirement may not be immediately obvious, but will be discussed briefly later in the chapter.

The complexities of the radiative-heat-transfer process in combustion and gasification systems are due mainly to the participating medium. In general, the system in combustors and gasifiers consists of nonuniform, emitting, absorbing, reflecting, solid surfaces that are in complex geometrical configurations and surround a multiphase, nonuniform, emitting, absorbing, transmitting, scattering particle or droplet-laden fluid which is also generating heat. The problem of finding the equilibrium conditions at every point in this system is complex. In addition to the mathematical difficulty, one of the more serious problems is the lack of knowledge of radiative properties of the materials involved in the system. The walls are typically coated with ash or slag of unknown composition and emittance. Fundamental radiative properties of the coal particles exist for only a very few selected coal types. Further, the properties of the particles are changing as they undergo chemical reaction, changing from mostly carbon particles to mostly ash.

The mathematical formulation of the radiation must provide heat-transfer rates to the appropriate equations of energy for the gas or particle/droplet phases. Depending on the form of the energy equation used (i.e., Eulerian or Lagrangian), the required radiative heat transfer will be in the form of a total radiative-heat-transfer rate to a single particle (i.e., energy per time) or a volumetric radiative-heat-transfer rate (i.e., energy per volume per time). In this chapter we will briefly present methods for computing these heat-transfer rates.

14.2. FUNDAMENTALS

There are several excellent textbooks on the topic of radiative heat transfer, these include Chandrasekhar (1960), Hottel and Sarofim (1967), and Siegel and Howell (1981). In this section, we will give a very brief review of some of the fundamental parameters involved in calculating radiative heat transfer. For a more detailed study, it is recommended that the aforementioned texts be studied.

14.2a. Intensity

The analysis of a radiative field is usually accomplished by the study of its intensity. To define intensity we must examine the energy flux in a pencil of radiation. In Figure 14.1 we consider a small area element dA with the unit normal vector **n**. Radiant energy is streaming through this area and in directions confined to an element of solid angle $d\Omega$. The angle θ is that angle between the outward normal **n** and the direction of the radiation considered. This defines a pencil of radiation. The energy transfer rate is proportional to dA and to $d\Omega$. The proportionality constant is called the intensity I of radiation with units of

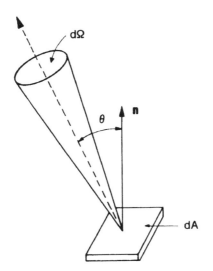

Figure 14.1.   Construction of a pencil of radiation.

energy per unit time per unit area per unit solid angle normal to the area:

$$dQ = I\,d\Omega\,\cos\theta\,dA \tag{14.1}$$

The energy flux q (i.e., energy/time/area) is the energy transfer rate Q (i.e., energy/time) per unit area; thus

$$dq = dQ/dA = I\cos\theta\,d\Omega \tag{14.2}$$

It is obvious that a pencil of rays with no divergence or with finite divergence but zero cross-sectional area (i.e., along a focal plane) contains no energy.

The intensity along a pencil of radiation is constant. In Figure 14.2, the two ends of the pencil are depicted by area dA_1 and area dA_2. They are located a distance R apart. The energy transfer rates through the two ends are equal and given by

$$dQ_1 = I_1\,dA_1\,d\Omega_1 = I_1\,dA_1\,dA_2/R^2 \tag{14.3}$$

$$dQ_2 = I_2\,dA_2\,d\Omega_2 = I_2\,dA_2\,dA_1/R^2 \tag{14.4}$$

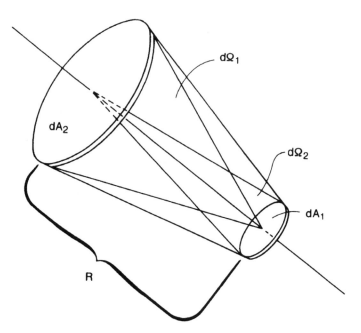

Figure 14.2. Invariance of intensity for a pencil of radiation

Since the two energy transfer rates are equal, then in the absence of absorption or refraction:

$$I_1 = I_2 \tag{14.5}$$

Thus, intensity is constant along a pencil. The increase in cross-sectional area normal to the pencil is exactly compensated for by the decrease in divergence angle.

The concept of intensity can be applied to the total radiation over the entire wavelength spectrum as well as to radiation at a given wavelength (monochromatic). It is often preferred to base the monochromatic intensity on frequency rather than wavelength since the later changes while passing from one medium to another due to changing indexes of refraction, while the frequency remains the same. The relationship between the total intensity I and the spectral intensity I_ν is simply

$$I = \int_0^\infty I_\nu \, d\nu \tag{14.6}$$

Although in theory, the intensity varies with the direction of the radiation, for most applications, the radiation field may be considered isotropic at a point and thus independent of the direction at that point. Surfaces which conform to this restriction are called diffuse surfaces.

14.2b. Emissive Power

A blackbody or ideal radiator is a body which emits and absorbs the maximum possible amount of radiation at any given wavelength and at any temperature. A blackbody is a theoretical concept which sets an upper limit to the emission of radiation in accordance with the second law of thermodynamics. It is a standard with which the radiation characteristics of other media may be compared. A black surface is an isotropic emitter and thus blackbody radiation is perfectly diffuse. For this isotropic radiation the energy flux at dA throughout the 2π steradians on one side of the surface is given by

$$q = I \int_{2\pi} \cos \theta \, d\Omega \tag{14.7}$$

This hemispherical energy flux is called the emissive power of the surface. The emissive power of a blackbody is designated by E_b and its intensity by I_b. By carrying out the integration of Eqn. (14.7) we see that

$$q = E_b = \pi I_b \tag{14.8}$$

TABLE 14.1. Stefan–Boltzmann Constant σ in Various Units

0.1713×10^{-8}	BTU ft^{-2} hr^{-1} (°R)$^{-4}$
4.88×10^{-8}	kcal m^{-2} hr^{-1} K^{-4}
1.356×10^{-12}	cal cm^{-2} sec^{-1} K^{-4}
5.67×10^{-12}	watts cm^{-2} K^{-4}

The famous Stefan–Boltsmann law relates the emissive power of a black surface to its absolute temperature:

$$E_b = \sigma T^4 \qquad (14.9)$$

where the proportionality constant σ is the Stefan–Boltzmann constant of Table 14.1.

The total emissive power given in Eqn. (14.9) represents the total thermal radiation emitted over the entire wavelength spectrum at a given temperature T. Of course, the blackbody emissive power is spectral in nature, meaning that it is a function of the wavelength of the radiation. The total emissive power and the monocromatic emissive power are related by

$$E_b = \int_0^x E_{b\lambda} \, d\lambda = \sigma T^4 \qquad (14.10)$$

A relationship which shows how the emissive power of a blackbody is distributed among the different wavelengths was derived by Max Planck in 1900 by means of quantum theory. The resulting hemispherical spectral emissive power as a function of wavelength can be found in any text on radiative heat transfer (i.e., Siegel and Howell, 1981).

With the radiative behavior of a blackbody serving as the standard we realize that the radiative behavior of real bodies depends on many factors, such as composition, surface finish, temperature, wavelength of the radiation, angle at which radiation is either being emitted or intercepted by the surface, and the spectral distribution of the radiation incident on the surface. Various emissive, absorptive, and reflective properties, both unaveraged and averaged, are used to describe the radiative behavior of real materials relative to blackbody behavior. The emissivity is a measure of how well a body can radiate energy as compared with a blackbody. Designating E_λ as the monochromatic emissive power of a real (nonblack) surface, the monochromatic hemispherical emittance or emissivity* of the surface ε_λ is defined by the relationship

$$E_\lambda = \varepsilon_\lambda E_{b\lambda} \tag{14.11}$$

Often it is more convenient to integrate the emissivity over all wavelengths to obtain the total hemispherical emissivity for the nonblack surface:

$$\varepsilon = E/E_b = \left(\int_0^\infty \varepsilon_\lambda E_{b\lambda}\, d\lambda \right) \Big/ \left(\int_0^\infty E_{b\lambda}\, d\lambda \right) \tag{14.12}$$

14.2c. Absorption

As a monochromatic pencil of radiation approaches the interface between two media 1 and 2 it will undergo different behavior at the interface. The radiation is partly reflected and the ratio of the reflected to the incident energy may be denoted by ρ_λ. The rest enters medium 2 in which a fraction α_λ is absorbed and the fraction τ_λ travels on and is finally absorbed in other media. In equilibrium the relationship

$$\rho_\lambda + \alpha_\lambda + \tau_\lambda = 1 \tag{14.13}$$

must hold. At opaque surfaces, the transmissivity (τ_λ) is 0 and the radiation is only altered by absorption and reflection. The reflected radiation is often characterized by scattering coefficients and the reflecting process referred to as scattering. Thus, as radiation traverses a continuous medium, it is weakened by its interactions with the continuum. The energy flux not transmitted is due to absorption and/or scattering and is sometimes referred to as attenuation, extinction, or interception.

The absorbed energy flux is that which does not re-emerge from the body as either a transmitted or scattered flux. The absorbed flux is converted into stored internal energy, measured by the body temperature. The ability to absorb energy is measured by the absorptivity α. It is obvious that if a surface will absorb radiant energy, then it will also emit radiation and thus there is a relationship between the absorptivity and the emissivity. Kirchhoff's law relates these two quantities and is usually stated for exchange between surfaces in radiative equilibrium. It states that the emissivity ε and the absorptivity α of such a surface are the same. Kirchhoff's law actually applies to monochromatic radiation, even though the surface may not be at equilibrium, and a net radiative

*Some authors distinguish between emissivity and emittance by reserving the use of emissivity for pure substances; throughout this book they will be used interchangeably.

flux may exist. The reason for this is that both α_λ and ε_λ are surface properties which depend solely on the condition of the surface. If that surface experiences no difficulty in maintaining equilibrium among the energy states associated with the absorption process at the given wavelength λ, then

$$\alpha_\lambda = \varepsilon_\lambda \qquad (14.14)$$

Since, in general, the emissivity and absorptivity of surfaces vary with the wavelength, a change in the temperature of the radiation source (and, consequently, in the spectral distribution of energy) incident on the surface changes the absorptivity.

In performing radiation calculations it can often be assumed that the surfaces are diffuse and grey. "Diffuse" signifies that the emission and thus the absorption of the surface is isotropic (i.e., independent of direction). "Grey" signifies that the spectral emissivity and absorptivity do not depend on wavelength. Thus, the total spectral absorptivity α equals the monochromatic absorptivity α_λ. These properties can, however, depend on temperature. Thus, at each surface temperature, the emitted radiation will be the same fraction of blackbody radiation for all wavelengths. Thus, even under conditions of net interchange, the emissivity and absorptivity of grey bodies are identical. It should be noted that the temperature effect on ε or α of bodies which are approximately grey is generally quite small.

For radiation in combustion chambers we are often concerned with radiative attenuation and emission from a volume of space. The effect of absorption on a pencil of radiation traversing and absorbing and scattering continuum is expressed by the Bouguer–Lambert Law. The change in intensity of the incident beam has been found experimentally to depend on the magnitude of the local intensity. The fractional decrease in intensity over a small distance dx is proportional to the distance:

$$dI = -IK_t\, dx \qquad (14.15)$$

Or upon integration:

$$I = I_0 \exp\left(-K_t x\right) \qquad (14.16)$$

For monochromatic radiation or for total radiation in a grey medium, the integration of Eqn. (14.15), resulting in Eqn. (14.16), is the Bouguer–Lambert Law, where I_0 is the incident intensity entering the volume of thickness x. It is obvious from our definition of transmissivity that

$$I/I_0 = \exp\left(-K_t x\right) = \tau \qquad (14.17)$$

The proportionality constant K_t is the total extinction coefficient of the continuum. It is a physical property of the material and it has units of reciprocal length. As we have already seen, it is composed of two parts—an absorption coefficient K_a and a scattering coefficient K_s, that is,

$$K_t = K_a + K_s \qquad (14.18)$$

For a volume of space occupied by absorbing and emitting gases and particles or droplets, the overall extinction, absorption, and scattering coefficients can be computed by adding the corresponding coefficients for the gases and particles or droplets present in the medium. It should be apparent that these overall coefficients are related to the individual emissivities, absorptivities, and reflectivities of the individual surfaces which make up the cloud (Hottel and Sarofim, 1967). This relationship is discussed further in Section 14.5.

14.2d. Scattering

In coal combustion and gasification applications, the presence of the particulate or droplet phases provides sources within the control volume for scattering or radiation. The interaction of a particle with radiation incident upon it depends on three dimensionless quantities: the complex refractive index $n(1 - \kappa i)$; the ratio of the characteristic particle dimension to wavelength of radiation which is sometimes called the particle size parameter $\pi d/\lambda$; and a particle shape. In principle, the relationship among these three particle parameters and the incident radiation may be determined by solving Maxwell's equations with the appropriate boundary conditions corresponding to the different particle parameters. In practice, such solutions are only available for a very limited number of particle parameters. Fortunately, limiting solutions may be obtained without recourse to Maxwell's equations. These limiting conditions include very large particles and particles corresponding with Rayleigh and Mie scattering. Detailed discussions of scattering are given by Van de Hulst (1957), Hottel and Sarofim (1967), and Siegel and Howell (1981).

When considering the effect of scattering on incident pencils of radiation, we must deal with two types of scattered light. The decrease in intensity due to scattered rays, which makes up a part of the extinction coefficient already discussed, is due to out-scattering. This is the part of the incident beam that is scattered in other directions, thereby reducing its intensity. The radiant intensity of the pencil may also be increased by scattering centers surrounding the volume element and thus scattering energy into the direction of the beam considered. This is known as in-scattering.

To formulate quantitatively the concept of scattering, we must specify the angular distribution of the scattered radiation. This is introduced by the phase function $P(\theta, \phi)$. This phase function has the physical interpretation of being the scattered intensity in a direction divided by the intensity that would be scattered in that direction if the scattering were isotropic. For isotropic scattering the phase function $P(\theta, \phi)$ equals 1. Since the sum of the probabilities over all directions must equal unity, the phase function is defined so that

$$(1/4\pi) \int_{\Omega = 4\pi} P(\theta, \phi)\, d\Omega = 1 \qquad (14.19)$$

When the scattered energy distribution is not uniform in all directions, but tends to intensify in one certain direction, the scattering is considered to be anisotropic. This phenomenon exists when the intersecting particles are larger in size than the wavelength of radiation.

The intensity of scattered radiation by a single particle may be computed by the solution of Maxwell's wave equation. This method developed by Gustove Mie is known as Mie theory (Chandrasekhar, 1960; Seigel and Howell, 1981). At low particle densities, the scattered intensity is equal to that from a single particle multiplied by the number of particles. Such an effect is called single scattering. As the particle density increases, the intensity of radiation, which has been scattered two or three times, becomes significant and this effect is known as multiple scattering. The dominant form of radiative heat transfer encountered in pulverized-coal combustion and gasification is both multiple and anisotropic.

Calculations which include contributions from scattering are usually based on the scattering cross section C_s. This is the apparent area that an object presents to an incident beam insofar as the ability of the object to deflect radiation from the beam is concerned. This apparent area may be quite different from the physical cross section of the scatterers. Table 14.2 summarizes scattering regimes for particles for some limiting cases. The scattering cross section not only depends upon the particle size but upon the material of the scattering body and state of the radiation (i.e., wavelength, polarization, and coherence).

The principles involved in scattering of radiation by particles are very involved. A thorough coverage of these principles is given by Van de Hulst

TABLE 14.2. Summary of Types of Scattering for Particles of Diameter d

Particle size	Scattering type	Scattering cross section C_s
$\lambda \gg d$, single scattering	Rayleigh	Proportional to V^2/λ^4
$\lambda \approx d$	Mie	Varies widely
$\lambda \ll d$	Fraunhofer and Fresnel diffraction plus reflection	$\sim 2(\pi d^2/4)$

(1957). For opaque coal particles, the scattering is caused by both reflection of the incident and defraction by the coal particle. The contribution from both sources makes up the total scattering cross section C_s. For example, for very large particles, as shown in Table 14.2, the scattering cross section is twice the projected particle area. Half of this contribution comes from the reflected radiation and the other half from the defracted radiation. The angle at which defraction rings occur is inversely proportional to particle size. Thus, for smaller and smaller particles, the defracted radiation is scattered into larger and larger angles. For particles of the same size range as the wavelength of the radiation, the contribution of the defraction must be calculated from the solution of Maxwell's equations. When the Mie equations or the Rayleigh equations are solved for scattering of smaller particles, the contribution due to defraction can be even more significant.

It must be realized that scattering by real coal particles is much more complex than can realistically be calculated. Scattering centers in real combustors and gasifiers consist of water droplets, char and coal particles of irregular shape and nonuniform composition, ash particles, soot particles, and particles consisting of mixtures of all of the above. The most complete solution to Maxwell's equations discussed so far (Mie theory) is only applicable to spherical particles of constant composition. It is obvious that many assumptions have to be made to include radiative heat transfer in a combustion or gasification application.

14.3   GENERAL EQUATION

Radiation in coal combustors and gasifiers is one example of a wide class of problems in which the radiative heat transfer is occurring in an absorbing medium where multiple scatter is important. Within such a scattering–absorbing medium, the intensity I of radiation is a function of position and direction. Now consider a pencil of radiation of intensity I passing normal to a surface element dA and traversing the distance dl. The change in intensity of the incident beam with respect to the path length dl is

$$dI/dl = -(K_a + K_s)I + K_a I_b + (K_s/4\pi) \int_{4\pi} I(\Omega)P(\theta, \phi)\, d\Omega \qquad (14.20)$$

The first term on the right-hand side of the Eqn. (14.20) represents the extinction of the radiation due to absorption and out-scattering. The second term shows how the beam is augmented due to emission of the material along the path length. The third term represents the augmentation of the incident beam due to in-scattering. This is the general radiative-heat-transfer equation for an absorbing–scattering medium. It is an integral–differential equation generally known as the transport equation. The integral term in Eqn. (14.20) introduces the contribution of multiple scattering.

14.4. SOLUTION METHODS

Exact solutions of the multiple scattering problem defined by Eqn. 14.20 have been discussed by Chandrasekhar (1960). These exact solutions are limited to isotropic scattering and infinite dispersion and they have only been obtained for simple geometries. Even in these cases, the numerical calculations are exceedingly detailed.

For the applications of interest to this book, the radiation undergoes attenuation and augmentation by several gas, particle, and droplet sources. These sources are continuously changing throughout the combustion or gasification process. Many simplifications have to be made in order to arrive at a viable computational strategy for the radiative heat transfer within the volume.

Due to the integral–differential nature of the equation of radiative transfer, the equation does not lend itself well to solution by numerical methods consistent with the methods used to solve for the fluid mechanics and chemistry of combustion and gasification. The main area of difficulty in Eqn. (14.20) is the last term on the right-hand side (in-scattering). This term is in integral form and must be reduced to a differential form in order to simplify the problem. Chandrasekhar (1960) has suggested that this be achieved by substituting the Gaussian quadrature formula for the integral term. Several methods for solving the equation have been presented over the past several years. We will discuss the most important of these in turn.

14.4a. The Hottel Zone Method

The zone method has been used with good success for many radiative-heat-transfer calculations for a number of years. The technique calculates the complete solution for a grey gas system and has been altered in some applications to treat nongrey effects. To date, the method has not been used for computations which have included three-dimensional calculations of the fluid mechanics and chemistry involved in combustion chambers. Most applications have used either assumed or measured heat release and flow patterns. In the limit of small control columns, the method is an exact formulation of radiative transfer. The technique breaks the computational volume into finite zones and solves for the interactions among zones. It results in a matrix system of equations which is not banded and for this reason does not couple easily to the differential equations for flow and chemical reaction. Excessive computer time and storage require-ments are demanded to accomplish any practical solution. For this reason, the method is not recommended for simultaneous heat-transfer and flow equations; however, the method offers good generality and high accuracy.

14.4b. The Monte Carlo Method

Monte Carlo methods for radiative heat transfer are potentially attractive because of the wide variety of geometries that can be considered. In this technique, bundles of intensities are generated within each control volume and radiated in many random directions. The individual rays are followed to extinction and allowed to undergo absorption, emission, and scattering as they pass through the continuum and interact with the walls. It has been proven to be a very powerful, rigorous, and flexible method to solve the general equations for radiative heat transfer. In order to accurately predict the overall performance of the combustion chamber, many random rays are required. The result is large computing time and large computer storage; again, probably too large to be included with simultaneous fluid mechanics solutions.

14.4c. The Diffusion Method

When considering an absorbing, scattering, emitting medium in dynamic radiative equilibrium, it is sometimes possible to consider the photon to be undergoing a diffusion process analogous to molecular transport processes. Obviously, the processes are somewhat different. Molecules collide with one another and travel with a Maxwell distribution of speeds. Photons collide with particles, not with one another, and their speed is constant. However, when systems are heavily loaded, and the mean free path of the photons is small relative to the system dimensions, then the diffusion approximation is an attractive simplification. Hottel and Sarofim (1967) have shown how the technique can be generalized to include effects of anisotropic scatter. In general, the mean free path of the photon in combustors and gasifiers is too long to be considered a diffusion process. Some densely loaded, high-pressure gasifiers have very short beam lengths. For such systems, the diffusion approximation might be an attractive alternative.

14.4d. Flux Methods

Flux methods are based on simplifying assumptions for angular variation of radiant intensities. This allows the exact integral–differential radiation transport equation to be reduced to a system of ordinary differential equations. These differential equations are ideally suited to simultaneous, finite-difference numerical solutions. They have been shown to be computationally economical and reasonably accurate. The method was originated and commonly used by astrophysicists for one-dimensional cases. Different models have been developed for four- and six-flux applications. Obviously, including more fluxes improves the accuracy. Some different flux models have been developed to include the effects of multiple and anisotropic scattering. All are relatively easy to use.

One flux method which has received increasing attention recently is the discrete ordinates method. A solution is found by solving the exact radiative transport equation for a set of discrete directions spanning the full range of solid angles. Angular integrals of intensity are evaluated by numerical quadrature. Unlike the Monte Carlo methods, the rays are not followed to extinction but only to interception by the wall. The number of ordinates chosen increases the accuracy of the technique.

The flux methods are recommended for incorporation into pulverized-coal combustion and gasification applications where the flow equations are also solved. The equations for one such method are given in the next section.

14.4e. Recommended Flux Method

The radiation submodel must provide the appropriate terms in the Lagrangian equation of energy and the Eulerian gas-phase energy equation. These terms are Q_{rp} for the particles and q'_{rg} for the gas. The two terms have different units due to the different Lagrangian and Eulerian approaches. The Lagrangian term Q_{rp} is the total radiative-heat-transfer rate to a single particle (i.e., energy/time). The Eulerian term q'_{rg} is the volumetric radiative-heat-transfer rate to the gas (i.e., energy/volume/time). With the aid of the local particle number density, the volumetric radiative-heat-transfer rate for the particles is obtained:

$$q'_{rp} = Q_{rp} n_p \tag{14.21}$$

The total volumetric radiative-heat-transfer rate is thus made up of particle and gas contributions:

$$q'_r = q'_{rg} + q'_{rp} \tag{14.22}$$

It is the task of the flux model to obtain these Eulerian rates.

In the flux approximation to the governing integral–differential radiative exchange equation, the anisotropic scattering term is approximated by assuming that all the radiation is scattered into one of six mutually orthogonal directions: forward, backward, and to the four sides. In this way, the integral term reduces to the sum of four algebraic terms and the governing equations become first-order, linear, differential equations (Chu and Churchill, 1955; Hottel and Sarofim, 1967).

The forward scattered component is

$$f = 2\pi \int_0^{\pi/2} P(\theta) \sin\theta \cos^2\theta \, d\theta \tag{14.23}$$

where $P(\theta)$ is the phase function for anisotropic particle scatter. The backward scattering component is

$$b = 2\pi \int_{\pi/2}^{\pi} P(\theta) \sin\theta \, d\theta \tag{14.24}$$

and the four sidewise scattering components are

$$s = \tfrac{1}{4}(1 - f - b) \tag{14.25}$$

For isotropic scattering, the phase function $P(\theta)$ is 1 and the above fractions degenerate to $\tfrac{1}{6}(f - b - s - 1)$.

The six transfer equations are then derived for the forward and backward directions of each of the three coordinate directions:

$$(1/K_t)(dI_x^+/dx) = -(1 - \omega_0 f)I_x^+ + \omega_0 b I_x^- + \omega_0 s(I_r^+ + I_r^- + I_\theta^+ + I_\theta^-)$$

$$+ \tfrac{1}{6}(1 - \omega_0)I_v \tag{14.26}$$

$$-(1/K_t)(dI_x^-/dx) = -(1 - \omega_0 f)I_x^- + \omega_0 b I_x^+ + \omega_0 s(I_r^+ + I_r^- + I_\theta^+ + I_\theta^-)$$

$$+ \tfrac{1}{6}(1 - \omega_0)I_b \tag{14.27}$$

$$(1/K_t)(dI_r^+/dr) = -(1 - \omega_0 f)I_r^+ + \omega_0 b I_r^- + \omega_0 s(I_x^+ + I_x^- + I_\theta^+ + I_\theta^-)$$

$$+ \tfrac{1}{6}(1 - \omega_0)I_b \tag{14.28}$$

$$-(1/K_t)(dI_r^-/dr) = -(1 - \omega_0 f)I_r^- + \omega_0 b I_r^+ + \omega_0 s(I_x^+ + I_x^- + I_\theta^+ + I_\theta^-)$$

$$+ \tfrac{1}{6}(1 - \omega_0)I_b \tag{14.29}$$

$$(1/K_t)(dI_\theta^+/d\theta) = -(1 - \omega_0 f)I_\theta^+ + \omega_0 b I_\theta^- + \omega_0 s(I_x^+ + I_x^- + I_r^+ + I_r^-)$$

$$+ \tfrac{1}{6}(1 - \omega_0)I_b \tag{14.30}$$

$$-(1/K_t)(dI_\theta^-/d\theta) = -(1 - \omega_0 f)I_\theta^- + \omega_0 b I_\theta^+ + \omega_0 s(I_x^+ + I_x^- + I_r^+ + I_r^-)$$

$$+ \tfrac{1}{6}(1 - \omega_0)I_b \tag{14.31}$$

TABLE 14.3.   Coefficients for Flux Sums in Eqns. 14.37–14.39 as Derived for the Six-Flux Method Shown in Eqns. 14.26–14.31

$$C_1 = -K_t(1 - \omega_0) + K_t[2\omega_0^2 s^2/(1 - \omega_0 f - \omega_0 b)]$$
$$C_2 = K_t\omega_0 b + K_t[2\omega_0^2 s^2/(1 - \omega_0 f - \omega_0 b)]$$
$$C_3 = K_t\omega_0 s + K_t[2\omega_0^2 s^2/(1 - \omega_0 f - \omega_0 b)]$$
$$C_4 = K_t[(1 - \omega_0)/6]\{1 + [2\omega_0 s/(1 - \omega_0 f - \omega_0 b)]\}$$
$$C_5 = \omega_0 s/(1 - \omega_0 f - \omega_0 b)$$
$$C_0 = (1 - \omega_0)/6(1 - \omega_0 f - \omega_0 b)$$

In each of these equations, I is a radiative flux, I_b is the blackbody emissive flux, and ω_0, which is the albedo for scatter, is expressed as

$$\omega_0 = K_s/(K_a + K_s) = K_s/K_t. \tag{14.32}$$

where K_a and K_s are the absorption and scatter coefficients.

The assumption of axial symmetry is invoked by letting $dI_\theta^+/dr = dI_\theta^-/dr = 0$. Equations (14.30) and (14.31) then become algebraic expressions for I_θ^+ and I_θ^-. This eliminates two of the six differential flux equations. The remaining four, first-order equations can be combined into two second-order equations by following the approach of Gosman and Lockwood (1973). The net radiation heat flux q is defined as

$$q = I^+ - I^- \tag{14.33}$$

and the flux sum F by

$$F = I^+ + I^- \tag{14.34}$$

The addition of each pair of radiative flux equations results in an expression for heat transfer. For the x component,

$$q_x = -\Gamma_F dF_x/dx \tag{14.35}$$

where

$$\Gamma_F = K_t[1 - \omega_0(f - b)]^{-1} \tag{14.36}$$

By subtracting each pair and combining with the expressions for the heat transfer, two second-order, working equations are obtained:

$$-\frac{d}{dx}\left(\Gamma_F \frac{dF_x}{dx}\right) = (C_1 + C_2)F_x + 2C_3 F_r + 2C_4 I_b \tag{14.37}$$

and

$$\frac{1}{r}\frac{d}{dr}\left(\Gamma_F\frac{dr\,F_r}{dr}\right) = (C_1 + C_2)F_r + 2C_3F_x + 2C_4I_b \qquad (14.38)$$

together with

$$F_\theta = 2C_5(F_x + F_r) + 2C_6I_b \qquad (14.39)$$

The necessary coefficients are defined in Table 14.3. The total volumetric heat-transfer rate is then derived:

$$q_r = K_a(F_x + F_\theta + F_r - I_b) \qquad (14.40)$$

Computing contributions of gases and particles is accomplished through the respective absorption coefficients:

$$K_a = K_{ag} + K_{ap} \qquad (14.41)$$

Then, from Eqns. (14.22) and (14.40),

$$q'_{rg} = K_{ag}(F_x + F_r + F_\theta - I_b) \qquad (14.42)$$

and

$$q'_{rp} = K_{ap}(F_x + F_r + F_\theta - I_b) \qquad (14.43)$$

It should also be mentioned that the six fluxes may be recovered from the solution procedure if desired. For example, for the x direction it can be shown that

$$I_x^+ = \tfrac{1}{2}(F_x + q_x) \qquad (14.44)$$

and

$$I_x^- = \tfrac{1}{2}(F_x - q_x) \qquad (14.45)$$

14.5. COEFFICIENTS

Before any of the suggested solution techniques can be used to calculate the radiative heat transfer in a combustor or gasifier, several coefficients and parameters must be defined. Some of these parameters are fundamental definitions. Others involve experimental coefficients which are obtained from data on various properties of the materials.

First, we will define a few parameters from more fundamental, measurable properties. The radiative medium of interest to coal combustors and gasifiers is composed of the transporting gas and the particulate or droplet phases. The absorption coefficient for these two phases is then

$$K_a = K_{ag} + K_{ap} \qquad (14.46)$$

If several particulate or gaseous phases are to be considered separately, they may be added directly into Eqn. (14.46). Often it is convenient to consider the continuum as a cloud of gases and particles that can emit and absorb radiation as a separate, single entity. In such cases it is desirable to relate the absorption coefficient to an emissivity of the cloud. The relationship between the cloud emissivity and the absorption coefficient can be calculated from the Bouguer–Lambert law, with contributions for both extinction and emission of radiation. Integration of the resulting differential equation and comparison to the definition of emissivity results in the following relationship between the emissivity and the absorption coefficient (see Problem 1):

$$\varepsilon = 1 - \exp(-K_a l) \qquad (14.47)$$

where l is a characteristic length or beam length for the cloud.

Since there is a full distribution of particle sizes and particle types within a combustor or gasifier, it is necessary to discretize the distribution (represent the continuous distribution by a finite number of discrete classes). In this way it is possible to calculate an absorption or scattering coefficient for a discrete number of particle sizes and types and thus compute an overall coefficient by summing over each of the individual particles. This method is simplified by the definition of efficiency factors for absorption scattering and extinction. For example, the efficiency factor for scattering (Q_s) is the ratio of the scattering cross section to the particle geometric cross section:

$$Q_s = C_s / \pi r^2 \qquad (14.48)$$

Similar efficiency factors can be defined for absorption and total extinction, extinction or attenuation being the additive effect of both absorption and scattering. The overall absorption and scattering coefficients for each of the particles in the jth classification (i.e., size and/or type) is

$$K_{aj} = (\pi/4)Q_a n_j d_j^2 \qquad (14.49)$$

$$K_{sj} = (\pi/4)Q_s n_j d_j^2 \qquad (14.50)$$

Summing all of the above absorption coefficients for gases, and over all particle classifications, the overall absorption coefficient for a small volume or cloud is

$$K_a = (1/l)\left[\ln(1 - \varepsilon_g) + (\pi/4)\sum_j (Q_a n_j d_j^2)\right] \tag{14.51}$$

The overall scattering coefficient is

$$K_s = (\pi/4)\sum_j (Q_s n_j d_j^2) \tag{14.52}$$

The problem of obtaining the proper coefficients for any of the models mentioned in this chapter is now reduced to obtaining absorption and scattering efficiencies for all particles and droplets of interest to the combustion and gasification process and to obtaining emissivities for the gases and the bounding surfaces in such reaction chambers. These coefficients are those which are based on properties of the surfaces and molecules themselves.

The optical properties for the particulates involved in combustion and gasification can be obtained from a solution of the Maxwell equations based on Mie theory if the refractive index of the particle is known. For example, char particles have been estimated to have an index of refraction of 1.93 $(1-0.53i)$ at a wavelength of $2\,\mu$m (Smoot and Pratt, 1979). For such a particle, Mie theory gives absorption, scattering, and extinction efficiencies as a function of the particle size as shown in Figure 14.3. In addition to giving the efficiencies as shown, the Mie theory also gives the angular distribution or phase function of the scattered radiation for each particle size.

Ash particles present a different type of problem. The uncertainties in evaluating ash radiation are due mainly to difficulties in selecting a mean ash particle size and in identifying the refractive indices for ash particles in order to obtain efficiency factors. Ash particles are known to behave as if they were opaque to radiation in the near-infrared region, but transparent to radiation in the visible region of the spectrum. Some of the mineral matter in pulverized coal is free and some is bound in molecular form in organic matter; but the majority is finely distributed as micron-size inclusions throughout the coal matrix. During combustion the mineral matter decomposes, fuses and agglomerates, as the carbonaceous matrix reacts. The number and size of ash particles resulting from any given initial coal particle size distribution will vary significantly, depending on the conditions of the combustion or gasification reactor and upon the types of mineral matter involved in the coal.

Some recent work has been done on obtaining the refractive indices for ash particles. Sarofim and Hottel (1978) and Smoot and Pratt (1979) have

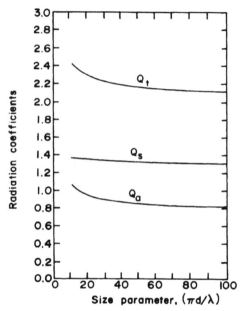

Figure 14.3. The absorption, scattering, and extinction efficiencies of char particles at wavelength of 2 μm and refractive index of 1.93 [1–0.53 i]. (Figure used with permission from Smoot and Pratt, 1979.)

presented various values for the complex refractive index, and for the resulting phase function and absorption efficiencies for ash particles in coal combustion and gasification environments.

Soot particles within these flames can also contribute significantly to the heat-transfer characteristics of the combustion process. Soot particles absorb highly, but scatter less significantly, due to their microscopic size. The particle size of the soot depends upon the fuel and combustion conditions. For all reported sizes, it is apparent that the size parameter is less than 0.4 and therefore sufficiently small to permit calculation of the absorption efficiency by the limiting form of the Mie equations known as the Rayleigh regime. The limiting factor in predictions of soot radiation is not uncertainty in optical properties, but uncertainty in soot concentrations in furnaces. Soot is undoubtedly formed in the fuel-rich regimes in flames in amounts that differ significantly with fuel type, mixing pattern, and temperature. The following approach is only an approximate method of estimating the soot loading of various flames.

To obtain the emittance of the soot flame, the theory used for ash and char particle clouds is also applicable. A mean diameter of 65 nm is typical of the range of values reported for mean diameters of soot particles in boilers. For particles of this size, Rayleigh scattering applies. In this case, the soot particle cloud emittance reduces to

$$\varepsilon_s = 1 - (1 + 350 C_1 C_2 C_3 C_4 f_{vh} T_F L)^{-4} \tag{14.53}$$

as shown by Sarofim and Hottel (1978). C_1 is the mass fraction of fuel which is carbon and C_2 is the fraction of that carbon which is volatile. The fraction of the volatile carbon which forms soot is C_3. C_4 represents the fraction of the soot formed which is present in the radiating gases;

$$C_4 = 1/(1 + t_{res}/t_{react}) \tag{14.54}$$

where t_{res} is the residence time for the particles and t_{react} is the chemical reaction rate time for the soot particles. Sarofim and Hottel indicate that at average furnace temperatures, the residence time is about 100 times the chemical reaction rate time for soot particles. In a mixing-limited process, the time ratio should be nearer 10 or 20. Thus, C_4 equals $\frac{1}{11}$ or $\frac{1}{21}$. Firm values of C_3 are not available, but furnace and laboratory measurements place the value between 0 and 0.2. Recent developments may lead to improvements in this approximate method.

The gas radiation in flames is due mainly to the transitions in the vibrational and rotational energy levels of molecules, resulting in nonluminous radiation at specific frequencies (i.e., spectral bands). The dominant contributors to the nonluminous radiation are CO_2 and H_2O. CO and hydrocarbons can contribute to the emission and attenuation of radiation within the flames in furnaces, but their contributions are localized and of secondary importance. The total emittance of carbon dioxide and water vapor clouds has been investiated thoroughly. Several sources are available to obtain the emissivity of carbon dioxide *vs.* temperature or the emissivity of water vapor *vs.* temperature. Use of these complicated charts, together with spectral-band correlations, provide a good basis for computing nonluminous radiation in furnaces; however, the method is somewhat tedious and, for purposes of coal flames, simpler approaches are justified.

One useful method of presenting such data in an approximate manner is suggested by Hadvig (1970). The emissivity–temperature product is plotted against the sum of the partial pressure–length products for different ratios of H_2O to CO_2 partial pressures, as shown in Figure 14.4. To use this chart, partial pressures of CO_2 and H_2O must be known from the fluid mechanics and chemistry calculation.

The emittances of the wall deposits are also important to furnace radiation. Wall *et al.* (1979) indicate that because of existing uncertainties, estimates of absorption or scattering coefficients of fly ash on walls can only be obtained by direct measurement in operating furnaces; such measurements are rare. Additional research is therefore necessary to obtain such measurements and to relate

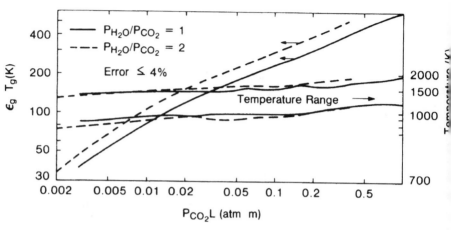

Figure 14.4. Gas emittance chart for restricted temperature range. (Figure used with permission from Hadvig, 1970.)

them to both laboratory measurements of particle size and the chemical and physical characteristics of fly ash. The most complete work of this type seems to have been done by Boow and Goard (1969). Materials composed mainly of oxides, magnesium, and silicon tend to be better emitters and absorbers than those containing high sodium or calcium. Data on emittance of various ash deposit surfaces gathered by Goetz *et al.* (1978) show the wide variability in total surface emittance as a function of coal type, surface temperature, etc. The variability in ash composition is here represented by the silica ratio (*Si*):

$$Si = \%SiO_2/(\%SiO_2 + \%CaO + \%Fe_2O_3 + \%MgO) \tag{14.55}$$

where the chemical symbols represent mass percentages of each compound in the ash. Figure 14.5 is a plot of total emittance of the ash surface *vs.* the temperature of the surface. The lines are plotted with constant ash composition and each is labeled with the corresponding Si ratio. The strong variation of total emittance with surface temperature is immediately obvious. At lower temperatures, the emittance of the deposit is much greater than at high temperatures. All of the curves indicate that the surface emittance decreases with increasing temperature until the ash sinters. Then, the coal ash shows a rapid increase in emittance and on cooling the emittance does not return to its original value. Godridge and Morgan (1971) have said that the effect of sintering on the emittance is to be expected, since the particle sizes increase with the onset of sintering (i.e., the particle swell slightly). It has already been noted that emittance tends to increase with increasing particle size.

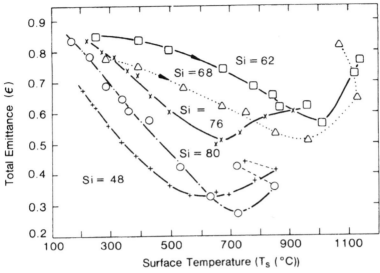

Figure 14.5. Ash deposit emittance as a function of surface temperature and mineral matter composition. (Figure used with permission from Mulcahy *et al.*, 1969.)

14.6. ILLUSTRATIVE PROBLEMS

1. Consider an absorbing emitting cloud of gases and particles with absorption coefficient K_{ac} and characteristic length l_c. Derive an expression for the emittance of the cloud (ε_c). (*Hint*: Start with the differential form of the Bouguer–Lambert law, then add a contribution for emission.)

2. Derive an expression for the radiative-heat-transfer flux q between a cloud of particles of diameter d, number density n, and temperature T_p, with a surrounding enclosure with wall temperature T_w. Allow for single scattering.

3. Using the phase function for char particles suggested by Smoot and Pratt (1979) $(P(\theta) = (8/3\pi)(\sin\theta - \theta\cos\theta)$, calculate the scatter coefficients f, b, and s for the flux model suggested in Section 14.4e.

4. Using recommendations made in the text surrounding Eqn. (14.53) and your own approximations, estimate the emittance of a soot cloud in a typical utility boiler. Compare this value to the typical gas cloud emittance in the same boiler (use the Hadvig chart of Figure 14.4).

Ever learning, and never able to come to the knowledge of the truth.

2 *Timothy 3:7*

NO$_x$ POLLUTANT FORMATION IN TURBULENT COAL SYSTEMS

15.1. INTRODUCTION

In Chapter 11 we introduced the concept and problems associated with chemical reactions in turbulent flows. We noted that among the most important questions in current combustion research are those regarding the interactions between the turbulent fluid mechanics and the chemical reactions. The apparent random fluctuations in the turbulent mixing of the reactants and products dramatically influences mean chemical kinetic rates that have reaction time scales on the order of the turbulent time scale or less. We saw that essentially all homogeneous gas combustion lies in this critical area, including the combustion of methane and higher-order hydrocarbon constituents of natural gas.

In Chapter 13 we extended these concepts to examine the effects of the chemical reaction and the turbulent fluid mechanics, when particles and or droplets were involved. In that chapter we saw that the gas-phase reactions were of significant importance, even when the source of the fuel was coming from the coal particles themselves. Techniques were presented for incorporating the effect of the fluctuating field on the mean chemical properties for coal combustion applications. We saw that the effect of these turbulent fluctuations was significant. We also saw that excluding these fluctuations caused significant error in the computational results.

In these chapters we have seen that one of the largest problems in understanding these interactions lies in the characterization of the time-mean net rate of formaton or destruction of the molecular species from the chemical reactions. Although the kinetic mechanisms are not always known, and kinetic constants difficult to identify, the major problem lies not in these areas but in obtaining the proper time-mean rate due to the presence of the turbulence.

In Chapter 11 we identified three different types of chemical reactions to help interpret the impact of the chemistry–turbulence interactions. These types were based upon the time scale of the chemical reactions and of the turbulent flow field. We discussed the methods for calculating the effects of the chemistry–turbulence interactions for the first two types of flames. We introduced the third type where the reaction time scale was on the same order of magnitude as the turbulent time scale, but we did not present any computational techniques for calculating this type of flame.

Although the fast-chemistry assumption presented in Chapter 11 and used in Chapter 13 for coal flames might be appropriate for some major species, it is not applicable to NO pollutant concentrations. The homogeneous NO pollutant reactions have a typical time scale on the order of 1 ms, while the typical integral time scale of turbulence is around 10 ms. Thus, the effect of turbulent fluctuations on the reaction rate cannot be ignored.

The NO formation process in coal flames is far from completely understood. Two aspects of the process must be addressed in order to formulate a model: the kinetic mechanism itself and the interactions of the kinetic mechanism with the turbulence. The kinetic mechanism for formation and reduction of NO involves hundreds of elementary reactions. Experimentally, observations have led to a sequence of global reactions which have been studied in some detail. The chemical reactions of interest take place in a turbulent environment. In order to calculate mean pollutant concentrations, the effects of the turbulence on the reaction process must be considered.

In this chapter we will examine briefly some possible alternatives for calculating NO formation in turbulent coal flames. We will focus on one technique particularly, show the importance of the turbulence on the calculation of the pollutant species, and present some computations and comparisons with major data. This type of computation involves a research area which is only in its infancy. We will identify the fundamental assumptions required to make the calculations tractable. The validity of these assumptions remains uncertain. Very few experimental observations have been made to evaluate the acceptability of these fundamental approximations. We will discuss some promising measurements which could be made in the next few years. The advent of optical measurement devices has presented some exciting new possibilities for evaluating these and other modeling assumptions. We will discuss briefly some of these experimental tools and present ideas of how they could be used to evaluate basic modeling approximations and thus provide computational techniques for incorporating the effect of chemically limited reactions in turbulent combustion calculations.

We will first examine the mechanisms involved in NO formation in coal flames. We will then examine how the turbulence fluctuations might affect the overall mean reaction rates, and how these effects might be incorporated into a computational strategy. We will then briefly examine some computations as compared with experimental data from laboratory reactors for NO pollutant formation. At this point we will review the critical assumptions necessary for such a model and discuss experiments and experimental methods which might help evaluate the critical assumptions.

Throughout this chapter keep it in mind that these techniques are specifically applicable only to the formation of nonequilibrium trace species. Trace species in chemically reacting environments have the distinct advantage of being able to be decoupled from the general thermochemical solution. Since pollutant species exist in only very small amounts, their presence seldom affects the other mean-field variables. For example, the presence of nitrogen oxide in a few hundred parts per million does not affect the local temperature, velocity, density, or mass fraction of major species, to any appreciable amount. We will see how this ability to decouple the trace species from the main field variables is a significant advantage in calculating out-of-equilibrium species. Generalizing the techniques of this chapter to major species that might be present in coal combustion is not immediately possible.

15.2. MECHANISM

To produce a mathematical description of the pollutant formation process it is necessary to identify a kinetic mechanism which satisfactorily explains the most significant experimental observations of the behavior of pollutants during coal reaction. Different pollutant reactions take place in all phases of the coal particle reaction sequence including coal devolatilization, homogeneous reactions of coal off gases, and heterogeneous reactions with coal char. We will examine each of these aspects of coal combustion or gasification in turn. Much has been written on each of these aspects of the pollutant formation process. In this chapter we only briefly present a summary of the major observations in order to formulate a simple overall NO formation mechanism. Such a mechanism must include major features but, by necessity, must be simple enough to include in a comprehensive combustion or gasification model.

15.2a. Coal Devolatilization

In the first step of devolatilization, the lower-molecular-weight fragments are driven from the particle. Some studies (Wendt and Pershing, 1977; Pohl and Sarofim, 1977), indicate that 0–20 % of the nitrogen in the coal is evolved in the early volatiles, primarily as HCN. The second step in devolatilization occurs at a higher temperature and evolves the heavier hydrocarbons and aromatics. It is postulated (Solomon and Colket, 1978; Wendt and Pershing,

1977; Wendt, 1980) that most of the nitrogen evolved from the coal particles is accomplished in this step.

The rate at which the nitrogen leaves the coal is still subject to controversy. Devolatilization studies (Pohl and Sarofim, 1977; Wendt, 1980) have shown that the nitrogen leaves the coal at a rate proportional to, but somewhat different from, the rate of coal weight loss (by factors of 1.25 to 1.5, (Pohl and Sarofim, 1977; Blair et al., 1977). However, in Chapter 13 it was shown that in a laboratory combustor at high temperatures the global rate of nitrogen release from coal was approximately equal to the rate of the coal weight loss.

We suggest that two basic assumptions be made relating to the devolatilization process (Smith et al., 1981). First, the assumption is made that nitrogen is devolatilized at a rate equal to the mean rate of coal weight loss. This assumption significantly reduces the number of equations required, since the devolatilization scheme used for coal weight loss can be used to predict the mean rate of nitrogen release from the coal. Second, the assumption is made that all of the nitrogen devolatilized from the coal is evolved as HCN or is rapidly converted to HCN in the gas phase. This assumption is justified by studies (Heap et al., 1978; Harding et al., 1982) which show that although HCN only forms a small portion of the nitrogen-containing compounds initially evolved from the coal, the remainder of the nitrogen compounds evolved initially are rapidly converted to HCN in the absence of oxygen. In fuel-rich gas flames, doped with nitrogen-containing compounds, HCN is the major product of the fuel nitrogen/hydrocarbon interactions, and fuel nitrogen exists mainly as HCN just downstream of the reaction zone (Heap et al., 1978; Malte et al., 1980). This assumption reduces the formation of NO from the coal nitrogen (fuel NO) to a mechanism which proceeds by decay of HCN in the gas phase.

15.2b. Homogeneous Formation of NO

Once the coal particle begins to devolatilize, the gaseous constituents react homogeneously in the gas phase. NO can be formed by three separate reaction processes in this phase: thermal NO, prompt NO, and fuel NO.

Thermal NO is formed by oxidation of atmospheric molecular nitrogen and is generally described by the Zeldovich or modified Zeldovich mechanism. These reactions are highly dependent on temperature and equivalence ratio, and NO formation by this mechanism is significantly reduced in fuel-rich systems and at temperatures below 1600–1800 K (Malte and Pratt, 1974; Malte and Rees, 1979). However, the temperature fluctuations in turbulent combustion systems can extend the importance of the Zeldovich mechanism to lower mean temperatures (Malte and Rees, 1979; Hayhurst and Vince, 1980). Nonetheless, because of the low temperatures of the fuel-rich pulverized-coal flames being considered (which lack oxygen atoms), the Zeldovich mechanism does not appear to be a significant source of NO (Wendt, 1980; Hayhurst and Vince, 1980; Corlett *et al.*, 1979). For this reason, this NO formation model has neglected the contributions of thermal NO. If the contributions of thermal NO are found to be significant, or it is desired to extend the model to fuel-lean systems, a method proposed by Fenimore and Fraenkel (1981) could be used to predict thermal NO concentrations.

Prompt NO, which has a weak temperature dependence, is formed by hydrocarbon fragments (resulting from the devolatilization process) attacking molecular nitrogen near the reaction zone of the flame (Hayhurst and Vince, 1980). Since the prompt NO mechanism requires a hydrocarbon to initiate the attack on the molecular nitrogen, the mechanism is much more prevalent in fuel-rich hydrocarbon flames. However, it has been estimated (Hayhurst and Vince, 1980) that if coal contains 1% nitrogen and 50% of that nitrogen is converted to NO, then prompt NO would account for less than 5% of the total NO formed. Since coal generally contains 1 to 2% nitrogen, prompt NO is likely to be overshadowed by fuel NO. Thus, this model has neglected the contributions of prompt NO. For fuel-rich pulverized-coal flames, neglecting prompt NO will probably result in greater error than neglecting thermal NO.

Fuel NO formed in the gas phase results from oxidation of devolatilized nitrogen constituents, and generally accounts for 60–80% of the total NO formed. Fuel NO has a weak temperature dependence in turbulent, diffusion-type, pulverized-coal flames (Wendt, 1980; Heap *et al.*, 1978; Pershing and Wendt, 1977), and is primarily sensitive to the equivalence ratio (Hayhurst and Vince, 1980). Further, nitrogen in the coal is first converted to HCN or remains in the char (Heap *et al.*, 1978). A proposed sequence for the decay of HCN is (Fenimore, 1979; Levy, 1980; Malte *et al.*, 1980; Rees *et al.*, 1981):

$$\text{HCN} \rightarrow \text{NH}_i \quad \begin{matrix} \nearrow \text{NO} \\ \\ \searrow \text{N}_2 \end{matrix} \tag{15.1}$$

Various elementary reactions are required to describe different aspects of the above sequence (Caretto, 1976; Takagi *et al.*, 1979; Levy, 1980).

It is assumed that the global reaction rate proposed by DeSoete (1975) for decay of HCN can adequately describe the formation of fuel NO. DeSoete's measurements were made with the addition of C_2N_2 to $C_2H_4/O_2/Ar$ in flat, premixed, laminar flames for fuel-rich or fuel-lean conditions. The proposed reactions apply specifically to NO formation following the formation of HCN:

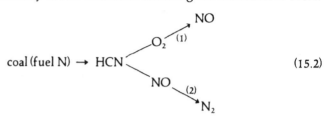

$$coal\,(fuel\,N) \rightarrow HCN \begin{array}{c} \nearrow O_2 \xrightarrow{(1)} NO \\ \\ \searrow NO \xrightarrow{(2)} N_2 \end{array} \tag{15.2}$$

The global reaction rates are reported in Table 15.1, Eqns. F and G. The order of reaction (1) is dependent on the O_2 mole fraction and is given by DeSoete (1975) as shown in Figure 15.1. The order is 0.0 for O_2 fractions greater than 18,000 ppm and 1.0 for O_2 less than 2500 ppm. These are instantaneous laminar rates, and it is necessary to time average these equations to obtain time-mean values of the rates in turbulent systems. The procedure for this is discussed in a subsequent section.

TABLE 15.1. Equation Set for NO Formation and Reduction Model in Turbulent, Coal-Laden, Reacting Flow Processes[a]

	Description		Equation
1.	Mass balance for O_2	(A)	$\tilde{Y}_{O_2} = \tilde{Y}^o_{O_2} - \tilde{Y}_{NO}M_{O_2}/M_{NO}$
2.	Species continuity for i = NO and HCN	(B)	$\bar{\rho}\tilde{u}(\partial \tilde{Y}_i/\partial x) + \bar{\rho}\tilde{v}(\partial \tilde{Y}_i/\partial r) - (\partial/\partial x)(\bar{D}_Y \partial \tilde{Y}_i/\partial x)$ $-(1/r)(\partial/\partial r)(r\bar{D}_Y \partial \tilde{Y}_i/\partial r) = \bar{W}_i$
3.	Overall mean reaction rates for each species	(C) (D)	$\bar{W}_{NO} = (\bar{w}_1 - \bar{w}_2 - \bar{w}_3)M_{NO}$ $\bar{W}_{HCN} = (\bar{w}_0 - \bar{w}_1 - \bar{w}_2)M_{HCN}$
4.	Rate of nitrogen release from the coal	(E)	$\bar{w}_0 = \omega_N \bar{S}_\rho/M_N$
5.	Instantaneous reaction rate terms	(F) (G)	$w_1 = \rho(1 \times 10^{11})X_{HCN}X^b_{O_2}\exp(-67.0\,kcal/RT)/M_m$ $w_2 = \rho(3 \times 10^{12})X_{HCN}X_{NO}\exp(-60.0\,kcal/RT)/M_m$
6.	Mean reaction rate term by convolution over probability density functions	(H)	$\bar{w}_i = \bar{\rho}\int_f \int_n [w_i(f,n)/\rho(f,n)]\tilde{P}(f)\tilde{P}(n)\,df\,dn$
7.	NO–Char reduction reaction	(I)	$\bar{w}_3 = \alpha_p n_p (4.18 \times 10^7)A_E \bar{\rho}_{NO}\exp(-34.7\,kcal/R\tilde{T})$
8.	Instantaneous mole fractions	(J)	$X_i = Y_i M_m/M_i$
9.	Instantaneous mass fractions	(K)	$Y_i = \pi_i Y'_i$
10.	Deviation from fully reacted mass fractions	(L)	$\pi_i = \tilde{Y}_i/\tilde{Y}'_i$

[a]From Smith *et al.* (1983).

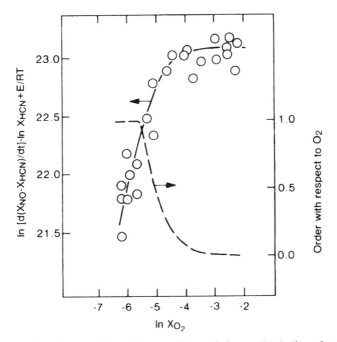

Figure 15.1. Order of reaction (1) as a function of O_2 mole fraction (X_{O_2}). The order is obtained from experimental data of HCN reaction with O_2 as shown. (Figure used with permission from DeSoete, 1975.)

15.2c. Heterogeneous Reactions of NO

The char remaining after devolatilization is primarily composed of carbon and mineral matter (ash), with small amounts of oxygen, hydrogen, sulfur, and nitrogen. A significant quantity of nitrogen still remains in the coal after devolatilization (Pershing and Wendt, 1979), and this can be heterogeneously oxidized to form NO. The conversion efficiency of char nitrogen to NO is much lower than the conversion of volatilized nitrogen (Pohl and Sarofim, 1977; Wendt, 1980; Pershing and Wendt, 1977). Char nitrogen contributes approximately 25% of the fuel NO in pulverized-coal flames (Pohl and Sarofim, 1977; Heap et al., 1976; Pershing and Wendt, 1977; Heap et al., 1978). The rate of formation of NO from heterogeneous reactions is incorporated by postulating that the nitrogen is evolved from the char at a mean rate proportional to the mean particle burnout. The evolved fuel nitrogen then follows the same reaction pathway as the nitrogen released from devolatilization.

NO may also be reduced by the char (Heap et al., 1978; Levy et al., 1981). The rate of NO reduction is described by Eqn. I shown in Table 15.1. This equation was obtained from measurements in an electrically heated, laminar flow furnace operated in the temperature range of 1250–1750 K with pulverized char in helium carrier gas. Other correlations are available in the literature for the heterogeneous reactions of NO (Wendt *et al.*, 1979; Wendt, 1980).

15.3. TURBULENCE FLUCTUATIONS

15.3a. Species Continuity

The combustion of pulverized coal usually occurs in highly turbulent systems. Coupling the NO kinetic mechanism with the turbulent fluid mechanics results in an example of the Type C flames discussed in Chapter 11.

The species continuity equations can be solved for the mean mass fraction of various species provided that the mean reaction rate can be obtained. The steady-state equation for Favre-averaged (Bilger, 1975) mass fractions in an axisymmetric reactor is shown in Table 15.1, Eqn. (B). The turbulent eddy diffusion coefficient $\tilde{D}_Y$ can be estimated from the local eddy diffusivity by choosing an appropriate turbulent Schmidt number. Turbulence closure is achieved through the conventional gradient-diffusion approximation.

One of the largest problems in solving the species continuity equation for Type C flames is obtaining the time-mean net rate of formation of the particular species from chemical reaction ($\bar{W}_1$) as discussed in Chapter 11. This rate is the difference between the net rates of formation and rates of depletion of species *i*. Methods for directly obtaining this time-mean reaction rate have been somewhat unproductive (Pratt, 1979). This has led to the fast-chemistry assumption (Pratt 1976) (i.e., local instantaneous equilibrium discussed in Chapters 11 and 13). In this approach, statistical probability density functions (PDFs) are used to obtain time-mean properties. Although this has proven somewhat satisfactory for some major species (Smith and Smoot, 1981), it is not applicable to NO pollutant concentrations. The homogeneous NO pollutant reactions have a typical time scale around 1 ms. The typical integral time scale of turbulence is around 10 ms. The effect of these fluctuations on the reaction rate cannot be ignored. However, the heterogeneous reactions of NO are thought to be much slower (i.e., on the order of 1 s). Hence, the time-mean reaction rate is calculated from mean properties by ignoring the effects of turbulence on T, A_E, and P_{NO} (see Table 15.1, Eqn. (I)).

Mean concentrations of pollutant species are obtained from solutions to the species continuity equation. Even though the NO pollutant species are of Type C flames, they can be decoupled from the other field equations since the small level of pollutants throughout the system will not affect the velocity, density, or temperature fields. However, the pollutant equation set is coupled since the reaction rates are functions of other pollutant concentrations. These equations can be solved simultaneously after all the main field variables have been converged. Recently, perturbation methods have been proposed for the analysis of pollutant formation processes. Methods proposed by Libby and Williams (1981) and Bilger (1979, 1980) deserve consideration. The main problem in adopting these techniques to pollutant species in coal flames is that perturbations from either the equilibrium or frozen concentrations are not necessarily small. Thus, ignoring the turbulent fluctuations in these perturbations for the coal flame causes unrealistic mass fractions to occur throughout the flow field.

15.3b. Time-Mean Reaction Rate

The time-mean net reaction rate $\bar{W}_i$ can be obtained from appropriate PDFs if the individual instantaneous reaction rates are only functions of the stoichiometry. The local equivalence ratio in a pulverized-fuel furnace can be calculated from the conserved scalar (mixture fraction f) and/or the local amount of coal off-gas (coal–gas mixture fraction η). The root-mean-square fluctuations in these two variables can be modeled, and appropriate probability density functions calculated (see Chapter 13). If $\bar{W}_i$ is a function of f and η, then it can be obtained as shown in Table 15.1, Eqn. (H). In most cases of pulverized-fuel combustion, the equivalence ratio is independent of f since the primary and secondary streams will have the same composition (air). Even in the few cases where f is required, the amount of coal off-gas at any point will be only weakly dependent on the amount of primary gas; thus, f and η can be considered statistically independent as shown in Table 15.1, Eqn. (H).

In 1978 Bilger reported a significant experimental finding. Over a broad region of a given gas-phase diffusion flame, the molecular species composition was only a function of the local equivalence ratio, even though the products were not in thermodynamic equilibrium. The conserved scalar, mixture fraction, is a measure of this equivalence ratio. Fenimore and Fraenkel (1981) tested this hypothesis for thermal NO concentrations in laminar, gaseous, diffusion flames. Their experiments confirmed this observation for thermal NO. Thus, the species reaction rate may be only a function of the mixture fraction and its rate of dissipation (Bilger, 1978, 1979), even where reactions are slow and thermodynamic equilibrium does not exist.

An example of this observed unique functional dependence of the composition on the mixture fraction for nonequilibrium flames is shown in Figure 15.2. Figure 15.3 shows the equilibrium composition as a function of mixture fraction. It can be seen that the measured data correlate well with the mixture fraction and yet are far from equilibrium. Apart from their significance for laminar flames, the observations hold great significance for the interaction between kinetically limited chemistry and the turbulent field. If the experimental data base could be extended to turbulent flames with similar relationships between local

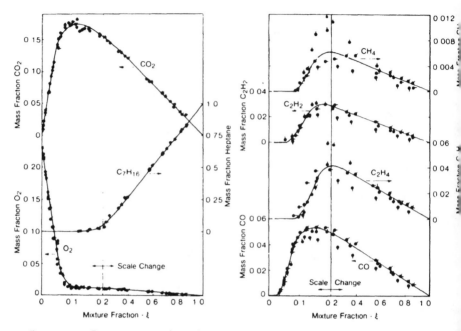

Figure 15.2. Composition correlation between (a) major and (b) minor species composition and mixture fraction in heptane diffusion flames. (Figure used with permission from Bilger, 1978.)

instantaneous chemical composition and the local instantaneous progress variables (such as mixture fraction), then current statistical techniques for calculating mean composition could be readily extended to nonequilibrium combustion.

If it can be assumed that the local mass fraction of any species is a function of the local equivalence ratio, then the time-mean reaction rates can be obtained by convolution over the local PDF. This approximation does not require that the relationship between mass fraction and equivalence ratio be unique for the whole flow field; it requires only that it be time independent with respect to the turbulent time scale. In other words, the reaction time might be different at different points throughout the reactor.

Bilger (1978, 1979, 1980) shows how the assumption that the reaction rate is only a function of equivalence ratio coupled with a perturbation analysis results in a simple expression for the reaction rate of out-of-equilibrium material. This method requires perturbations to be small or the quasi-equilibrium function to be known.

For the approach of this chapter, the local instantaneous mass fraction is taken to be a linear function of its fully reacted mass fraction:

$$Y_i = \pi_i Y_i' \qquad (15.3)$$

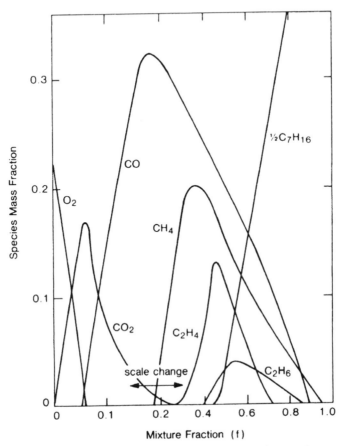

Figure 15.3. Equilibrium composition for the heptane/air flame data shown in Figure 15.2. These equilibrium predictions were performed without solid carbon as an allowed species. (Figure used with permission from Bilger, 1978.)

where Y_i' is the mass fraction of species i that would result if all the original reactant formed only species i (i.e., if $i =$ HCN then Y_i' is the mass fraction of HCN if all fuel nitrogen formed only HCN). Y_i' is only a function of stoichiometry. The variable π_i is a linearization constant for species i and must be independent of the turbulent fluctuations, although it may be spatially variable. It can be shown that π_i is obtained from the time-mean mass fraction field:

$$\pi_i = \tilde{Y}_i / \tilde{Y}_i' \tag{15.4}$$

In this equation, π_i is the fractional deviation of the Favre-mean mass fraction i from the mean fully reacted value $\tilde{Y}_i'$. Thus, the Y_i's are only functions of the local equivalence ratio and hence only functions of f and η. The time-mean reaction rate is then obtained for each pollutant species by convolution over the PDF as shown in Table 15.1, Eqn. (H). All of the equations to be solved for the desired Favre-mean mass fraction are summarized in Table 15.1.

The implied relationship between the mixture fractions and the original amount of reactant, assumes that the histories of fluid elements at a given location in the reactor are not important. Because of the assumed time independence of π_i, fluid elements at a given location with mixture fractions f and η have the same local species mass fractions Y_i. This assumption is based on the experimental observations that the local composition is a unique function of the stoichiometry.

The assumed relationship still allows for the history effects of the fluid elements at different locations in the reactor. This is a result of each π_i being dependent on the local mean velocity, density, etc. Fluid elements with the same mixture fractions f and η may have different local mass fractions Y_i at different locations.

The implications of the assumed relationship between the mixture fractions and the original amount of reactant need to be studied further. Both experimental and computational studies need to be conducted to evaluate this assumption, particularly with regards to second-order reactions.

15.4. COMPUTATIONS

It is apparent from the description of the NO pollutant formation mechanism in turbulent coal systems, as described in the preceding sections, that such a scheme although decoupled from the main field, is highly dependent upon the values of the main field variables. The interactive nature of this submodel with the several other submodels in a comprehensive model makes independent evaluation difficult. The detail included in such models is limited by uncertainties in the kinetic mechanism and turbulence interactions. In this section we will review briefly a few comparisons of calculations made with the approach outlined in the last section with experimental data from laboratory research furnaces.

Before comparing predictions with experimental data, it is useful to examine the effect of the turbulent fluctuations on the NO predictions. Figure 15.4 shows two contour maps for predicted NO concentrations in a pulverized-coal combustor. The combustor is axisymmetric and the contour plots show predictions from the centerline to the wall and from the inlet to the outlet. The two maps shown in this figure were generated from the same input conditions with the only difference being that Figure 15.4a was calculated from reaction rates using mean temperature and composition only. All turbulent fluctuations were ignored. In the plot of Figure 15.4b the mean reaction rate was calculated by

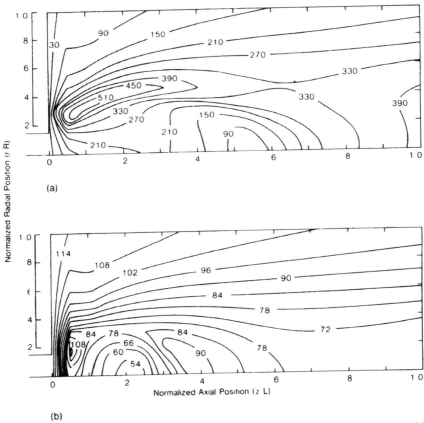

Figure 15.4. NO concentration contours (in ppm) as calculated from the NO formation model described in the text. (a) As calculated from mean values only (no fluctuations). (b) As calculated including turbulent fluctuations. The predictions are for the axisymmetric lifted coal flame of Thurgood and Smoot (1979). Conditions for this flame were discussed in Chapter 8.

integration over the probability density function as outlined in the preceding sections. In this figure it is apparent that the incorporation of the effects of the turbulence on the mean reaction rate gives a dramatically different result from that obtained when ignoring these effects. It is obvious that mean values are not sufficient to obtain the mean reaction rate.

The comparisons of calculations from the predictive method with experimental data will be discussed next. For these calculations, the NO calculational procedure outlined in this chapter was incorporated into the two-dimensional pulverized-coal gasification or combustion code (PCGC-2) discussed in Chapter 7. This computer program incorporates most of the suggested calculational procedures of this book.

The first method used to evaluate the NO model involved comparisons of the predicted and measured effects of key variables on nitrogen pollutant emissions. These comparisons indicate the extent to which the NO model predicts the effects of two experimental operating parameters on NO reactions in pulverized-coal systems. This demonstrates one way of evaluating modeling utility. Does the model predict proper trends?

The stoichiometric ratio (SR) is a measure of the overall air to fuel ratio in a combustion system. The SR is defined as the ratio of inlet air to the air required for stoichiometric combustion. Increased stoichiometric ratio increases NO concentrations in coal systems (Lee *et al.*, 1979; Asay, 1982; Rees *et al.*, 1981; Brown *et al.*, 1977; Harding *et al.*, 1982). The local stoichiometry affects NO formation and reduction in various ways, but a primary effect is the amount of oxygen available for NO formation. A fuel-rich environment reduces NO formation from volatile nitrogen. Figure 15.5 shows a comparison of predicted and measured outlet NO concentrations as a function of stoichiometric ratio. The combustor is a single burner, axisymmetric, laboratory furnace. The two cases shown are for similar runs with different coal types. Both a high-moisture subbituminous coal and a Utah bituminous coal were examined.

The lower stoichiometric ratios result in less oxygen available for oxidation of the volatile nitrogen. Nitrogen is devolatilized into a fuel-rich zone which reduces NO formation and enhances reduction of the NO formed. Particle

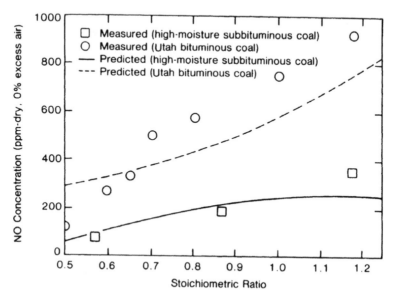

Figure 15.5. Comparison of predicted and observed effect of stoichiometric ratio on outlet NO concentrations. (Predictions used with permission from Hill, 1983.)

temperatures will increase at higher stoichiometric ratios which results in greater nitrogen devolatilization, and increased NO formation (Rees *et al.*, 1981; Pershing and Wendt, 1979).

For both the swirling and nonswirling diffusion flames, the measured and predicted values indicate that the exit NO levels increase with stoichiometric ratios. However, the measured effect of stoichiometric ratio is greater than the predicted effect at higher *SR* values. This could partially result from some assumptions in the coal combustion code, but it probably results primarily from the underprediction of NO emissions at these values. Underprediction of the NO emissions could result from underestimation of the percent nitrogen conversion to HCN during devolatilization. In addition, at higher stoichiometric ratios, thermal NO formation increases, which is not predicted by the NO model. This would result in underprediction of the total NO emissions at these values of stoichiometric ratio. In addition, coal burnout increases at these values, but is usually underpredicted, which results in lower NO concentrations.

The swirl number is a measure of the angular momentum imparted to the inlet streams, and is defined as the flux of angular momentum divided by the product of the axial momentum flux and the burner radius. In coal systems, swirl is often imparted to only the secondary inlet stream which is the situation with the measurements and predictions presented here. The increased angular momentum of the secondary stream generally results in faster mixing of the coal and oxidizer. Increased mixing enhances ignition, and often causes attachment of the flame to the coal burner.

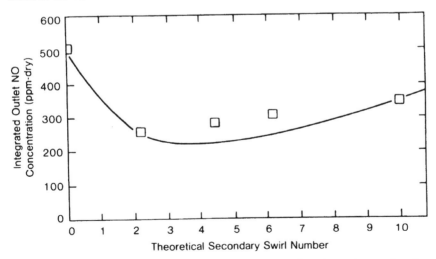

Figure 15.6. Comparison of the predicted (———) and observed (□) effect of secondary swirl number on outlet NO concentrations for a subbituminous coal. (Predictions used with permission from Hill, 1983.)

Figure 15.6 compares predicted and measured NO concentrations as a function of the theoretical swirl number for a subbituminous coal in a laboratory-scale, single-burner, axisymmetric combustor. The observed NO concentrations are the average of integrated outlet values measured at $SR = 0.87$ and 1.17 and adjusted to stoichiometric conditions. Figure 15.7 compares predicted and measured centerline NO concentrations as a function of the theoretical swirl number for a bituminous coal in the same laboratory furnace. At low primary velocities (i.e., the two cases of Figures 15.6 and 15.7, where $u_p = 10 \text{ ms}^{-1}$), the mixing between the primary and swirling secondary streams initially occurs only at the edge of the primary jet, in the region of high shear. This mixing promotes early combustion in this annular mixing layer with the flame attached to the burner. This hot annular region provides a radiative heat source which promotes devolatilization of the coal particles in the fuel-rich central core. This fuel-rich environment is where the early nitrogen is being evolved from the coal and thus minimizes NO formation. As the swirl number of the secondary stream is increased, mixing of the streams is enhanced. This increases contact of the volatiles and oxidizer, and, subsequently, increases NO formation.

The second method used to evaluate the NO model is by comparison of detailed local measurements with overall model predictions. A comparison was made with data measured at the International Flame Research Foundation (IFRF) in Imjuiden, Holland. The coal used was a West German Saar coal whose characteristics are shown in Table 15.2. Predictions were made for both attached and detached coal flames.

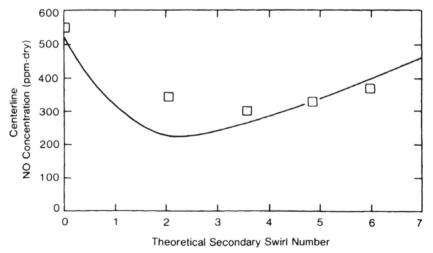

Figure 15.7. Comparison of the predicted (——) and observed (□) effect of secondary swirl number on centerline NO concentration for a bituminous coal. (Predictions from Hill, 1983.)

TABLE 15.2. Characteristics of Typical West German Saar Coal in IFRF Tests

Elemental analysis	Weight percent (dry)
C	74.65
H	4.70
N	1.12
S	0.85
O	11.08
Percent volatiles (dry)	32.0
Percent ash (dry)	7.6
High calorific value	32×10^6 J/kg

Figures 15.8 to 15.12 show comparisons of predicted and measured center-line and radial profiles for the attached IFRF pulverized-coal flame. The reactor operating conditions for this attached flame are summarized in Table 15.3.

Figures 15.8–15.10 show comparisons of experimental and predicted radial profiles for both oxygen and temperature. These predictions are made from the pulverized-coal combustion/gasification code (PCGC-2) without contribution from the NO submodel. However, these predictions have a direct effect on the NO model predictions, and in fact the pollutant model predictions can be only as good as the predictions of the coal combustion processes. Figures 15.8–15.10 are included to show the extent to which PCGC-2 can predict some of the general coal combustion features. Only those features most pertinent to the prediction of the NO pollutant formation process are shown.

Figure 15.8a shows comparison of observed and predicted centerline oxygen profiles. The oxygen profile is further illustrated in Figure 15.9 which shows radial profiles of oxygen at axial locations of 0.5, 1.0, 1.5, and 3.5 m. Figure 15.8b shows comparison of predicted and measured centerline temperature profiles. Figure 15.10 shows radial profiles of temperature at axial locations of 0.1, 0.75, 1.9, and 3.5 m.

Figures 15.8a and 15.9 show that only general trends are in agreement between measured and predicted oxygen profiles at most axial locations. One noticeably large discrepancy in the oxygen profiles, as shown in Figures 15.8a and 15.9, indicates that the calculated oxygen profiles overpredict the measured profiles at most axial positions in the reactor. Figure 15.8a indicates that the observed buildup of oxygen occurs more rapidly, but to a lesser extent, than the predicted centerline oxygen levels. Lower predicted oxygen levels are especially noticeable near the wall of the reactor (as shown by Figure 15.9) in the large recirculation zone characteristic of this reactor.

Figures 15.8b and 15.10 show fair agreement between measured and predicted temperature profiles at most locations in the reactor. The measured

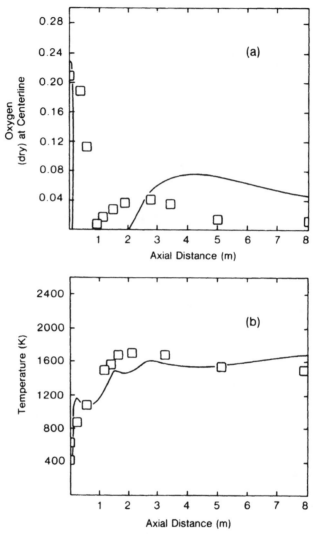

Figure 15.8. Comparison of centerline oxygen (a) and temperature (b) profiles as measured in IFRF attached flame and predicted with conditions shown in Table 15.3. (Predictions used with permission from Hill, 1983.)

temperature profile appears to have essentially the same ignition point, but a slightly slower temperature rise to higher levels in the early region of the reactor. The measured temperature profile decreases more rapidly, and results in a slightly lower reactor outlet temperature. It is clear that quantitative differences exist between the calculated and measured results, but that general overall trends are represented.

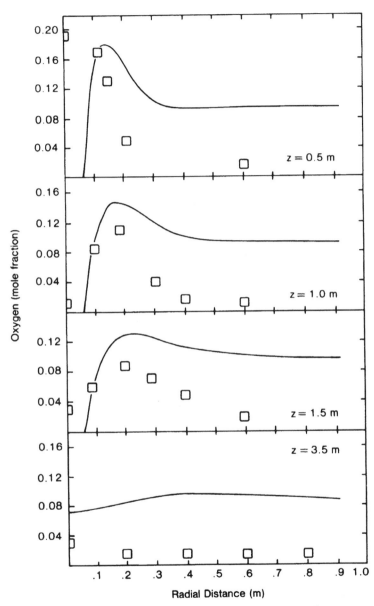

Figure 15.9. Comparison of radial oxygen mole fraction at 0.5, 1.0, 1.5, and 3.5 m as measured in IFRF attached flame and as predicted by PCGC-2. Conditions shown in Table 15.3. (Predictions used with permission from Hill, 1983.)

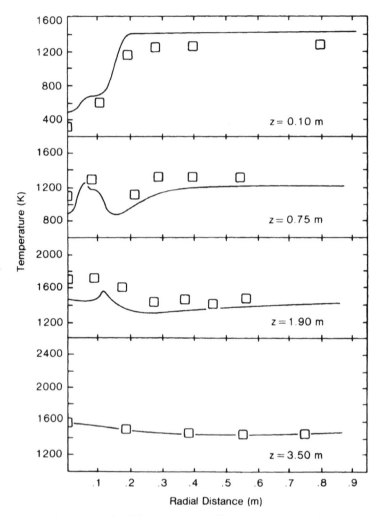

Figure 15.10. Comparison of radial temperature profile at 0.1, 0.75, 1.9, and 3.5 m as measured in IFRF attached flame as predicted by PCGC-2. Conditions shown in Table 15.3. (Figure used with permission from Hill, 1983.)

Figure 15.11 shows comparison of experimental and predicted centerline pollutant profiles for nitrogen oxide (NO) and hydrogen cyanide (HCN). The nitrogen oxide profiles are further illustrated in Figure 15.12 at axial locations 0.10, 0.75, 1.00, 1.9, and 3.50 m. These figures show that the NO model predicts the general trends and overall magnitude of the measured NO pollutant levels. Figure 15.11a shows that observed nitrogen oxide formaton begins earlier; but

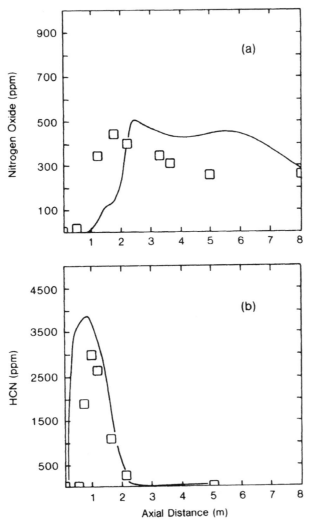

Figure 15.11. Comparison of centerline (a) NO concentration and (b) HCN concentration as measured in IFRF attached flame and as predicted by PCGC-2. Conditions shown in Table 15.3. (Figure used with permission from Hill, 1983.)

at about the same rate and reaches similar peak values as predicted pollutant levels. The observed NO decay also begins earlier, and decay occurs to generally lower centerline values. Figure 15.12 illustrates similar trends between observed and predicted centerline NO levels. Figure 15.12 shows that observed NO values just off-centerline continue at higher levels in early regions of the reactor.

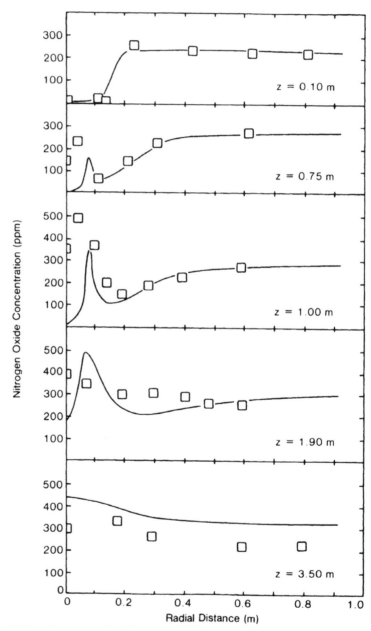

Figure 15.12. Comparison of radial measured (IFRF) nitrogen oxide concentrations at several axial locations with predictions (with permission from Hill, 1983). Conditions shown in Table 15.3.

TABLE 15.3. Conditions used in Laboratory Reactor[a]
(West German Saar coal, attached coal flame)

Primary air feed rate (g/s)	70.6
Secondary air feed rate (g/s)	573.3
Coal feed rate (g/s)	58.9
Mean particle diameter (μm)	67.0
Solids loading (wt. %)	80
Primary velocity (m/s)	21.9
Secondary velocity (m/s)	25.1
Overall stoichiometric ratio (SR)	1.10
Primary preheat temperature (K)	400
Secondary preheat temperature (K)	725
Swirl number (experimental)	0.0
Visible ignition distance (m)	0.0
Reactor width (square, m)	1.9

[a]For IFRF test data, see Chapter 8.

In aft-regions of the reactor, off-centerline NO values are slightly lower than predicted NO values. This agrees with trends observed from centerline NO values in Figure 15.11a. Figure 15.12 also shows good agreement between observed and predicted NO levels in the large recirculation zone at all axial locations, especially at early regions in the reactor. The largest discrepancy between observed and predicted NO levels occurs at 3.50 m, and could be the result of the slower decay of NO in the aft-regions of the reactor as shown in Figure 15.11.

Figure 15.11b shows that observed centerline HCN formation occurs later in the reactor, and to a lesser extent than the predicted centerline HCN formation. Observed and predicted centerline HCN decay occurs at essentially the same rate and location in the reactor. Sufficient measured radial HCN profiles were not available for comparison with predicted radial HCN profiles, and any conclusions drawn from such a figure would be of limited validity.

A delay in the predicted centerline NO formation is shown in Figure 15.11a, and a high predicted centerline HCN level is shown in Figure 15.11b. These discrepancies could be attributed to the completely depleted centerline oxygen levels as shown in Figure 15.8a. Figure 15.9 shows a total depletion of predicted oxygen levels immediately off the centerline in the early regions of the reactor. This could be the main factor resulting in the discrepancy between observed and predicted NO levels near the centerline in these regions of the reactor. Figure 15.9 also shows high predicted oxygen levels in the large recirculation zone of the reactor. These high predicted oxygen levels do not appear to have a strong influence on the predicted NO levels shown in Figure 15.12.

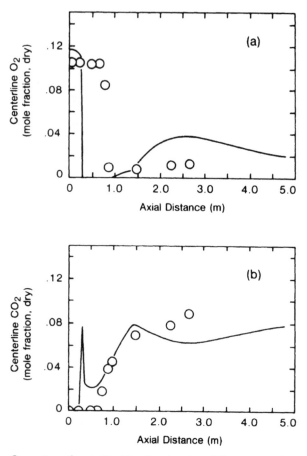

Figure 15.13. Comparison of centerline (a) carbon dioxide and (b) oxygen profiles as measured in IFRF lifted flame and as predicted by Hill (1983, with permission). Conditions shown in Table 15.4.

Figures 15.13–15.18 show comparisons of predicted and measured centerline and radial profiles for a detached IFRF pulverized coal flame from the same furnace as the previous IFRF case. The inlet conditions for this case are summarized in Table 15.4.

Again, Figures 15.13–15.15 are included to show the extent to which PCGC-2 represents some of the coal combustion processes most essential for prediction of the pollutant formation processes. Discrepancies between theory and measurement in this case are mostly attributable to the differences in the location of the ignition point. Again it is clear that some detail is predicted but, clearly, some quantitative differences exist.

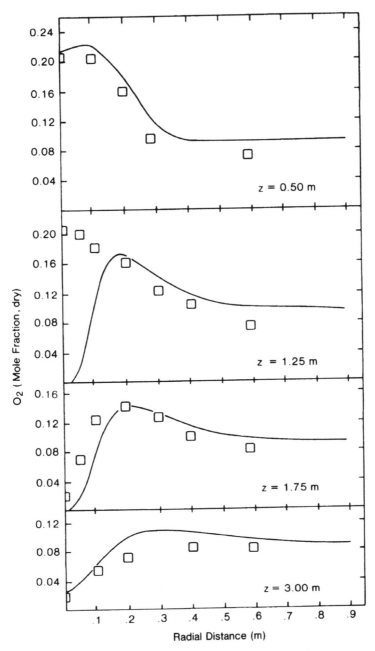

Figure 15.14. Comparison of radial oxygen mole fraction at several axial locations as measured in IFRF lifted flame and as predicted by Hill (1983, with permission). Conditions from Table 15.4.

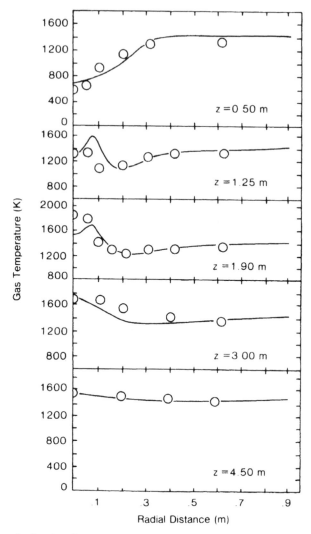

Figure 15.15. Predicted and measured radial gas temperature profiles for the IFRF detached bituminous coal flame case. Conditions shown in Table 15.4. (Prediction from Hill, 1983, with permission.)

15.5. NEW MEASUREMENTS NEEDED

The theory for NO formation outlined in this chapter relies on the species concentrations being functions only of the local equivalence ratio. Comparisons between predicted and measured profiles does not satisfactorily resolve the

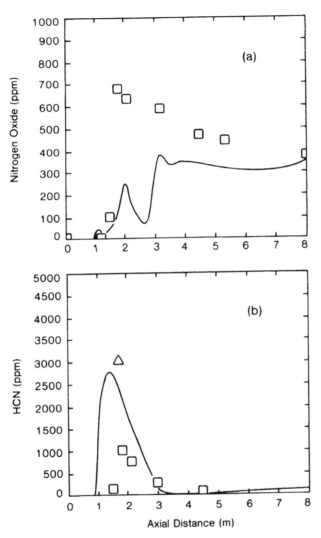

Figure 15.16. Comparison of centerline (a) NO and (b) HCN profiles as measured in IFRF lifted flame and predicted by Hill (1983, with permission). Conditions shown in Table 15.4.

validity of the assumptions used in the NO model. Before those methods can be confidently extended to general Type C flames, evaluation of the crucial assumption with more direct experimental data must be explored. We have discussed some measurements that have been taken in laminar flames for gaseous reactants. These measurements must be extended to local instantaneous measurements in both turbulent gas and coal-laden flames.

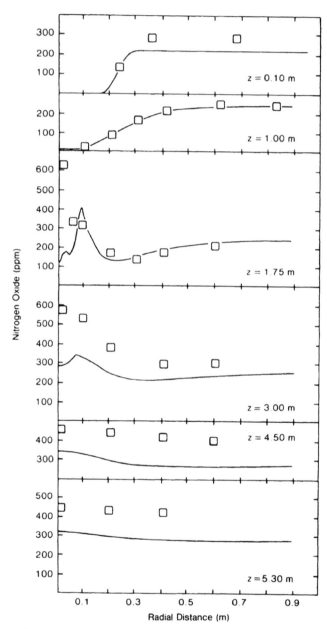

Figure 15.17. Comparison of radial NO concentration profiles at several axial locations as measured in IFRF lifted flame and predicted by Hill (1983, with permission). Conditions shown in Table 15.4.

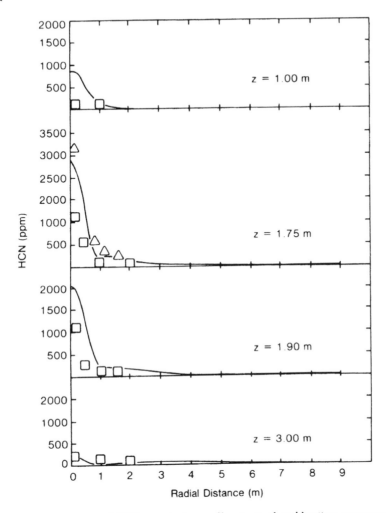

Figure 15.18. Comparison of HCN concentration profiles at several axial locations as measured in IFRF lifted flame and predicted by Hill (1983, with permission). Conditions shown in Table 15.4.

Optical diagnostic techniques for reliably resolving the local instantaneous composition in practical turbulent combustors have recently become applicable to practical gas flames (Drake *et al.*, 1982). The nonintrusive nature of these measurements will be beneficial in characterizing the interactions between chemistry and turbulence, independent of specific theoretical models. Laser-based instrumentation provides the means to obtain well-constructed, detailed, pertinent data. Several complementary techniques are capable of providing measurements of local composition and temperature. Among the methods

TABLE 15.4. Conditions Used in laboratory Reactor[a] (West German Saar coal, detached coal flame)

Primary air feed rate (g/s)	120.6
Secondary air feed rate (g/s)	527.8
Coal feed rate (g/s)	58.9
Mean particle diameter (μm)	67.0
Solids loading (wt %)	47.5
Primary velocity (m/s)	40.7
Secondary velocity (m/s)	9.6
Overall stoichiometric ratio (SR)	1.10
Primary preheat temperature (K)	440
Secondary preheat temperature (K)	735
Swirl number (experimental)	0.0
Visible ignition distance (m)	0.80–1.00
Reactor width (square, m)	1.9

[a]For IFRF test data, see Chapter 8.

which appear to be of most general utility are Raman spectroscopy (particularly, coherent Raman techniques), laser-induced fluorescence (LIF), and multiple-photon ionization (MPI). These methods have been shown to work satisfactorily in gas flames but the interference caused by high particle loadings in pulverized-fuel flames is a serious obstacle which must be surmounted.

Optical measurements of simultaneous species composition in turbulent gaseous flames can be extremely useful in evaluating the functional dependence of the chemistry in Type C flames on the local instantaneous equivalence ratio. A comparable experimental research effort must be expended on coal flames. The ability to use the optical tools in particle- or droplet-laden systems has not yet been demonstrated. Such a capability will undoubtedly be achieved in the next several years; however, in the meantime, gaseous composition measurements in laminar coal-laden flames could lead the research effort. Even these studies in laminar coal flames would give insight into the dependence of the chemistry on the local equivalence ratio. Extensions to measurements of instantaneous gaseous properties in turbulent coal flames must be explored before the techniques outlined in this chapter can be rigorously evaluated.

15.6. ILLUSTRATIVE PROBLEMS

1. From the IFRF lifted flame (discussed in Section 15.4) on the centerline at an axial position of 3.0 m, calculate the reaction rate of HCN conversion to NO by reaction 1 from Eqn. (15.2). Use the experimental data and mean values only; ignore turbulent fluctuations. Use the rate constants reported in Table 15.1.

2. Repeat Problem 1 for the reaction rate of HCN conversion to N$_2$ at the same location. Compare the two rates.

3. From the data given for the IFRF lifted flame of Problems 1 and 2, estimate a value for $\tilde{f}$ and $\tilde{\eta}$ on the centerline at an axial location of 3.0 m.

4. From Figure 15.12 calculate the centerline value for π_{NO} at $z = 1.75$, 3.0, 4.5, and 5.3 m. Use the measured data points.

5. From the estimated $\tilde{f}$ and $\tilde{\eta}$ of Problem 3 and assuming $g_f = \tilde{f}(1 - \tilde{f})$ and $g_\eta = \tilde{\eta}(1 - \tilde{\eta})$, repeat Problem 1 for the mean reaction rate $\bar{W}_1$.

6. Explain as many reasons as possible for the differences between Figure 15.4a and 15.4b.

APPENDIX

CONVERSION FACTORS

This appendix was abstracted from the government publication by E. A. Mechtly, "The International Systems of Units—Physical Constants and Conversion Factors," second revision, NASA SP-7012, Washington, D.C. 1973. Permission to include this material in this publication was obtained from the Scientific and Technical Information Office, NASA, Washington, D.C.

The following table expresses the definitions of miscellaneous units of measure as exact numerical multiples of coherent SI units, and provides multiplying factors for converting numbers and miscellaneous units to corresponding new numbers and SI units.

The digits of each numerical entry following "E" represent a power of 10. An asterisk leading each number expresses an exact definition. For example, the entry "*2.54E − 2" expresses the fact that 1 in. = 2.54×10^{-2} m, exactly, by definition. Most of the definitions are extracted from National Bureau of Standards documents. Numbers not preceded by an asterisk are only approximate representations of definitions, or are the results of physical measurements.

Conversion Factors to SI Units

To convert from	To	Multiply by
atmosphere	newton/square meter (N/m²)	*1.013E5
bar	newton/square meter (N/m²)	*1.00E5
British thermal unit (mean)	joule (J)	1.055 87E3
British thermal unit (thermochemical)	joule (J)	1.054 350E3
British thermal unit (39°F)	joule (J)	1.059 67E3
British thermal unit (60°F)	joule (J)	1.054 68E3

405

To convert from	To	Multiply by
calorie (International Steam Table)	joule (J)	4.1868
calorie (mean)	joule (J)	4.190 02
calorie (thermochemical)	joule (J)	*4.184
calorie (15°C)	joule (J)	4.185 80
calorie (20°C)	joule (J)	4.181 90
calorie (kg, International Steam Table)	joule (J)	4.1868E3
calorie (kg, mean)	joule (J)	4.190 02E3
calorie (kg, thermochemical)	joule (J)	*4.184E3
Celsius (temperature)	kelvin (K)	$t_K = t_C + 273.15$
centimeter of mercury (0°C)	newton/square meter (N/m²)	1.333 22E3
centimeter of water (4°C)	newton/square meter (N/m²)	9.806 38E1
electron volt	joule (J)	1.602 191 7E − 19
erg	joule (J)	*1.00E − 7
Fahrenheit (temperature)	kelvin (K)	$t_K = \frac{5}{9}(t_F + 459.67)$
Fahrenheit (temperature)	Celsius (°C)	$t_C = \frac{5}{9}(t_F − 32)$
fluid ounce (U.S.)	cubic meter (m³)	*2.957 352 956 25E5
foot	meter (m)	*3.048E − 1
foot of water (39.2°F)	newton/square meter (N/m²)	2.988 98E3
gallon (U.S. dry)	cubic meter (m³)	*4.404 883 770 86E − 3
gallon (U.S. liquid)	cubic meter (m³)	*3.785 411 784E − 3
horsepower (550 ft · lbf/s)	watt (W)	7.456 998 7E2
inch	meter	*2.54 E − 2
kayser	1/meter (1/m)	*1.00 E2
kilocalorie (International Steam Table)	joule (J)	4.186 8 E3
kilocalorie (mean)	joule (J)	4.190 02 E3
kilocalorie (thermochemical)	joule (J)	4.184 E3
kilogram mass	kilogram (kg)	*1.00
kilogram force (kgf)	newton (N)	*9.806 65
lbf (pound force, avoirdupois)	newton (N)	*4.448 221 615 260 5
lbm (pound mass, avoirdupois)	kilogram (kg)	*4.535 923 7 E − 1
liter	cubic meter (m³)	*1.00 E − 3
meter	wavelengths Kr 86	*1.650 763 73 E6
micron	meter (m)	*1.00 E − 6
mile (U.S. statute)	meter (m)	*1.609 344 E3
pascal	newton/square meter (N/m²)	*1.00
poise	newton second/square meter (N · s/m²)	*1.00 E − 1
pound force (lbf avoirdupois)	newton (N)	*4.448 221 615 260 5
pound mass (lbm avoirdupois)	kilogram (kg)	*4.535 923 7 E − 1
quart (U.S. dry)	cubic meter (m³)	*1.101 220 942 715E − 3
quart (U.S. liquid)	cubic meter (m³)	9.463 592 5E − 4
Rankine (temperature)	kelvin (K)	$t_K = (\frac{5}{9})t_R$
slug	kilogram (kg)	1.459 390 29E1

To convert from	To	Multiply by
ton (long)	kilogram (kg)	*1.016 046 908 8E3
ton (metric)	kilogram (kg)	*1.00E3
ton (short, 2000 lb)	kilogram (kg)	*9.071 847 4E2
torr (0°C)	newton/square meter (N/m^2)	1.333 22E2

NOMENCLATURE

Symbol	Units	Definition
A	m^2	area
A	s^{-1}	frequency factor in Arrhenius rate constant
b	kg-atoms kg^{-1}	atom or mole number
b	—	backward scatter component
B	arbitrary	intensive property
B	—	blowing parameter
B	$m^{-3} s^{-1}$	birth rate of pore
B	—	rate parameter
C	—	constant
C	—	coal
C	$J kg^{-1} K^{-1}$	heat capacity
C	$kg m^{-3}$	concentration of unreacted coal in solid particles
C	$kmol m^{-3}$	molar concentration
d	m	diameter
D	m	diameter
D	$m^{-3} s^{-1}$	death rate of pores
D	$m^2 s^{-1}$	diffusivity
D	$kg s^{-1}$	numerical coefficient for diffusion
E	J	energy
E	$J kmol^{-1}$	activation energy
E	—	log-law constant
E	$J m^{-2} s^{-1}$	emissive power
f	—	mixture fraction
f	—	volume of space occupied by particles
f	—	forward scatter component
F	N	force
F	—	transformed mean

Symbol	Units	Definition
g	$m\,s^{-2}$	gravitational acceleration
g	—	root-mean-square fluctuation
G	—	transformed variance
h	$J\,kg^{-1}$	mass enthalpy
H	$J\,kmol^{-1}$	molar enthalpy
i	$J\,kg^{-1}$	specific internal energy
I	—	turbulent intensity
I	$J\,m^{-2}\,s^{-1}\,st^{-1}$	intensity of radiation
j	$kg\,m^{-2}\,s^{-1}$	mass flux
J	$N\,s$	impulse
J	—	total number of particles
k	$J\,m^{-1}\,s^{-1}\,K^{-1}$	thermal conductivity
k	$m^2\,s^{-2}$	turbulent kinetic energy
k	$m\,s^{-1}$, other	mass-transfer coefficient
k	$m\,s^{-1}, s^{-1}$	reaction rate constant
K	—	radiation coefficient
K	—	total number of elements
K	$s\,m^{-2}$	burning rate constant
l	m	length scale
L	m	length
m	kg	mass
m	—	true order of reaction
m	$g\,cm^{-1}\,s^{-1}$	mass addition rate
M	$kg\,kmol^{-1}$	molecular weight
n	m^{-3}	number density
n	—	unit normal
n	—	apparent reaction order
P	Pa	pressure
P	$m^2\,s^{-3}$	production
P	—	probability density
P	—	phase function
q	$J\,m^{-2}\,s^{-1}$	heat-transfer flux
Q	$m^3\,s^{-1}$	volumetric flow rate
Q	$J\,s^{-1}$	heat-transfer rate
Q	—	constant relating quantity of high-rate volatiles to proximate volatiles
r	$kg\,m^{-3}\,s^{-1}$	reaction rate
r	m	radius
R	m	total radius
R	$kg\,m^{-2}\,s^{-1}$	specific reaction rate
R	$J\,kmol^{-1}\,K^{-1}$	universal gas constant

Symbol	Units	Definition
$\underline{R}$	$\mathrm{cal\,gmol^{-1}\,K^{-1}}$	gas constant
Re	—	Reynolds number
Ri	—	Richardson number
s	variable	conserved scalar
s	—	side scatter component
s	$\mathrm{s^{-1}}$	dilatation rate of particle per unit volume
S	arbitrary	source or sink term
S	—	swirl number
S	—	solids
S	$\mathrm{kg\,m^{-3}}$	concentration of unreactive char in solid particles produced by devolatilization reaction
Si	—	silica ratio
t	s	time
T	s	time (total)
T	K	temperature
u	$\mathrm{m\,s^{-1}}$	axial velocity
v	$\mathrm{m\,s^{-1}}$	velocity
v	$\mathrm{m^3}$	particle volume
v	—	fraction of solid lost as volatiles
V	$\mathrm{m^3}$	volume
V	$\mathrm{kg\,m^{-3}}$	concentration of volatiles
w	$\mathrm{m\,s^{-1}}$	azimuthal velocity
w	$\mathrm{m\,s^{-1}}$	gas velocity at control surface
w	$\mathrm{kmol\,m^{-3}\,s^{-1}}$	individual reaction rate
W	$\mathrm{N\,m}$	work
W	$\mathrm{kmol\,m^{-3}\,s^{-1}}$	overall reaction rate
W	—	fraction of functional coal group reacted during devolatilization process
x	m	coordinate direction
X	—	particle size fraction
X	—	mole fraction
y	m	coordinate direction
Y	—	starting location fraction
Y	—	mass fraction
Y	—	devolatilization coefficient
Y	—	stoichiometric coefficient
z	m	coordinate direction
α	—	fractional intermittency
α	kg	component mass
β	arbitrary	intensive property per unit mass

Symbol	Units	Definition
$\dot{\gamma}$	$N\,m^{-2}$	strain rate
γ	—	fractional heat loss
γ	—	particle swelling parameter
Γ	—	Rabinowitsch–Mooney consistency variable
δ	—	Kroenecker delta
Δ	—	increment, change
ε	$m^2\,s^{-3}$	dissipation rate
ξ	—	internal active particle surface area/external equivalent sphere surface
η	—	coal–gas mixture fraction
η	—	effectiveness factor
H	—	transformation variable
θ	—	void fraction
θ	—	angle
θ	—	porosity, $1 - \rho_{ap}/\rho_s$
κ	—	constant
λ	m	wavelength
μ	$kg\,m^{-1}\,s^{-1}$	viscosity
ν	$m^2\,s^{-1}$	kinematic viscosity or diffusivity
ν	—	stoichiometric coefficient
ν	—	surface product stoichiometric factor
π	—	proportionality constant
ρ	$kg\,m^{-3}$	density
ρ_c	$g\,m^{-2}\,s^{-1}$	specific char reaction rate (surface area basis)
σ	$J\,m^{-2}\,s^{-1}\,K^{-4}$	Stefan–Boltzmann constant
σ	—	Gaussian coefficient for distribution of activation energies
τ	$kg\,m^{-1}\,s^{-2}$	shear stress
τ	—	tortuosity factor
ϕ	arbitrary	arbitrary variable
ϕ	—	Thiele modulus
ϕ	—	initial equivalence ratio
χ	—	consumption rate/maximum consumption rate
ω	—	albedo for scatter
ω	—	mass fraction

Subscripts
a   average, ash, adiabatic, absorption, apparent
b   black, boundary

c	convection, raw coal, coal off-gas, char, concentration
cs	control surface
cv	control volume
d	drag, diffusion
D	diffusive
daf	dry ash-free
e	effective
f	mixture fraction, furnace, fluid, fuel
g	gas, root-mean-square fluctuation, gas phase
h	enthalpy, char
H	hydrocarbon
in	inlet
i	coordinate direction, species, inlet, intrinsic, coal functional group
j	coordinate direction, particle
k	turbulent kinetic energy, particle, kinetic
K	Knudsen
l	laminar, heterogeneous oxidation reaction index, loss
m	mass transfer, devolatilization reaction index, mean, mixture, mass
n	reaction order
o	initial value, without blowing, oxidizer, oxidation, central, original
O	oxygen
O_2	oxygen
p	particle, primary, proximate
r	radiation, reaction
R	reaction
s	surface, secondary, scattering, solid
S	char fraction
t	turbulent, total
v	per volume
w	wall, water
ε	dissipation
η	coal–gas mixture fraction
ν	spectral
ϕ	arbitrary variable
∞	maximum value

Superscripts

$'$	rate, bulk property, fluctuating property
$-$	conventional Reynolds average
$\sim$	Favre average
$+$	forward direction

$-$ backward direction
h enthalpy
m mass, mixing
N temperature exponent
o initial value, standard state
t turbulent
u axial momentum
v radial momentum
α coal component
σ solid, bulk
ρ recirculation

REFERENCES

Abramovich, G. N., "Effect of Solid-Particle or Droplet Admixture on the Structure of a Turbulent Gas Jet," *Int. J. Heat Mass Transfer* **14**, 1039 (1971).

Alden, M., Edner, H., Grafstrom, P., and Svanberg, S., "Two-Photon Excitation of Atomic Oxygen in a Flame," *Opt. Comm.* **42**, 244 (1982).

Altenkirch, R. A., Peck, R. E., and Chen, S. L., "The Appearance of Nitric Oxide and Cyanide in One-Dimensional Coal Dust/Oxidizer Flames," *Combust. Sci. Technol.* **20**, 49 (1979).

Amundson, N. R., and Arri, L. E., "Char Gasification in a Counter Current Reactor" *AIChE J.* **24**, 87 (1978).

Amundson and co-workers (i.e., Amundson, 1981; Srinivas and Amundson 1980, 1981; and Zygourakis *et al.*, 1982).

Amundson and co-workers (i.e., Srinivas and Amundson, 1981; Zygourakis *et al.*, 1982).

Annamalai, K., "Critical Regimes of Coal Ignition," *Trans. ASME* **101**, 576 (1979).

Annamalai, K., and Durbetaki, P., "A Theory on Transition of Ignition Phase of Coal Particles," *Combust. Flame* **29**, 193 (1977).

Anson, D., "Fluidized Bed Combustion of Coal for Power Generation," *Progr. Energy Combust. Sci.* **2**, 61 (1976).

Anthony, D. B., and Howard, J. B., "Coal Devolatilization and Hydrogasification," *AIChE J.* **22**, 625 (1976).

Anthony, D. B., Howard, J. B., Hottel, H. C., and Meissner, H. P., "Rapid Devolatilization of Pulverized Coal," *15th Symposium (International) on Combustion*, The Combustion Institute, Pittsburgh, PA, 1303 (1975).

Asay, B. W., "Effects of Coal Type and Moisture Content on Burnout and Nitrogenous Pollutant Formation," Ph.D. dissertation, Department of Chemical Engineering, Brigham Young University, Provo, UT (1982).

Asay, B. W., Smoot, L. D., and Hedman, P. O., "Effect of Coal Moisture on Burnout and Nitrogen Oxide Formation," *Combust. Sci. Technol.* **35**, 15 (1983).

Avedesian, M. M., and Davidson, J. F., "Combustion of Carbon Particles in a Fluidized Bed," *Trans. Instn. Chem. Engrs.* **51**, 121 (1973).

Ayling, A. B., and Smith, I. W., "Measured Temperatures of Burning Pulverized-Fuel Particles, and the Nature of the Primary Reaction Product," *Combust. Flame* **18**, 173, (1972).

Badzioch, S., and Hawksley, P. G. W., "Kinetics of Thermal Decomposition of Pulverized Coal Particles," *Ind. Eng. Chem. Process Des. Dev.* **9**, 521 (1970).

Bailey, G. H., Slater, I. W., and Eisenblam, P., "Dynamics, Equations, and Solutions for Particles Undergoing Mass Transfer," *British Chem. Engineering* **15**, 912 (1970).

Baronavski, A. P., and McDonald, J. R., "Measurement of C_2 Concentrations in an Oxygen-Acetylene Flame: An Application of Saturation Spectroscopy," *J. Chem. Phys.* **66**, 3300 (1977a).

Baronavski, A. P., and McDonald, J. R., "Application of Saturation Spectroscopy to the Measurement of C_2, $^3\Pi_u$ Concentrations in Oxy-acetylene Flames," *Appl. Opt.* **16**, 1897 (1977b).

Barnhart, J. S., Thomas, J. F., and Laurendeau, N. M., *Pulverized Coal Combustion and Gasification in a Cyclone Reactor, Experiment and Models*, DOE/E-49-18, Purdue University, West Lafayette, IN (1979).

Barriga, A., and Essenhigh, R. H., "A Mathematical Model of a Combustion Pot," Western States Section, The Combustion Institute, Pittsburgh, PA (April, 1979).

Batchelder, H. R., and Sternberg, J. C., "Thermodynamic Study of Coal Gasification," *Ind. Eng. Chem.* **42**, 877 (1950).

Batchelder, H. R., Busche, R. M., and Armstrong, W. P., "Kinetics of Coal Gasification," *Ind. Eng. Chem.* **45**, 1856 (1953).

Beck, V. J., "Extension and Verification of a One-Dimensional Computer Model of Coal Combustion and Gasification," M.S. thesis, Department of Chemical Engineering, Brigham Young University, Provo, UT (1980).

Beér, J. M., "The Fluidised Combustion of Coal," *16th Symposium (International) on Combustion*, The Combustion Institute, Pittsburgh, PA, 439 (1976).

Beér, J. M., Lee, K. B., Marsden, C., and Thring, M. W., "Flames de Charbon Pulverize a Melange Controle," *Conference on Combustion and Energy Conversion*, Paris, Paper 1.9, Institute Francais des Combustibles et de l'Energie (1964).

Belt, R. J., and Bissett, L. A., *An Assessment of Flash Pyrolysis and Hydropyrolysis*, U.S. Department of Energy Report METC/RI-70/2, Morgantown Energy Technology Center, Morgantown, WV (1978).

Bergman, P. D., and George, T. J., "Coal Slurries: State of the Art," Unpublished Paper, U.S. Department of Energy, Pittsburgh Energy Technology Center, Pittsburgh, PA (1983).

Berry, R. J., "An Ancient Fuel Provides Energy for Modern Times," *Chem. Eng.* **87**, 73 (April 21, 1980).

Bhatia, S. K., and Perlmutter, D. D., "A Random Pore Model for Fluid-Solid Reactions: I. Isothermal, Kinetic Control," *AIChE J.* **26**, 379 (1980).

Bhatia, S. K., and Perlmutter, D. D., "A Random Pore Model for Fluid-Solid Reactions: II. Diffusion and Transport Effects," *AIChE J.* **27**, 247 (1981).

Bilger, R. W., "A Note on Favre Averaging in Variable Density Flows," *Combust. Sci. Technol.* **11**, 215 (1975).

Bilger, R. W., "Turbulent Jet Diffusion Flames," *Progr. Energy Combust. Sci.* **1**, 87 (1976).

Bilger, R. W., "Probe Measurements in Turbulent Combustion," *Progr. Astro. Aero.* **53**, 49 (1977).

Bilger, R. W., "Reaction Rates in Diffusion Flames," *Combust. Flame* **30**, 277 (1978).

Bilger, R. W., "Effects of Kinetics and Mixing in Turbulent Combustion," *Combust. Sci. Technol.* **19**, 89 (1979).

Bilger, R. W., *Topics in Applied Physics: Turbulent Reacting Fows* (P. A. Libby and F. A. Williams, eds.), Vol. 44, Springer-Verlag, New York, NY, 65 (1980).

Bird, R. B., Stewart, W. E., and Lightfoot, E. N., *Transport Phenomena*, Wiley, New York, NY (1960).

Bissett, L. A., "An Engineering Assessment of Entrainment Gasification," U.S. Department of Energy Report MERC/RI-78/2, Morgantown, WV (April 1978).

Blair, D. W., Wendt, J. O. L., and Bartok, W., "Evolution of Nitrogen and Other Species During Controlled Pyrolysis of Coal," *16th Symposium (International) on Combustion*, The Combustion Institute, Pittsburgh, PA, 475 (1977).

Blake, T. R., Brownell, D. H., Jr., Garg, S. K., Herline, W. E., Pritchett, J. W., and Schneyer, G. P., *Computer Modeling of Coal Gasification Reactors, Year 2*, DOE/FE-1770-32, Systems, Science and Software, La Jolla, CA (1977).

Blake, T. R., Herline, W. E., and Schneyer, G. P., "Numerical Simulation of Coal Gasification Processes," 87th National Meeting of the AIChE, Boston, MA (1979).

Bodle, W. W., and Schora, F. C., "Coal Gasification Technology Overview," Paper presented at *Advances in Coal Utilization Technology Symposium*, Louisville, KY (May 14–18, 1979).

Bodle, W. W., Punwami, D. V., and Mensinger, M. C., "Re: Peat," *CHEMTECH* 8, 559, (September 1978).

Boni, A. A., and Penner, R. C., "Sensitivity Analysis of a Mechanism for Methane Oxidation Kinetics," *Combust. Sci. Technol.* 15, 99 (1977).

Bonczyk, P. A., and Shirley, J. A., "Measurement of CH and CN Concentration in Flames by Laser-Induced Saturated Fluorescence," *Combust. Flame* 34, 253 (1979).

Boow, J., and Goard, P. R. C., "Fireside Deposits and Their Effect on Heat Transfer in a Pulverized Fuel-Fired Boiler: Part III. The Influence of the Physical Characteristics of the Deposits on its Radiant Emittance and Effective Thermal Conductance," *J. Inst. Fuel* 42, 346 (1969).

Boussinesq, J., "Theorie de l'Écoulment Tourbillant," *Memoirs Am. Acad. Art Sci.* 23, No. 46 (1877).

Bradshaw, P., Ferriss, D. H., and Atwell, N. P., "Calculation of Boundary-Layer Development Using the Turbulent Energy Equation," *J. Fluid Mech.* 28, 593 (1967).

Breen, B. P., "Combustion in Large Boilers: Design and Operating Effects on Efficiency and Emissions," *16th Symposium (International) on Combustion*, The Combustion Institute, Pittsburgh, PA, 19 (1977).

Brown, B., "Effect of Coal Type on Entrained Gasification," Ph.D. dissertation, (in progress), Department of Chemical Engineering, Brigham Young University, Provo, UT (1985).

Brown, R. A., Mason, H. B., Pershing, O. W., and Wendt, J. O. L., "Investigation of First and Second Variables on Control of NO_x Emissions in a Pulverized Coal Furnace," 83rd National Meeting, AIChE, Houston, TX (1977).

Bryers, R. W., ed., *Ash Deposits and Corrosion Due to Impurities in Combustion Gases*, Hemisphere Publishing Corp., Livingston, NJ (1978).

Bueters, K. A., Cogoli, J. G., and Habelt, W. W., "Performance Prediction of Tangentially Fired Utility Furnaces by Computer Model," *15th Symposium (International) on Combustion*, The Combustion Institute, Pittsburgh, PA, 1245 (1974).

Burgess, D., Hertzberg, J., Richmond, J. K., Liebman, I., Cashdollar, K. L., and Lazzara, C. P., "Combustion, Extinguishment and Devolatilization in Coal Dust Explosions," Western States Section, The Combustion Institute, Provo, UT (April 1979).

Caretto, L. S., "Modeling the Gas Phase Kinetics of Fuel Nitrogen," Western States Section, The Combustion Institute, University of Utah, Salt Lake City, UT (1976).

Cashdollar, K. L., and Hertzberg, N., *SPIE, 253, Modern Utilization of Infrared Technology VI* (1980).

Ceely, F. J., and Daman, E. L., "Combustion Process Technology," *Chemistry of Coal Utilization*, second supplementary volume (M. A. Elliott, ed.), Wiley–Interscience, New York, NY, 1313 (1981).

Chan, C., "Measurement of OH in Flames Using Laser-Induced Fluorescence Spectroscopy," Ph.D. dissertation, Department of Mechanical Engineering, University of California, Berkeley, CA (1979).

Chan, C., and Daily, J. W., "Measurement of OH Quenching Cross-Sections in Flames Using Laser-Induced Fluorescence Spectroscopy," Paper No. 79–20, Western States Section, The Combustion Institute, Provo, UT (1979).

Chan, R. K. C., Meister, C. A., Scharff, M. F., Dietrich, D. E., Goldman, S. R., Levine, H. B., Scharff, M. F., and Ubhayakar, S. K., "A Computer Model for the Bigas Gasifier: Formulation of the Model," U.S. Department of Energy Final Report J-570-80-008A/2183, Jaycor Corp., Del Mar, CA (October 31, 1980).

Chan, R. K.-C., Chiou, M. J., Dietrich, D. E., Dion, D. R., Klein, H. H., Laird, D. H., Levine, H. B., Meister, C. A., Scharff, M. R., and Srinivas, F., "Computer Modeling of Mixing and Agglomeration in Coal Conversion Processes, Vol. 1., Model Formulation; Vol. 2, User's Manual; U.S. Department of Energy Final Report DOE/ET/10329-1211, Jaycor Corp., Del Mar, CA (February 1982).

Chandrasekhar, S., *Radiative Transfer*, Dover, New York, NY (1960).

Chanhon, S. P., and Longenbach, J. R., *ACS Div. Fuel Chem. Preprints* 23, 73 (1978).

Chapyak, E. J., Blewett, P. J., and Cagliostro, D. J., "Verification Studies of Entrained Flow Gasification and Combustion Systems with the SIMMER-2 Code," 10th International Association of Mathematics and Computers in Simulation (IMACS), World Conference, Montreal, Canada (August 1982).

Chaurey, K. H., and Sharma, K. C., "Coal-Based Ammonia Plants—Preliminary Operating Experience of Coal Gasification at Talcher and Ramagundam Fertilizer Plants of the Fertilizer Corporation of India," *TVA Ammonia from Coal Symposium*, Muscle Shoals, AL (May 1979).

Chen, P. J., Schneyer, G. P., Peterson, E. W., Blake, T. R., Cook, J. L., and Brownell, D. H., Jr., "Computer Modeling of Coal Gasification Reactors," U.S. Department of Energy Final Report DOE/ET/10232-T1, Vols. 1–3, S-Cubed Co., San Diego, CA (April 1981).

Chu, C. T., and Churchill, S. W., "Numerical Solution of Problems in Multiple Scattering of Electromagnetic Radiation," *J. Phys. Chem.* **59**, 955 (1955).

Corlett, R. C., Monteith, L. E., Halgren, C. A., and Malte, P. C., "Molecular Nitrogen Yields from Fuel-Nitrogen in Backmixed Combustion," *Combust. Sci. Technol.* **19**, 95 (1979).

Cranfield, R. R., and Gelhart, D., "Large Particle Fluidisation," *Chem. Eng. Sci.* **29**, 935 (1974).

Crosley, D. R. (ed.), "Laser Probes for Combustion Chemistry," ACS Symposium Series 61, American Chemical Society, Washington, DC, 1980.

Crowe, C. T., Sharma, M. P., and Stock, D. E., "The Particle-Source-In Cell (PSI-CELL) Model for Gas-Droplet Flows," *Trans. ASME, J. Fluids Eng.* **99**, 325 (June 1977).

Cukier, R. I., Levine, H. B., and Shuler, K. E., "Nonlinear Sensitivity Analysis of Multiparameter Model Systems," *J. Comp. Phys.* **26**, 1 (1978).

Daily, J. W., "Pulsed Resonance Spectroscopy Applied to Turbulent Combustion Flows," *Appl. Opt.* **15**, 955 (1976).

Daily, J. W., "Saturation Effects in Laser Induced Fluorescence Spectroscopy," *Appl. Opt.* **16**, 568 (1977).

Daily, J. W., "Saturation of Fluorescence in Flames with a Gaussian Laser Beam," *Appl. Opt.* **17**, 225 (1978).

Davidson, J. F., and Harrison, D., *Fluidised Particles*, Cambridge University Press, Cambridge, MA (1963).

Deguingand, B., and Galant, S., "Upper Flammability Limits of Coal Dust-Air Mixtures," *18th Symposium (International) on Combustion*, The Combustion Institute, Pittsburgh, PA, 705 (1980).

deLesdernier, D. L., Johnson, S. A., and Engleman, V. S., "Conceptual Design and Economic Analysis for CWM Utilization in an Oil-Designed Utility Boiler," Fourth International Symposium on Coal Slurry Combustion, Vol. 2, PETC, Pittsburgh, PA (May 10–12, 1982).

DeSai, P. R., and Wen, C. Y., "Computer Modeling of the MERC Fixed Bed Gasifier, U.S. Department of Energy Report, MERC/CR-78/3, Morgantown, WV (March, 1978).

DeSoete, G. G., "Overall Reaction Rates of NO and N_2 Formation from Fuel Nitrogen," *15th Symposium (International) on Combustion*, The Combustion Institute, Pittsburgh, PA, 1093 (1975).

Digiuseppe, T. G., Hudgens, J. W., and Lin, M. C., "New Electronic States in CH_3, Observed Using Multiphoton Ionization," *J. Chem. Phys.* **76**, 3337 (1982).

Donaldson, C. duP., "A Computer Study of Boundary Layer Transition," *AIAA J.* **7**, 271 (1969).

Drake, M. C., Bilger, R. W., and Storner, S. H., "Raman Measurements and Conserved Scalar Modeling in Turbulent Diffusion Flames," *19th Symposium (International) on Combustion*, The Combustion Institute, Pittsburgh, PA, 459 (1982).

Drew, D. A., "Averaged Field Equations for Two-Phase Media," *Studies Appl. Math* **50**, 133 (1971).

Dryer, F. L., and Glassman, I., "Fundamental and Semi-Global Kinetic Mechanisms of Hydrocarbon Combustion," DOE/COO-4272-3, Princeton University, Princeton, NJ (1978).

Dutta, S., and Wen, C. Y., "Reactivity of Coal and Char. 2. In Oxygen-Nitrogen Atmosphere," *Ind. Eng. Chem. Process Des. Dev.* **16**, 31 (1977).

Dutta, S., Wen, C. Y., and Belt, R. J., "Reactivity of Coal and Char. 1. In Carbon Dioxide Atmosphere," *Ind. Eng. Chem. Process Des. Dev.* **16**, 20 (1977).

Eagen, T., Blackadar, R., and Essenhigh, R. H., "Kinetics of Gasification in a Combustion Pot: A Comparison of Theory and Experiment," *16th Symposium (International) On Combustion*, The Combustion Institute, Pittsburgh, PA, 515 (1977).

Edelman, R. B., and Fortune, O. F., "A Quasi-Global Chemical Kinetic Model for the Finite Rate Combustion of Hydrocarbon Fuels with Application to Turbulent Burning and Mixing in Hypersonic Engines and Nozzles," AIAA Paper No. 69–86, New York, NY (January 1969).

Edelman, R. B., and Harsha, P. T., "Laminar and Turbulent Gas Dynamics in Combustors—Current Status," *Progr. Energy Combust. Sci.* **4**, 1 (1978).

Edmister, W. C., Perry, H., Correy, R. C., Elliott, M. A., "Thermodynamics of Gasification of Coal with Oxygen and Steam," *ASME J.* **74**, 621 (July 1952).

Ekmann, J. M. "Transport and Handling Characteristics of Coal-Water Mixtures," Fourth International Symposium on Coal Slurry Combustion, Vol. 4, DOE/PETC, Pittsburgh, PA (May 10–12, 1982).

Elder, J. L., Schidt, L. D., Steiner, W. A., and Davis, J. D., "Relative Spontaneous Heating Tendencies of Coals," U.S. Bureau of Mines R. I. 8206, Washington, DC (1977).

Elghobashi, S. E. and Abou–Arab, T. W., "A Two-Equation Turbulence Model for Two-Phase Flows," *Phys. Fluids* **26**(4), 931 (1983).

Elliot, M. A. (ed.), *Chemistry of Coal Utilization*, second supplementary volume, Wiley, New York, NY (1981).

Elliott, M. A., and Yoke, G. R., "The Coal Industry and Coal Research and Developments in Perspective," *Chemistry of Coal Utilization*, second supplementary volume 1 (M. A. Elliott, ed.), Wiley, New York, NY (1981).

Engleman, U. S., "Survey and Evaluation of Kinetic Data of Reactions in Methane/Air Combustion," EPA-600/2-76-003, Exxon Research and Engineering Co., Linden, NJ (January, 1976).

EPRI, Coal Gasification for Electric Utilities, *EPRI Journal*, 6 (April 1979).

Essenhigh, R. H., "Combustion and Flame Propagation in Coal Systems: A Review," *16th Symposium (International) on Combustion*, The Combustion Institute, Pittsburgh, PA, 372 (1977).

Essenhigh, R. H., "Fundamentals of Coal Combustion," second supplementary volume, *Chemistry of Coal Utilization* (M. A. Elliott, ed.), Wiley, New York, NY, 1153 (1981).

Essenhigh, R. H., Froberg, R., and Howard, J. B., "Predicted Burning Rates of Single Carbon Particles," *Ind. Eng. Chem.* **57**, 33 (1965).

Essenhigh, R. H., Wen, C. Y., and Lee, E. S., *Coal Conversion Technology*, Addision-Wesley, Reading, MA (1979).

Farthing, G. A., Jr., Johnson, S. A., and Vecci, S. J., "Combustion Tests of Coal-Water Slurry," EPRI Report CS-2286, Babcock & Wilcox Co., Alliance, OH (March 1982).

Feistel, P. P., Van Heek, K. H., and Juentgen, H., *Ger. Chem. Eng.* (English translation) **1**, 294 (1978).

Fenimore, C. P., "Studies of Fuel-Nitrogen Species in Rich Flame Gases," *17th Symposium (International) on Combustion*, The Combustion Institute, Pittsburgh, PA, 661 (1979).

Fenimore, C. P., and Fraenkel, H. A., "Formation and Interconversion of Fixed-Nitrogen Species in Laminar Diffusion Flames," *18th Symposium (International) on Combustion*, The Combustion Institute, Pittsburgh, PA, 143 (1981).

Field, M. A., "Predicting the Burning Time of the Coke Residue of Pulverized Fuel," *Brit. Coal Util. Res. Assoc. Monogr. Bull.* **28**, 61 (1964).

Field, M. A., "Rate of Combustion of Size-Graded Fractions of Char from a Low-Rank Coal Between 1200°K and 2000°K," *Combust. Flame* **13**, 237 (1969).

Field, M. A., "Measurements of the Effect of Rank on Combustion Rates of Pulverized Coal," *Combust. Flame* **14**, 237 (1970).

Field, M. A., and Roberts, R. A., "Measurement of the Rate of Reaction of Carbon Particles with Oxygen in the Pulverized Coal Size Range for Gas Temperatures Between 1400°K and 1800°K," *BCURA Memb. Circ. No. 325*, Leatherhead, England (1967).

Field, M. A., Gill, D. W., Morgan, B. B., and Hawksley, P. G. W., "Combustion of Pulverized Fuel: Part 6. Reaction Rate of Carbon Particles," *Brit. Coal Util. Res. Assoc. Monogr. Bull.* **31**, 285 (1967).

Finson, M. L., Kothandaraman, G., Lewis, P. F., Simons, G. A., Wilemski, G., and Wray, K. L., *Modeling of Coal Gasification for Fuel Cell Utilization*, U.S. Department of Energy Final Report SAN-1254-2, Physical Sciences Inc., Woburn, MA (1978).

Fletcher, T. H., "A Two-Dimensional Model for Coal Gasification and Combustion," Ph.D. dissertation, Department of Chemical Engineering, Brigham Young University, Provo, UT (1983).

Frank-Kamenetski, *Diffusion and Heat Exchange in Chemical Kinetics* (translated from Russian by N. Thonel), Princeton University Press, Princeton, NJ (1955).

Froberg, R. W., and Essenhigh, R., "Reaction Order and Activation Energy on Carbon Oxidation During Internal Burning," *17th Symposium (International) on Combustion*, The Combustion Institute, Pittsburgh, PA, 179 (1979).

Gavalas, G. R., "A Random Capillary Model with Application to Char Gasification at Chemically Controlled Rates," *AIChE J.* **26**, 577 (1980).

Gavalas, G. R., "Analysis of Char Combustion Including the Effect of Pore Enlargement," *Combust. Sci. Technol.* **24**, 197 (1981).

Germane, G. J., Smoot, L. D., Diehl, S. P., Richardson, K. H., and Rawlins, D. C., "Coal-Water Mixture Laboratory Combustion Studies," Fifth International Symposium on Coal Slurry Combustion, Vol. II, U.S. Department of Energy, Pittsburgh, PA, 973 (April 1983).

Ghassemzadeh, M. R., Sommer, T. M., Farthing, G. A., and Vecci, S. J., "Rheology and Combustion Characteristics of Coal-Water Mixtures," Fourth International Symposium of Coal Slurry Combustion, Vol. II, Session 3, PETC/DOE, Pittsburgh, PA (May 1982).

Gibbs, B. M., *Inst. Fuel (Symp. Series No. 1)* **1**, A5–1 (1975).

Gibson, J., "The Constitution of Coal and Its Relevance to Coal Conversion Processes", *J. Inst. Fuel* **51**, 67 (1978).

Gibson, M. M., and Launder, B. E., "Ground Effects on Pressure Fluctuations in the Atmospheric Boundary Layer," *J. Fluid Mech.* **86**, 491 (1978).

Gibson, M. M., and Morgan, B. B., "Mathematical Model of Combustion of Solid Particles in a Turbulent Stream with Recirculation,," *J. Inst. Fuel* **43**, 517 (1970).

Gidaspow, D., and Ettehadieh, B., "Fluidization in Two-Dimensional Beds with a Jet,' *Ind. Eng. Chem. Fundam.* **22**, 193 (1983a).

Gidaspow, D., Lin, C., and Seo, Y. C., "Fluidization in Two-Dimensional Beds with a Jet: 1. Experimental Porosity Distributions," *Ind. Eng. Chem. Fundam.* **22**, 187 (1983b).

Given, P. H., "The Distribution of Hydrogen in Coals and Its Relation to Coal Structures," *Fuel* **39**, 147 (1960).

Given, P. H. (ed.), *American Conference on Coal Science*, Adv. Chem. Ser. **55**, American Chemical Society, Washington, DC (1964).

Given, P. H. and Biswas, B., "Dependence of Coal Liquefaction Behavior on Coal Characteristics. 2. Role of Petrographic Composition," *Fuel* **54**, 40 (1979).

Glassett, J. H., "Elevated Pressure Mapping of the Brigham Young University Coal Gasification Reactor," M. E. Project, Chemical Engineering Department, Brigham Young University, Provo, UT (1984).

Glassman, I., *Combustion*, Academic Press, New York, NY, 194 (1977).

Glownia, J. H., and Sander, R. K., *Appl. Phys. Lett.* **40**, 648 (1982).

Gluskoter, H. J., Shimp, N. F., and Rich, R. R., "Coal Analysis Trace Elements and Mineral Matter," *Chemistry of Coal Utilization*, Second Supplementary Volume, (M. A. Elliott, ed.), John Wiley, New York, NT, 369 (1981).

Godridge, A. M., and Morgan, E. S., 'Emmissivities of Materials from Coal and Oil-Fired Boilers," *J. Inst. Fuel* **44**, 207 (1971).

Godridge, A. M., and Read, A. W., "Combustion and Heat Transfer in Large Boiler Furnaces," *Progr. Energy Combust. Sci.* **2**, 83 (1976).

Goetz, G. J., Nsakala, N. Y., and Borio, R. W., "Development of a Method for Determining Emissivities and Absorptivities of Coal Ash Deposits," ASME Paper No. 78-WA/Ru-6, Combustion Engineering Publication TIS-5890, Combustion Engineering Co., Windsor, CN (1978).

Goetz, G. J., Nsakala, N. Y., Patel, K. L., Lao, T. C., "Combustion and Gasification Kinetics of Chars from Four Commercially Significant Coals of Varying Rank", Second Annual Conference on Coal Gasification, EPRI, Palo Alto, Ca (October 1982).

Goldschmidt, V. W., Householder, M. K., Ahmadi, G., and Chuang, S. C., "Turbulent Diffusion of Small Particles Suspended in Turbulent Jets," *Progr. Heat Mass Transfer, Vol. VI*, 487 (1971).

Gordon, A. L., Caram, H. S., and Admundson, N. R., 'Modelling of Fluidized Bed Reactors-V," *Chem. Eng. Sci.* **33**, 713 (1978).

Gordon, S., and McBride, B., "Computer Program for Calculation of Complex Chemical Equilibrium Compositions," NASA SP-273 (1971).

Gosman, A. D., and Lockwood, F. C., "Incorporation of a Flux Model for Radiation into a Finite-Difference Procedure for Furnace Calculations," *14th Symposium (International) on Combustion*, The Combustion Institute, Pittsburgh, PA, 661 (1973).

Gosman, A. D., Pun, W. M., Ruchal, A. K., Spalding, D. B., and Wolfshstein, R., *Heat and Mass Transfer in Recirculating Flow*, Academic Press, New York, NY (1969).

Gosman, A. D., Lockwood, F. C. and Salooja, A. P., "The Prediction of Cylindrical Furnaces Gaseous Fueled with Premixed and Diffusion Burners," *17th Symposium (International) on Combustion*, The Combustion Institute, Pittsburgh, PA 747 (1979).

Gosman, A. D., Lockwood, F. C., Megahed, I. E. A., and Shah, N. G., "The Prediction of the Flow, Reaction and Heat Transfer in the Combustion Chamber of a Glass Furnace," 18th Aerospace Sciences Meeting, Pasadena, CA (January 1980).

Gouldin, F. C., Depsky, J. S., and Lee, S. L., "Velocity Field Characteristics of a Swirling Flow Combustor," AIAA Paper No. AIAA-83-0314 (1983).

Granoff, B., and Nuttall, H. E., Jr., "Pyrolysis Kinetics for Oil-Shale Particles," *Fuel* **56**, 234 (1977).

Gray, D., Cogoli, J. G., and Essenhigh, R. H., "Problems in Pulverized Coal and Char Combustion," No. 6, *Adv. Chem. Ser.* **131**, 72, American Chemical Society, Washington, DC (1976).

Grieser, D. R., and Barnes, R. H., "Laser Probes for Combustion Chemistry," (D. R. Crosley, ed.), ACS Symposium Series, 143, American Chemical Society, Washington, DC (1980).

Grumer, J., "Recent Research Concerning Extinguishment of Coal Dust Explosions," *15th Symposium (International) on Combustion*, The Combustion Institute, Pittsburgh, PA, 103 (1975).

Hadvig, S., "Gas Emissivity and Absorptivity: A Thermodynamic Study," *J. Inst. Fuel* **42**, 129 (1970).

Halpern, J. B., Hancock, G., Lenzi, M., and Welge, K. H., "Laser Induced Fluorescence from NH_2 ($^2A_1$). State Selected Radiative Lifetimes and Collisional De-Excitation Rates," *J. Chem. Phys.* **63**, 4808 (1975).

Hamblen, D. G., Solomon, P. R., and Hobbs, R. H., "Physical and Chemical Characterization of Coal." U.S. Environmental Protection Agency EPA-600/7-80-106, United Technologies Research Center, Hartford, CN (May 1980).

Harding, N. S., "Effects of Secondary Swirl and Other Burner Parameters on Nitrogen Pollution Formation in a Pulverized Coal Combustor," Ph.D. Dissertation, Chemical Engineering Department, Brigham Young University, Provo, UT (1980).

Hamor, R. J., Smith, I. W., and Tyler, R. J., 'Kinetics of Combustion of a Pulverized Brown Coal Char Between 630 and 2200 K," *Combust. Flame* **21**, 153 (1973).

Hanks, R., "Advanced Fluid Mechanics," Department of Chemical Engineering, Brigham Young University, Provo, UT (September 1, 1980).

Hansen, L. D., Phillips, L. R., Ahlgren, R. B., Mangelson, N. F., Eatough, D. J., and Lee, M. L., Proceedings of U.S. Department of Energy Environmental Control Symposium, Washington, DC (1978).

Harding, N. S., Jr., Smoot, L. D. and Hedman, P. O., "Nitrogen Pollutant Formation in a Pulverized Coal Combustor: Effect of Secondary Stream Swirl," *AIChE J.* **28**, 573 (1982).

Hashimoto, K., and Silveston, P. L., "Gasification: Part I. Isothermal, Kinetic Control Model for a Solid with a Pore Size Distribution," *AIChE J.* **19**, 259 (1973a).

Hashimoto, K., and Silveston, P. L., "Gasification: Part II. Extension of Diffusion Control," *AIChE J.* **19**, 268 (1973b).

Hassan, M. M., Lockwood, F. C., and Moneib, H. A., "Measurements in a Gas-Fired Cylindrical Furnace," *Combust. Flame* **51**, 249 (1983).

Haurman, D. J., Dryer, F. L., Schug., and Glassman, I., "A Multiple-Step Overall Kinetic Mechanism for the Oxidation of Hydrocarbons," *Combust. Sci. Technol.* **25**, 219 (1981).

Hawthorne, W. R., Weddell, D. S., and Hottel, H. C., "Mixing and Combustion in Turbulent Gas Jets," Third Symposium on Combustion, Flame and Explosion Phenomena, Williams & Wilkin, Baltimore, MD (1949).

Hayhurst, A. N., and Vince, I. M., "Nitric Oxide Formation from N_2 in Flames: The Importance of 'Prompt' NO," *Progr. Energy Combust. Sci.* **6**, 35 (1980).

Heap, M. P., Corley, T. L., Dau, C. J. and Tyson, T. J., "The Fate of Fuel Nitrogen—Implications for Combustor Design and Operation,' Energy and Environmental Research Corp., Hemisphere Publishing Corp., Washington, DC (1978).

Heap, M. P., Lowes, T. M., Walmsley, R., Bartledo, H., and LeVaguerese, P., "Burner Criteria from NO_x Control: Vol. I., Influence of Burner Variables on NO_x in Pulverized Coal Flames," U.S. E.F.T.S. Publ. No. EPA-600/2-76-061a (1976).

Hebden, D., and Stroud, H. J. F., "Coal Gasification Processes," *Chemistry of Coal Utilization*, second supplementary volume (M.A. Elliott, ed.), Wiley–Interscience, New York, NY, 1599 (1981).

Hedman, P. O., Highsmith, J. R., Soelberg, N. R., and Smoot, L. D., "Detailed Local Measurements in the BYU Entrained Gasifier," IFRF International Symposium on Conversion of Solid Fuels, Newport Beach, CA (October 1982).

Hein, K., "Preliminary Results of C.15 Trials," International Flame Research Foundation, P. F. Panel Meetings (August and December, 1970).

Hendrickson, T. A. (ed.), *Synthetic Fuels Data Handbook*, Cameron Engineers, Inc., Denver, CO (1975).

Heredy, L. A., and Wender, I., *ACS Div. Fuel Chem. Preprints* **25**, 4 (1980).

Herring, J. R., "Subgrid Scale Modeling—An Introduction and Overview," *Turbulent Shear Flows I* (F. Dorst, B. E. Launder, F. W. Schmidt, and J. H. Whitelaw, eds.), Springer-Verlag, Berlin (1979).

Hertzberg, M., Litton, C. D., and Garloff, R., "Studies of Incipient Combustion and Its Detection," U.S. Bureau of Mines R. I. 8206, Pittsburgh, PA (1977).

Hertzberg, M., Cashdollar, K. L., Ng, D. L., and Conti, R. S., "Domains of Flammability and Thermal Ignitability for Pulverized Coals and Other Dusts: Particle Size Dependences and Microscopic Residue Analyses," *19th Symposium (International) on Combustion*, The Combustion Institute, Pittsburgh, PA, 1169 (1982).

Highly, J., and Merrick, D., *AIChE Symposium Series* **70** (137), 336 (1974).

Hill, S. C., "Modeling of Nitrogen Pollutants in Turbulent Pulverized Coal Flames," Ph.D. Dissertation, Department of Chemical Engineering, Brigham Young University, Provo, UT (1983).

Hinze, J. O., *Turbulence*, 2nd ed., McGraw-Hill, New York, NY (1967).

Hinze, J. O., "Turbulent Fluid and Particle Interaction," *Progr. Heat Mass Transfer, Vol. VI* 943 (1971).

Horio, M. Rengarajan, P., Krishnan, R., and Wen, C. Y., "Fluidized Bed Combustor Modeling," Report NAS3-19725, West Virginia University, Morgantown, WV (1977).

Horton, M. D., "Fast Pyrolysis," *Pulverized-Coal Combustion and Gasification* (L.D. Smoot and D. T. Pratt, eds.), Plenum, New York, NY (1979).

Horton, M. D., Goodson, F. P., and Smoot, L. D., "Characteristics of Flat, Laminar Coal-Dust Flames," *Combust. Flame* **28**, 187 (1977).

Hottel, H. C., and Sarofim, A. F., *Radiative Transfer*, McGraw-Hill, New York, NY (1967).

Howard, J. B., "Fundamentals of Coal Pyrolysis and Hydropyrolysis," second supplementary volume, *Chemistry of Coal Utilization* (M. A. Elliott, ed.), Wiley, New York, NY, 665 (1981).

Howard, J. B., and Essenhigh, R. H., "Mechanism of Solid-Particle Combustion with Simultaneous Gas-Phase Volatiles Combustion," *11th Symposium (International) on Combustion*, The Combustion Institute, Pittsburgh, PA, 399 (1967).

Howard, J. B., Peters, W. A., and Serio, M. A., "Coal Devolatilization Information for Reactor Modelling," EPRI AP-1803, Massachusetts Institute of Technology, Cambridge, MA (April 1981).

Hubbard, E. H., *J. Inst. Fuel* **33**, 386 (1960).

Hulburt, H. M., and Katz, S., "Some Problems in Particle Technology: A Statistical Mechanical Formulation," *Chem. Eng. Sci.* **19**, 555 (1964).

Hutchinson, P., Khalil, E. E., Whitelaw, J. H., and Wigley, G., "The Calculation of Furnace Flow Properties and Their Experimental Verification," *ASME, J. Heat Transfer* **81**, 276 (1976).

IGT, *Preparation of a Coal Conversion Systems Technical Data Book*, Institute of Gas Technology, U.S. Department of Energy Report FE 2286-32, Chicago, IL (February 1979).

Johnson, J. L., "Fundamentals of Coal Gasification, *Chemistry of Coal Utilization*, second supplementary volume (M. A. Elliott, ed.), Wiley–Interscience, New York, NY, 1491 (1981).

Johnson, S. A., and Sommer, T. M., "Commercial Evaluation of a Low NO_x Combustion System as Applied to Coal-Fired Utility Boilers," Joint Symposium on Stationary NO_x Control, Denver, CO (October 1980).

Jones, P. G., "Fluid Dynamics Measurements in a Simulated Entrained Coal Gasifier," M.S. Thesis, Chemical Engineering Department, Brigham Young University, Provo, UT (1983).

Jones, W. P., and Launder, B. E., "The Prediction of Laminarisation with a Two-Equation Model of Turbulence," *Int. J. Heat Mass Transfer* **15**, 301 (1972).

Kanury, A. M., *Introduction to Combustion Phenomena*, Gordon & Breach, New York, NY, (1975), p. 90.

Kennedy, I. M., and Kent, J. H., "Laser Scattering Measurements in Turbulent Diffusion Flames," AIAA 18th Aerospace Sciences Meeting, Pasadena, CA (January 14–16, 1980).

Kent, J. H., and Bilger, R. W., "Turbulent Diffusion Flames," Kolling Report F-37, Department of Mechanical Engineering, University of Sydney, Sydney, Australia (1971).

Kim, A. G., "Laboratory Studies on Spontaneous Heating of Coal," U.S. Bureau of Mines Information Circular No. 8756, U.S. Department of the Interior, Pittsburgh, PA (1977).

Kimber, G. M., and Gray, M. D., "Rapid Devolatilization of Small Coal Particles," *Combust. Flame* **11**, 360 (1967).

Kobayashi, H., Howard, J. B., and Sarofim, A. F., "Coal Devolatilization at High Temperatures," *18th Symposium (International) on Combustion*, The Combustion Institute, Pittsburgh, PA, 411 (1977).

Krazinski, J. C., Buckins, R. O., and Krier, H., "Coal Dust Flames: A Review and Development of a Model of Flame Propagation," *Progr. Energy Combust. Sci.* **3**, 35 (1977).

Krazinski, J. C., Buckins, R. O., and Krier, H., "Coal Dust Flames: A Review and Development of a Model in Flame Propagation," *Progr. Energy Combust. Sci.* **5**, 31 (1979).

Kriegbaum, R. A., and Laurendeau, N. M., "Pore Model for the Gasification of a Single Coal Char Particle," Eastern States Section, The Combustion Institute, Hartford, CN (1977).

Kuchta, J. M., Rowe, V. R. and Burgess, D. S., "Spontaneous Combustion Susceptibility of U.S. Coals, "U.S. Bureau of Mines Report 8474, Pittsburgh, PA (1980).

Kurtzrock, R., personal communication, U.S. Department of Energy, Pittsburgh Energy Technology Center, Pittsburgh, PA (May, 1982).

LaFollette, R., "Temperature and Heat Flux Measurements in a Laboratory Scale Pulverized Coal Combustor," M.S. thesis, Department of Chemical Engineering, Brigham Young University, Provo, UT (1983).

Launder, B. E., "Stress Transport Closures—Into the Third Generation," *Turbulent Shear Flows I* (F. Dorst, B. E. Launder, F. W. Schmidt, and J. H. Whitelaw, eds)., Springer-Verlag, Berlin (1979).

Launder, B. E., and Spalding, D. B., *Mathematical Models of Turbulence*, Academic Press, London (1972).

Launder, B. E., Reece, G. J., and Rodi, W., "Progress in the Development of a Reynolds-Stress Turbulence Closure," *J. Fluid Mech.* **68**, 537 (1975).

Laurendeau, N. M., "Heterogeneous Kinetics of Coal Char Gasification and Combustion," *Progr. Energy Combust. Sci.* **4**, 221–270 (1978).

Lee, J. W., Chen, S. L., Pershing, M. P., and Heap, M. P., "Pollutant Production in Pulverized Coal Flames and the Effect of Coal Characteristics," Western States Section/Combustion Institute, Brigham Young University, Provo, UT (1979).

Leavitt, D. R. "Coal Dust and Swirl Effects on Gas and Particle Mixing Rates in Confined Jets," M.S. thesis, Department of Chemical Engineering, Brigham Young University, Provo, UT (August 1980).

Leonard, B. P., "Stable and Accurate Convective Modeling Procedure Based on Quadratic Upstream Interpolation," *Computer Methods Appl. Mech. Eng.* **19**, 15 (1979).

Leslie, D. C., "Analysis of a Strongly Sheared, Nearly Homogeneous Turbulent Shear Flow," *J. Fluid Mech.* **98**, 435 (1980).

Levy, J. M., "Modeling of Fuel-Nitrogen Chemistry in Combustion: The Influence of Hydrocarbons," Fifth EPA Fundamental Combustion Research Workshop, Newport Beach, CA (1980).

Levy, J. M., Pohl, J. H., Sarofim, A. F., and Song, Y. H., "Combustion Research on the Fate of Fuel Nitrogen under Conditions of Pulverized Coal Combustion," EPA-600/7-78-165, Massachusetts Institute of Technology, Cambridge, MA (1978).

Levy, M. R., Chan, L. K., Sarofim, A. F., and Beér, J. M., "NO/Char Reactions at Pulverized Coal Flame Conditions," *18th Symposium (International) on Combustion,* The Combustion Institute, Pittsburgh, PA, 111 (1981).

Lewis, M. H., "Local Measurements in Turbulent Natural Gas Combustion," M.S. thesis, Department of Chemical Engineering, Brigham Young University, Provo, UT (1979).

Lewellen, W. S., Teske, M. E., and Donaldson, C. du P., "Variable Density Flows Computed by a Second-Order Closure Description of Turbulence," *AIAA J.* **14**, 382 (1976).

Lewellen, W. E., Segur, H., and Varma, A. K., *Modeling Two Phase Flow in a Swirl Combustor,* Final Report, ERDA Contract No. EY-76-C-024062 (1977).

Lewis, G. H., "Carbon Conversion in an Entrained Coal Gasifier," M.S. thesis, Department of Chemical Engineering, Brigham Young University, Provo, UT (1981).

Lewis, J. R., and Tung, S. E. (with Park, D., and/or Lee, D., in selected volumes), *Modeling of Fluidized Bed Combustion of Coal,* Final Report, Vols. I–VI, DOE/MC/16000-1294, Massachusetts Institute of Technology, Cambridge, MA (May 1982).

Lewis, M. H., and Smoot, L. D., "Turbulent Gaseous Combustion Part I: Local Species Concentration Measurements," *Combust. Flame* **42**, 183 (1981).

Lewis, P. F., and Simons, G. A., "Char Gasification: Part II. Oxidation Results," *Combust. Sci. Technol.* **20**, 117 (1979).

Libby, P. A., and Williams, F. A., "Some Implications of Recent Theoretical Studies in Turbulence Combustion," *AIAA J.* **19**, 261 (1981).

Libby, P. A., and Williams, F. A. (eds.), *Turbulent Reacting Flows,* Springer-Verlag, New York, NY (1980).

Lilley, D. G., "Flowfield Modeling in Practical Combustors: A Review," *J. Energy* **3**, 193 (1979).

Lilly, G. P., "Effect of Particle Size on Particle Eddy Diffusivity," *Ind. Eng. Chem. Fundam.* **12**, 268 (1973).

Linares-Solano, A., Mahajan, O. P., and Walker, P. L., Jr., "Reactivity of Heat-Treated Coals in Steam," *Fuel* **58**, 327 (1979).

Littlewood, K., "Gasification Theory and Application," *Progr. Energy Combust. Sci.* **3**, 35 (1977).

Liu, Y. A. (ed.), "Physical Cleaning of Coal," Marcel Dekker, New York, NY (1982).

Lockwood, F. C., Salooja, A. P., and Syed, S. A., "A Prediction Method for Coal-Fired Furnaces," *Combust. Flame* **38**, 1 (1980).

Longwell, J. P., and Weiss, M. A., "Mixing and Distribution of Liquids in High-Velocity Air Streams," *Ind. Eng. Chem.* **45**, 667 (1953).

Love, M. D., and Leslie, D. C., "Studies of Subgrid Modeling with Classical Closures and Burger's Equation," *Turbulent Shear Flows I* (F. Dorst, B. E. Launder, F. W. Schmidt, and J. H. Whitelaw, eds)., Springer-Verlag, Berlin (1979).

Lowe, A., Wall, T. F., and Stewart, I. McC., "A Zoned Heat Transfer Model of a Large Tangentially Fired Pulverized Coal Boiler," *15th Symposium (International) on Combustion,* The Combustion Institute, Pittsburgh, PA, 1261 (1974).

Lowry, H. H. (ed.), *Chemistry of Coal Utilization,* Vol. I, Wiley, New York, NY (1945).

Lowry, H. H. (ed.), *Chemistry of Coal Utilization,* supplementary volume, Wiley, New York (1963).

Lumley, J. L., and Khajeh-Nouri, B., "Computation of Turbulent Transport," *Adv. Geophys.* **A18** 169 (1974).

Macek, A., "Coal Combustion in Boilers: A Mature Technology Facing New Constraints," *17th Symposium (International) on Combustion,* The Combustion Institute, Pittsburgh, PA, 65 (1978).

Mackowski, D. W., Altenkirch, R. A., Peck, R. E., and Tong, T. W., "Infrared Plyrometer Measurement of Particle and Gas Temperatures in Pulverized-Coal Flames," 1982 Spring Meeting Combustion Institute/Western States Section, Salt Lake City, UT (1982).

Mallard, W. G., Miller, J. H., and Smyth, K. C., "Resonantly Enhanced Two-Photon Photoionization of NO in an Atmospheric Flame," *J. Chem. Phys.* **76**, 3483 (1982).

Malte, P. C., and Pratt, D. T., "The Role of Energy-Releasing Kinetics in NO_x Formation: Fuel-Lean, Jet-Stirred CO-Air Combustion," *Combust. Sci. Technol.* **9**, 221 (1974).

Malte, P. C., and Rees, D. P., "Mechanisms and Kinetics of Pollutant Formation During Reaction of Pulverized Coal" (L. D. Smoot and D. T. Pratt, eds.), *Pulverized Coal Combustion and Gasification,* Plenum Press, New York, 183 (1979).

Malte, P. C., Schmidt, S. C., Kramlich, J. C., Spitzer, K. D., Yee, D., and Singh, S., "The Influence of Finite Mixing on OH and NO_x Concentrations in a Jet-Stirred Reactor," AIAA 18th Aerospace Sciences Meeting, Pasadena, CA (1980).

Mandel, G., "Gasification of Coal Char in Oxygen and Carbon Dioxide at High Temperature," M.S. thesis, Massachusetts Institute of Technology, Cambridge, MA (1977).

Manfred, R. K., and Ehrlich, S., "Combustion of Clean Coal-Water Slurry," 44th Annual Meeting of American Power Conference, Chicago, IL (April 1982).

Marinero, E. E., Rettner, C. T., and Zare, R. N., "Quantum-State-Specific Detection of Molecular Hydrogen by Three-Photon Ionization," *Phys. Rev. Lett.* **48**, 1323 (1982).

Marsden, C., "Anthracite Dust Cloud Combustion," Ph.D. thesis, University of Sheffield, Sheffield, England (1964).

Mayers, A. M., "The Rate of Reduction of Carbon Dioxide by Graphite," *Am. Chem. Soc. J.* **56**, 70 (1934a).

Mayers, A. M., "The Rate of Oxidation of Graphite by Steam," *Am. Chem. Soc. J.* **56**, 1879 (1934b).

McCann, C. R., Demeter, J. J., and Bienstock, D., "Combustion of Pulverized Solvent-Refined Coal," *J. Eng. Power, Trans. ASME* **99**, 305 (July 1977).

McHale, E. G., Scheffee, R. S., and Rossmeissl, N. P., "Combustion of Coal-Water Slurry," *Combust. Flame* **45**, 121 (1982).

McIntosh, M. J., and Coates, R. L., "Experimental and Process Design Study of a High Rate Entrained Coal Gasification Process," Final Report, DOE Contract EX-76-C-01-1548, Eyring Research Institute, Provo, UT (1978).

McLean, W. J., Hardesty, D. R., and Pohl, J. H., "Direct Observations of Devolatilizating Pulverized Coal Particles in a Combustion Environment," *18th Symposium (International) on Combustion,* The Combustion Institute, Pittsburgh, PA 1239 (1981).

McNair, M. B., "Energy Data Report: Coal Distribution," DOE/EIA-0125 (80/2Q) (1980).

McNeil, D., "High Temperature Coal Tar," Chapter 17, *Chemistry of Coal Utilization,* second supplementary volume (M. A. Elliott, ed.), Wiley, New York (1981).

McRanie, R. D., "Full-Scale Utility Boiler Test with Solvent Refined Coal (SRC)," Final Report to U.S. Department of Energy, Contract No. EX-76-C-01-2222, Southern Company Services, Inc., Birmingham, AL (July 1979).

Mehta, A. K., "Mathematical Modeling of Chemical Processes for Low BTU Gasification of Coal for Electric Power Generation," Final Report, ERDA Contract No. E(49-18)-1545, Combustion Engineering, Windsor, MA (1976).

Mehta, B. N., and Aris, R., "Communications on the Theory of Diffusion and Reaction-VII the Isothermal pth Order Reaction," *Chem. Engr. Sci.* **26**, 1699 (1971).

Mellor, G. L., and Herring, H. J., "A Survey of the Mean Turbulent Field Closure Models," *AIAA J.* **11**, 590 (1973).

Melville, E. K., and Bray, N. C., "A Model of the Two-Phase Turbulent Jet," *Int. J. Heat Mass Transfer* **22**, 647 (1979).

Memmott, V. J., "Rates of Mixing of Particles and Gases in Confined Jets," M.S. thesis, Department of Chemical Engineering, Brigham Young University, Provo, UT (1977).

Memmott, V. J., and Smoot, L. D., "Cold Flow Mixing Rate Data for Pulverized Coal Reactors," *AIChE J.* **24**, 466 (1978).

Michel, J. B., and Payne, R., "Detailed Measurement of Long Pulverized Coal Flames for the Characterization of Pollutant Formation," IFRF Document Number F 09/a/23, Ijmuiden, The Netherlands (1980).

Michelfelder, S., and Lowes, T. M., "Report on the M-2 Trails," IFRF Document F 36/a/4, Ijmuiden, The Netherlands (1974).

Meyer, J. P., Wells, J. W., Cox, J. R., Belk, J. P., Frazier, G. C., and Wham, R. M., ORNL-5475, Oak Ridge National Laboratory, Oak Ridge, TN (November 1980).

Mie, G., *Ann. Phys.* **25**, 377 (1908).

Milne, T. A., and Beachey, J. E., "The Microstructure of Pulverized Coal-Air Flames. II. Gaseous Species, Particulate and Temperature Profiles," *Combust. Sci. Technol.* **16**, 139 (1977).

Mims, C. A., Neville, M., Quann, R. J., and Sarofim, A. F., "Laboratory Studies of Trace Element Transformations During Coal Combustion," AIChE 87th Annual Meeting, Boston, MA (1979).

Mitchell, R. E., and McLean, W. J., "On the Temperature and Reaction Rate of Burning Pulverized Fuels," 1982 Spring Meeting Combustion Institute/Western States Section, Salt Lake City, UT (1982).

Mitchell, R. E., Sarofim, A. F., and Clomburg, L. A., "Experimental and Numerical Investigation of Confined Laminar Diffusion Flames," *Combust. Flame* **37**, 227 (1980).

Modarress, D., Wuerer, J., and Elghobashi, S., "An Experimental Study of a Turbulent Round Two-Phase Jet," AIAA Paper No. AIAA-82-0964 (1982).

Mulcahy, M. F. R., Boow, J., and Goard, P. R. C., "Fireside Deposits and Their Effect on Heat Transfer on a Pulverized Fuel-Fired Boiler. Part III: The Influence of Physical Characteristics of the Deposits on its Radiant Emmittance and Effective Thermal Conductance," *J. Inst. Fuel.* **42**, 346 (1969).

Mulcahy, M. F. R., and Smith, I. W., "The Kinetics of Combustion of Pulverized Coke, Anthracite and Coal Chars," Proceedings of "CHEMECA 70" Conference, Session 2, 101 Butterworth, Australia (1971).

Nagy, J., Dorsett, J. G., and Cooper, A. R., "Explosibility of Carbonaceous Dusts," U.S. Bureau of Mines R.I. 6597, Pittsburgh, PA (1965).

Neavil, R. C., "Coal Origin, Analysis, and Classification," Paper No. 111b, AIChE 72nd Annual Meeting, New York, NY (1979).

Nettleton, M. A., "Temperature Measurements on Burning Coal Particles in a Radiating Enclosure," *Combust. Flame* **9**, 311 (1965).

Nettleton, M. A., "Burning Rates of Devolatilized Coal Particles," *Ind. Eng. Chem. Fundam.* **6**, 20 (1967).

Nettleton, M. A., "Particulate Formation in Power Station Boiler Furnaces," *Progr. Energy Combust. Sci.* **5**, 223 (1979).

Neville, M., Quann, R. J., Haynes, B. S., and Sarofim, A. F., "Vaporization and Condensation of Mineral Matter During Pulverized Coal Combustion," *18th Symposium (International) on Combustion,* The Combustion Institute, Pittsburgh, PA, 126 (1980).

Nowacki, P., *Coal Liquefaction Processes*, Noyes Data Corporation, Park Ridge, NJ (1979).

Nuttel, H. E., and Roach, G. F., "An Interdisciplinary Investigation of Coal Gasification Mechanics and Kinetics for the Optimal Development of New Mexico's Energy Resources," Final Report 75-119, University of New Mexico, Albuquerque, NM (1978).

Oberjohn, W. J., Cornelius, D. K., Fiveland, W. A., Schnipke, R. J., and Wang, J. H., "Computational Tools for Pulverized Coal Combustion," DOE/PC/40265-3, Babcock and Wilcox Co., Alliance, Ohio (Jan. 1982).

O'Brien, E. E., "Turbulent Reacting Flows" (P. A. Libby and F. A. Williams eds.), *Topics in Applied Physics*, **44**, 185 (1980).

O'Brien, T., and Pierce, T., "Preliminary Sensitivity Analysis of Devolatilization Reactions in 1-DICOG," U.S. Department of Energy preprint, Morgantown Energy Technology Center, Morgantown, WV (1981).

Olofsson, J., "Mathematical Modelling of Fluidized Bed Combustors," No. ICTIS/TR14, IEA Coal Research, London (1980).

Omenetto, N., Benetti, P., and Rossi, G., "Flame Temperature Measurements by Means of Atomic Fluorescence Spectrometry," *Spectrochem. Acta.* **27**, 453 (1972).

OTA, "The Direct Uses of Coal," Office of Technology Assessment, Washington, DC (1979).

Otto, K., Bartosiewicz, L., and Shelef, M., "Catalysis of Carbon-Steam Gasification by Ash Components from Two Lignites," *Fuel* **58**, 85 (1979a).

Otto, K., Barosiewicz, L., and Shelef, M., "Effects of Calcium, Strontium, and Barium as Catalysts and Sulphur Scavengers in the Steam Gasification of Coal Chars," *Fuel* **58**, 565 (1979b).

Owen, P. R., "Pneumatic Transport," *J. Fluid Mech.* **39**, 407 (1969).

Padia, A. S., Sarofim, A. F., and Howard, J. B., "Behavior of Ash in Pulverized Coal Under Simulated Combustion Conditions," Spring Meeting of the Central States Section, The Combustion Institute, Pittsburgh, PA (April 1976).

Pan, Y. S., Bella, G. T., Snedden, R. B., Wieczenski, D. E., and Joubert, J. I., "Exploratory Coal-Water and Coal-Methonal Mixture Combustion Tests in Oil-Designed Boilers," Fourth International Symposium on Coal Slurry Combustion, Vol. 2, Session 3, PETC, Pittsburgh, PA (May 1982).

Patankar, S. V., *Studies in Convection* (B. W. Launder, ed.), Vol. 1, Academic Press, New York, NY (1975).

Patankar, S. V., and Spalding, D. B., *Heat and Mass Transfer in Boundary Layers*, second edition, Intertext Books, London (1970).

Patel, J. G., "U-Gas Technology Status," Paper presented at *Advances in Coal Utilization Technology Symposium*, Louisville, KY (May 14–18, 1979).

Perry, H., "The Gasification of Coal," *Sci. Am.* **230**, 19 (1974).

Pershing, D. W., and Wendt, J. O. L., "Relative Contributions of Volatile Nitrogen and Char Nitrogen to NO_x Emissions from Pulverized Coal Flames," *Ind. Eng. Chem. Process Des. Dev.* **18**, 60 (1979).

Pershing, D. W., and Wendt, J. O. L., "Pulverized Coal Combustion: The Influence of Flame Temperature and Coal Composition on Thermal and Fuel NO_x," *16th Symposium (International) on Combustion*, The Combustion Institute, Pittsburgh, PA (1977).

Pitt, G. J., and Millward, G. R., *Coal and Modern Coal Processing: An Introduction*, Academic Press, New York, NY (1979).

Pohl, J. H., and Sarofim, A. F., "Devolatilization and Oxidation of Coal Nitrogen," *16th Symposium (International) on Combustion*, The Combustion Institute, Pittsburgh PA, 491 (1977).

Pratt, D. T., "Mixing and Chemical Reaction in Continuous Combustion," *Prog. Energy Comb. Sci.* **1**, 73 (1976).

Pratt, D. T., "Gas-Phase Chemical Kinetics" (L. D. Smoot and D. T. Pratt eds.), *Pulverized Coal Combustion and Gasification*, Plenum, New York, NY, 65 (1979).

Pratt, D. T., and Wormeck, J. J., "CREK, A Computer Program for Calculation of Combustion Reaction Equilibrium and Kinetics in Laminar or Turbulent Flows," Thermal Energy Laboratory, Washington State University, Pullman, WA (May 1976).

Radovic, L. R., and Walker, P. L., Jr., "Reactivities of Chars Obtained as Residues in Selected Coal Conversion Processes," *Fuel Processing Technology* **8**, 149 (1984).

Rees, D. P., "Pollutant Formation During Pulverized Coal Combustion," Ph.D. Dissertation, Chemical Engineering Department, Brigham Young University, Provo, UT (1980).

Rees, D. P., Smoot, L. D., and Hedman, P. O., "Nitrogen Oxide Formation Inside a Laboratory, Pulverized Coal Combustor," *18th Symposium (International) on Combustion*, The Combustion Institute, Pittsburgh, PA, 1305 (1981).

Regnier, P. R., and Taran, J. P. E., *Appl. Phys. Lett.* **23**, 240 (1973).

Reid, W. T., "Coal Ash, Its Effect on Combustion Systems," *Chemistry of Coal Utilization*, Second Supplementary Volume, M. A. Elliott, ed., John Wiley & Sons, New York, NY 1389 (1981).

Remenyi, K., "Combustion Stability," Akademiai Kiado, Budapest (1980).

Richter, W., and Quack, R., "A Mathematical Model of a Low-Volatile Pulverized Fuel Flame," *Heat Transfer in Flames* (N. H. Afgan and J. M. Beér, eds.), Scripta Technica, 95 (1974).

Rockney, B. H., Cool, T. A., and Grant, E. R., "Detection of Nascent NO in a Methane/Air Flame by Multiphoton Ionization," *Chem. Phys. Lett.* **87**, 141 (1982).

Rodi, W., "A New Algebraic Relation for Calculating the Reynolds Stresses," *ZAMM* **56** (1976).

Rosner, D. E., "A Course in Combustion Science and Technology," *Chem. Eng. Ed.* **193**, 95 (Fall, 1980).

Samuelsen, G. S. and Brum, R. D., "Two-Component Laser Anemometry Measurements in a Non-Reacting and Reacting Complex Flow Model Combustor," Western States Section of the Combustion Institute, Sandia National Laboratories, Sandia, CA (1982).

Samuelsen, G. S., Trolinger, J. D., Heap, M. P., and Seeker, W. R., "Observation of the Behavior of Coal Particles During Thermal Decomposition," *Combust. Flame* **40**, 7 (1981).

Sarofim, A. F., and Beér, J. M., "Modelling of Fluidized Bed Combustion," *17th Symposium (International) on Combustion*, The Combustion Institute, Pittsburgh, PA, 189 (1978).

Sarofim, A. F., and Flagen, R. C., "NO Control for Stationary Combustion Sources," *Progr. Energy Combust. Sci.* **2**, 1 (1976).

Sarofim, A. F., and Hottel, H. C., "Radiative Transfer in Combustion Chambers: Influence of Alternate Fuels," Sixth International Heat Transfer Conference, Toronto, Canada, 199 (1978).

Sarofim, A. F., Howard, J. B., and Padia, A. S., "The Physical Transformation of Mineral Matter in Pulverized Coal Under Simulated Combustion Conditions," *Combust. Sci. Technol.* **16**, 187 (1977).

Saxena, S. C., Grewal, N. S., and Venhatoramana, M., U.S. Department of Energy Report FE-1787-10, Argonne National Laboratory, Argonne, IL (1978a).

Saxena, S. C., Chen, T. P., and Jonke, A. A., "A Plug Flow Model for Coal Combustion and Desulfurization in Fluidized Beds: Theoretical Formulation," ANL/CEN/FE-78-11, Argonne National Laboratory, Argonne, IL (1978b).

Scheffee, R. S., Skolnik, E. G., Rossmeissl, N. P., Heaton, H. L., and McHale, E. T., "Further Development and Evaluation of Coal-Water Mixture Technology," Fourth International Symposium on Coal Slurry Combustion, Vol. 3, Session 6, PETC/DOE, Pittsburgh, PA (May 1982).

Schneyer, G. P., Peterson, E. W., Chen, P. J., Cook, J. L., Brownell, D. H., Jr., and Blake, T. R., U.S. Department of Energy, Final Report, Vol. 2, DOE/ET/10242-T1, Systems, Science and Software, San Diego, CA (April 1981).

Schweiger, R. G., "Burning Tomorrow's Fuels," *Power* **123**, 1 (February 1979).

Seeker, W. R., Samuelsen, G. S., Heap, M. P., and Trolinger, J. D., "The Thermal Decomposition of Pulverized Coal Particles," *18th Symposium (International) on Combustion*, The Combustion Institute, Pittsburgh, PA, 1213 (1981).

Siegel, R., and Howell, J. R., *Thermal Radiation Heat Transfer*, McGraw-Hill, New York, NY (1981).

Selcuk, N., Siddall, R. G., and Beér, J. M., "A Comparison of Mathematical Models of the Radiative Behavior of A Large-Scale Experimental Furnace," *16th Symposium (International) on Combustion*, The Combustion Institute, Pittsburgh, PA, 53 (1976).

Selph, C., "Generalized Thermochemical Equilibrium Program for Complex Mixtures," Rocket Propulsion Laboratory, Edwards Air Force Base, CA (1965).

Semenov, N. N. *Chemical Kinetics and Chain Reactions*, Oxford University Press, London (1935). •

Sergeant, G. D., and Smith, I. W., "Combustion Rates of Bituminous Coal Char in the Temperature Range 800 to 1700 K," *Fuel* **52**, 52 (1973).

Sharp, J. L., "Particle and Gas Mixing in Confined, Recirculating Coaxial Jets with Angular Injection," M.S. thesis, Department of Chemical Engineering, Brigham Young University, Provo, UT (1981).

Siminski, V. J., Wright, F. J., Edelman, R. B., Economos, C., and Fortune, O. F., "Research on Methods of Improving the Combustion Characteristics of Liquid Hydrocarbon Fuels," AFAPL TR 72-74, Vols. I and II, Air Force Aeropropulsion Laboratory, Wright Patterson Air Force Base, OH (February 1972).

Simons, G. A., "The Structure of Coal Char: Part II. Pore Combination," *Combust. Sci. Technol.* **19**, 227 (1979a).

Simons, G. A., "Char Gasification: Part I. Transport Model," *Combust. Sci. Technol.* **20**, 107 (1979b).

Simons, G. A., and Finson, M. L., "The Structure of Coal Char: Part I. Pore Branching," *Combust. Sci. Technol.* **19**, 217 (1979).

Singer, J. G. (ed.), *Combustion: Fossil Power Systems*, Chapter 13, "Fuel Burning Systems," Combustion Engineering, Inc., Winsor, CN (1981).

Singh, S. P. N., Moyers, J. C., and Carr, K. R., "Coal Beneficiation—The Cinderella Technology," Paper for CCAWG Meeting, Oak Ridge National Laboratory, Oak Ridge, TN (December 9, 1982).

Skinner, F. D., "Mixing and Gasification of Pulverized Coal," Ph.D. dissertation, Chemical Engineering Department, Brigham Young University, Provo, UT (1980).

Skinner, F. D., and Smoot, L. D., "Heterogeneous Reactions of Char and Carbon," *Pulverized Coal Combustion and Gasification* (L. D. Smoot and D. P. Pratt, eds.), Plenum, New York, NY, 149 (1979).

Skinner, F. D., Smoot, L. D., and Hedman, P. O., "Mixing and Gasification of Coal in an Entrained Flow Gasifier," ASME 80-WA/HT-30, New York, NY (1980).

Sloan, D., personal communication, Brigham Young University, Provo, UT (1982). See also Germane, G. J., and Smoot, L. D., "Basic Combustion and Pollutant Formation Processes for Pulverized Fuels," Quarterly Progress Report #7, DOE Contract No. DE-FG22-80PC-30306 (July 15, 1982).

Sloan, D., "Modeling of Turbulent, Swirling Systems and Single-Particle Char Combustion," Doctoral Dissertation (In preparation), Brigham Young University, Provo, Utah (1984).

Smith, I. W., "Kinetics of Combustion of Size-Graded Pulverized Fuels in the Temperature Range 1200–2270°K," *Combust. Flame* **17**, 303 (1971a).

Smith, I. W., "The Kinetics of Combustion of Pulverized Semi-Anthracite in the Temperature Range 1400–2200°K," *Combust. Flame* **17**, 421 (1971b).

Smith, I. W., "The Combustion Rates of Pulverized Coal Char Particles," Conference on Coal Combustion Technology and Emission Control, California Institute of Technology, Pasadena, CA (February 1979).

Smith, I. W., "The Intrinsic Reactivity of Carbons to Oxygen," *Fuel* **57**, 409 (1978).

Smith, I. W., and Tyler, R. J., "Internal Burning of Pulverized Semi-Anthracite: The Relation Between Particle Structure and Reactivity," *Fuel* **51**, 312 (1971).

Smith, I. W., and Tyler, R. J., "The Reactivity of a Porous Brown Coal Char to Oxygen Between 630 and 1812°K," *Combust. Sci. Tech.* **9**, 87 (1974).

Smith, I. W., "The Combustion Rates of Coal Chars: A Review," *19th Symposium (International) on Combustion*, The Combustion Institute, Pittsburgh, PA 1045 (1982).

Smith, P. J., "Theoretical Modeling of Coal or Gas Fired Turbulent Combustion or Gasification," Ph.D. dissertation, Department of Chemical Engineering, Brigham Young University, Provo, UT (1979).

Smith, P. J., and Smoot, L. D., "Mixing and Kinetic Processes in Pulverized Coal Combustors," Volume 2: User's Manual for Computer Program for 1-DICOG, EPRI, Final Report 364-1-3, Brigham Young University, Provo, UT (October 1979).

Smith, P. J., and Smoot, L. D., "One Dimensional Model for Pulverized Coal Combustion and Gasification," *Combust. Sci. Technol.* **23**, 17 (1980).

Smith, P. J., and Smoot, L. D., "Turbulent Gaseous Combustion Part II: Theory and Evaluation for Local Properties," *Combust. Flame* **42**, 277 (1981).

Smith, P. J., Fletcher, T. J., and Smoot, L. D., "Model for Pulverized Coal-Fired Reactors," *18th Symposium (International) on Combustion,* The Combustion Institute, Pittsburgh, PA, 1285 (1981).

Smith, R. D., "The Trace Element Chemistry of Coal During Combustion and the Emissions from Coal-Fired Plants," *Progr. Energy Combust. Sci.* **6**, 201 (1980).

Smoot, L. D., "Pulverized Coal Diffusion Flames—A Perspective Through Modeling," *18th Symposium (International) on Combustion,* The Combustion Institute, Pittsburgh, PA, 1185 (1980).

Smoot, L. D., and Hill, S. C., "Critical Requirements in Combustion Research," *Progr. Energy Combust. Sci.* **9**, 77 (1983).

Smoot, L. D., and Horton, M. D., "Propagation of Laminar Pulverized Coal-Air Flames," *Progr. Energy Combust. Sci.* **3**, 235 (1977).

Smoot, L. D., and Pratt, D. T. (eds.), *Pulverized Coal Combustion and Gasification,* Plenum, New York, NY (1979).

Smoot, L. D., and Smith, P. J., "Modeling Pulverized Coal Reaction Processes" (L. D. Smoot and D. T. Pratt eds.), *Pulverized Coal Combustion and Gasification,* Plenum, New York, NY, 217 (1979).

Smoot, L. D., Horton, M. D., and Williams, G. A., "Propagation of Laminar Pulverized Coal-Air Flames," *16th Symposium (International) on Combustion,* The Combustion Institute, Pittsburgh, Pa, 375 (1976).

Smoot, L. D., Hedman, P. O., and Smith, P. J., "Pulverized Coal Combustion Research at Brigham Young University," accepted for publication in *Progr. Eng. Comb. Sci.* (1984).

Smyth, K. C., and Mallard, W. G., "Two-Photon Ionization Processes of PO in a C_2H_2/Air Flame," *J. Chem. Phys.* **77**, 1779 (1982).

Soelberg, N. R., "Local Measurements in an Entrained Coal Gasifier," M.S. thesis, Chemical Engineering Department, Brigham Young University, Provo, UT (1983).

Solomon, P. R., "Characterization of Coal and Coal Thermal Decomposition," Chapter III Report, Advanced Fuel Research, Inc., East Hartford, CN (1980).

Solomon, P. R., personal communication, Advanced Fuel Research, East Hartford, CN (June 25, 1982).

Solomon, P. R., and Colket, M. B., "Evolution of Fuel Nitrogen in Coal Devolatilization," *Fuel* **57**, 749 (1978).

Solomon, P. R., and Colket, M. B., "Coal Devolatilization," *17th Symposium (International) on Combustion,* The Combustion Institute, Pittsburgh, PA 131 (1979).

Solomon, P. R., Hamblen, D. G., Krause, J. L., and Sickle, A., "Vaporization and Pyrolysis of Coal-Water Mixtures," *1983 International Conference on Coal Science,* Pittsburgh, PA, 687 (1983).

Soo, S. L., *Fluid Dynamics of Multiphase Systems,* Blaisdell, Waltham, MA (1967).

Spackman, W., "The Characteristics of American Coals in Relation to Their Conversion into Clean Energy Fuels," Pennsylvania State University, DDE Report FE-2030-13, University Park, PA (May 1980).

Spackman, W., "The Characteristics of American Coals in Relation to Their Conversion into Clean Energy Fuels," U.S. DOE Final Report, Contract No. AC01-76ET10615, Pennsylvania State University, University Park, PA (1982).

Spalding, D. B., "Turbulence Modeling: Solved and Unsolved Problems," *Turbulent Mixing in Nonreactive and Reactive Flows* (S. N. B. Murthy, ed.), Plenum, New York, NY (1975).

Sprouse, K. M., *Theory of Pulverized Coal Conversion in Entrained Flows,* Technical Memorandum for DOE/EZ-77-C-01-2518, Rockwell International, Canoga Park, CA (1977).

Sprouse, K. M., and Schuman, M. D., "Predicting Lignite Devolatilization with the Multiple Parallel and Two-Competing Reaction Models," *Combust. Flame* **43**, 265 (1981).

Srinivas, B., and Amundson, N. R., "A Single-Particle Char Gasification Model," *AIChE J.* **26**, 487 (1980).

Srinivas, B., and Amundson, N. R., "Intraparticle Effects in Char Combustion: III. Transient Studies," *Can. J. Chem. Eng.* **59**, 728 (1981).

Starley, G. P., Manis, S. C., Bradshaw, F. W., and Pershing, D. W., "Formation and Control of NO_x Emissions in Fixed-Bed Coal Combustion," 1982 Spring Meeting Combustion Institute/Western States Section, Salt Lake City, UT (1982).

Stickler, D. B., and Gannon, R. E., "Modeling of Three Entrained Flow Coal Processes," 89th National Meeting of AIChE, New York, NY (August 1979).

Stott, J. B., "The Spontaneous Heating of Coal and the Role of Moisture Transfer," prepared for the U.S. Bureau of Mines by the Department of Chemical Engineering, University of Canterbury, Christchurch, 1 New Zealand, Cont. 146 (July 1980).

Strehlow, R. A., Savage, L. D., and Sorenson, S. C., "Coal Dust Combustion and Suppression," 10th AIAA/SAE Propulsion Conference, Paper No. 74-1112, San Diego, CA (October 1974).

Strimbeck, G. R., Holden, J. H., Bonar, F., Plants, K. D., Pears, C. D., and Hirst, L. L., "Gasification of Pulverized Coal at Atmospheric Pressure: Discussion of Pilot-Plant Development, Study of Process Variables, and Relative Gasification Characteristics of Coal of Different Rank," U.S. Bureau of Mines Report of Investigation 5559, Pittsburgh, PA (1960).

Stull, D. R., and Prophet, H., JANAF Thermochemical Tables, second edition, National Bureau of Standards, Washington, DC (1971).

Suuberg, E. M., Peters, W. A., and Howard, J. B., "Product Composition and Kinetics of Lignite Pyrolysis," Symposium of Coal Gasification Kinetics, *ACS Chem. Soc. Preprints* **22**, 112 (1977).

Suuberg, E. M., Peters, W. A., and Howard, J. B., "Product Compositions and Formation Kinetics in Rapid Pyrolysis of Pulverized Coal—Implications for Combustion," *17th Symposium (International) on Combustion,* The Combustion Institute, Pittsburgh, PA, 177 (1979).

Takagi, T., Tatsumi, T., and Ogasawara, M., "Nitric Oxide Formation from Fuel Nitrogen in Staged Combustion: Roles of HCN and NH_i," *Combust. Flame* **35**, 17 (1979).

Takagi, T., Shin, H., and Ishio, A., "Properties of Turbulence in Turbulent Diffusion Flames," *Combust. Flame* **40**, 121 (1981).

Tamanini, F., "On the Numerical Prediction of Turbulent Diffusion Flames," Central and Western States Section of the Combustion Institute, San Antonio, TX (April 21–22, 1975).

Taylor, D. D., and Flagan, R. C., "Laboratory Studies of Submicron Particles from Coal Combustion," *18th Symposium (International) on Combustion,* The Combustion Institute, Pittsburgh, PA, 1227 (1980).

Tennekes, H., and Lumley, J. L., *A First Course in Turbulence*, MIT Press, Cambridge, MA (1972).

Thomas, W. J., "Effect of Oxidation on the Pore Structure of Some Graphitized Carbon Blacks," *Carbon* **3**, 435 (1977).

Thorsness, C. B., and Sherwood, A. E., "Moving Equilibrium Front Model for In-Situ Gasification," VCRL-52524, Lawrence Livermore Laboratory, Livermore, CA (July 1978).

Thurgood, J. R., "Mixing and Combustion of Pulverized Coal," Ph.D. dissertation, Chemical Engineering Department, Brigham Young University, Provo, UT (1979).

Thurgood, J. R., and Smoot, L. D., "Volatiles Combustion," *Pulverized Coal Combustion and Gasification* (L. D. Smoot and D. T. Pratt, eds.), Plenum, New York, NY, 169 (1979).

Thurgood, J. R., Smoot, L. D., and Hedman, P. O., "Rate Measurements in a Laboratory–Scale Pulverized Coal Combustor," *Combust. Sci. Technol.* **21**, 213 (1980).

Tice, C. L., "Particle and Gas Mixing Rates in Confined, Coaxial Jets with Recirculation," M.S. thesis, Department of Chemical Engineering, Brigham Young University, Provo, UT (1979).

Tillman, D. A. *Wood as an Energy Resource*, Academic Press, New York, NY (1978).

Timothy, L. D., Sarofim, A. F., and Beér, J. M., "Characteristics of Single Particle Combustion," *19th Symposium (International) on Combustion,* The Combustion Institute, Pittsburgh, PA, 1123 (1982).

Tomita, A., Mahajan, O. P., and Walker, Jr., P. L., "Catalysis of Char Gasification by Minerals," *ACS Div. Fuel Chem. Preprints* **22**, 4 (1977).

Trusdell, C., and Toupin, R., "The Classical Field Theories," *Handbuch der Physik*, Springer-Verlag, Berlin (1960).

Ubhayakar, S. K., Stickler, D. B., Von Rosenberg, C. W., and Gannon, R. E., "Rapid Devolatilization of Pulverized Coal in Hot Combustion Gases," *16th Symposium (International) on Combustion,* The Combustion Institute, Pittsburgh, PA, 427 (1976).

Ubhayakar, S. K., Stickler, D. B., and Gannon, R. E., "Modelling of Entrained-Bed Pulverized Coal Gasifiers," *Fuel* **56**, 281 (1977).

Ulrich, G. D., "... The Mechanism of Fly-Ash Formation in Coal-Fired Utility Boilers," DOE Report FE-2205-16, University of New Hampshire, Durham, NH (1979).

Van de Hulst, H. C., *Light Scattering by Small Particles*, Wiley, New York, NY (1957).

van Dijk, C. A., Curran, F. M., Lin, K. C., and Crouch, S. R., "Two-Step Laser-Assisted Ionization of Sodium in a Hydrogen-Oxygen-Argon Flame," *Anal. Chem.* **53**, 1275 (1981).

von Fredersdorff, C. G., and Elliot, M. A., *Chemistry of Coal Utilization, Supplementary Vol.* (H. H. Lowry, ed.), Wiley-Interscience, New York, NY (1963), p. 893.

Vu, B. T., and Gouldin, F. C., "Flow Measurements in a Model Swirl Combustor," *AIAA J.* **20**, 642 (1982).

Walker, P. L., Shelef, M., and Anderson, R. T., "Catalysis of Carbon Gasification," *Chemistry and Physics of Carbon*, Vol. 4 (Walker, P. L., ed.), Marcel Dekker, New York, NY, 287 (1968).

Walker, P. L., Jr., Foresti, R. J., Jr., and Wright, C. C., "Surface Area Studies of Carbon–Carbon Dioxide Reaction," *Ind. Eng. Chem.* **45**, 1703 (1953).

Walker, P. L., Jr., Rusinko, F., Jr., and Austin, L. G., "Gas Reactions of Carbon," *Advan. Catal.* **11**, 135 (1959).

Wall, T. F., Lowe, A., Wibberly, L. J., and Stewart, M. C., "Mineral Matter in Coal and the Thermal Performance of Large Boilers," *Progr. Energy Combust. Sci.* **5**, 1 (1979).

Wall, T. F., Lowe, A., Wibberley, L. J., Mai–Viet, T., and Gupta, R. P., "Fly Ash Characteristics and Radiative Heat Transfer in Pulverized-Coal-Fired Furnaces," *Combust. Sci. Technol.* **26**, 107 (1981).

Walsh, P. M., Zhang, M., Farmayan, W. F., and Beér, J. M., "Ignition and Combustion of Coal-Water Slurry in a Confined Turbulent Diffusion Flame," Accepted for publication *20th Symposium (International) on Combustion,* Ann Arbor, MI (August 1984).

Webb, B. W., "LDV Measurements in Confined Coaxial Jets with and without Swirl," M.S. thesis, Department of Mechanical Engineering, Brigham Young University, Provo, UT (1982).

Wegener, D. C., unpublished data from Phillips Petroleum Co., Bartlesville, OK. (1982). Chars provided and characterized by Combustion Laboratory, Brigham Young University, Provo, UT (1982).

Wells, W. F., Kramer, S. K., Smoot, L. D., and Blackham, A. U., "Reactivity and Combustion of Coal Chars," to be presented at the *20th Symposium (International) on Combustion,* Ann Arbor, MI (August 1984).

Wells, J. R., and Krishnan, R. P., Interim Annual Report for 1979, ORNL/TM-7398, Oak Ridge National Laboratory, Oak Ridge, TN (October 1980).

Wen, C. Y., *Optimization of Coal Gasification Processes*, R & D Report No. 66, Interim Report No. 1, Office of Coal Research Contract No. 14-01-0001-497, West Virginia University, Morgantown, WV (1972).

Wen, C. Y., and Chuang, T. Z., *Entrained Coal Gasification Modeling*, DOE/FE-2274-T1, West Virginia University, Morgantown, WV (1978).

Wen, C. Y., Chen, H., Onozaki, M., "User's Manual for Computer Simulation and Design of the Moving Bed Coal Gasifier," DOE/MC/16474-1390, West Virginia University, Morgantown, WV (January 1982).

Wendt, J. O. L., "Fundamental Coal Combustion Mechanisms and Pollutant Formation in Furnaces," *Progr. Energy Combust. Sci.* **6**, 201 (1980).

Wendt, J. O. L., and Pershing, D. W., Physical Mechanisms Governing the Oxidation of Volatile Fuel Nitrogen in Pulverized Coal Flames," *Combust. Sci. Technol.* **16**, 111 (1977).

Wendt, J. O. L., Pershing, D. W., Lee, J. W., and Glass, J. W., "Pulverized Coal Combustion: NO$_x$ Formation Mechanisms under Fuel Rich and Staged Combustion Conditions," *17th Symposium (International) on Combustion,* The Combustion Institute, Pittsburgh, PA, 77 (1979).

White, W. B., Johnson, S. M., and Dantzig, G. B., "Chemical Equilibrium in Complex Mixtures," *J. Chem. Phys.* **28**, 751 (1958).

Williams, F. A., *Combustion Theory,* Addison-Wesley, MA (1965).

Williams, F. A., and Libby, P. A., AIAA Paper 80-0012, 18th Aerospace Sciences Meeting, Pasadena, CA (1980).

Wilson, C. L., *Coal—Bridge to the Future,* Ballinger, Cambridge, MA, 244 (1980).

Winslow, A. M., "Numerical Model of Coal Gasification in a Packed Bed," *16th Symposium (International) on Combustion,* The Combustion Institute, Pittsburgh, PA, 503 (1977).

Wiser, W. H., *ACS, Div. Fuel Chem. Preprints* **20**, 122 (1975).

Yang, R. T., and Steinburg, M., "A Diffusion Cell Method for Studying Heterogeneous Kinetics in the Chemical Reaction/Diffusion Controlled Region. Kinetics of $C + CO_2 \rightarrow 2CO$ at 1200–1600°C," *Ind. Eng. Chem. Fundam.* **16**, 235 (1977).

Yen, T. F. (ed.), and Chilingarian, G. V., *Oil Shale: Developments in Petroleum Science 5,* Elsevier, New York, NY (1976).

Yoon, H., Wei, J., and Denn, M. M., "A Model for Moving-Bed Coal Gasification Reactors," *AIChE J.* **24**, 885 (1978).

Young, B. C., and Smith, I. W., "The Kinetics of Combustion of Petroleum Coke Particles at 1000 to 1800 K: The Reaction Order," *18th Symposium (International) on Combustion,* The Combustion Institute, Pittsburgh, PA, 1249 (1981).

Zabetakis, M. G., "Flammability Characteristics of Combustible Gases and Vapors," U.S. Bureau of Mines Bulletin **627**, 22 (1965).

Zakkay, V., Miller, G., and Brenton, A., "Heat Transfer Characteristics in a Fluidized Bed with Heat Exchangers," NYU/DAS 77/17, New York University, New York, NY (August 1977).

Zygourakis, K., Arri, L., and Amundson, N. R., "Studies on the Gasification of a Single Char Particle," *Ind. Engr. Chem. Fundam.* **21**, 1 (1982).

———, Fourth International Symposium on Coal Slurry Combustion, (Vols. 1–4), DOE/PETC, Pittsburgh Energy Technology Center, Pittsburgh, PA (May 10–12, 1982).

———, Fifth International Symposium on Coal Slurry Combustion (Vols. I–II), Pittsburgh Energy Technology Center, Pittsburgh, PA (April, 1983).

INDEX